D1259482

SCIENTISTS AT WAR

SCIENTISTS AT WAR

THE ETHICS OF COLD WAR WEAPONS RESEARCH

Sarah Bridger

 Harvard University Press

CAMBRIDGE, MASSACHUSETTS
LONDON, ENGLAND
2015

First printing

Library of Congress Cataloging-in-Publication Data
Bridger, Sarah.
   Scientists at war : the ethics of Cold War weapons research / Sarah Bridger.
      pages cm
   Includes bibliographical references and index.
   ISBN 978-0-674-73682-5
   1. Science—Social aspects.   2. Science—Moral and ethical aspects.
3. Military weapons—Technological innovations—Moral and ethical
aspects.   4. Military research—United States—History—20th century.
5. Military research—United States—Moral and ethical aspects.
6. Nuclear weapons—Moral and ethical aspects—United States.
7. Cold War.   I. Title.
   Q125.B728 2015
   174'.96234—dc23   2014035595

For my parents

Contents

Abbreviations Used in Text .... *ix*

Prologue: The Conscience of a Physicist. 1

1. The Sputnik Opportunity 13

2. The Moral Case for a Test Ban 30

3. The Science of Nonnuclear War 63

4. Into the Ethical Hot Pot 88

5. Disaster and Disillusionment in Vietnam. 115

6. Institutional Reckonings at MIT. 155

7. The New Left Assault on Neutrality. 194

8. Collapse of the Sputnik Order 222

9. A United Front against Star Wars. 245

Epilogue: Science and Ethics after the Cold War 270

Notes *275*

Acknowledgments *329*

Index *331*

Abbreviations Used in Text

AAAS	American Association for the Advancement of Science
ABM	antiballistic missile
ACDA	Arms Control and Disarmament Agency
AEC	Atomic Energy Commission
AFB	Air Force Base
AFSAB	U.S. Air Force Scientific Advisory Board
APS	American Physical Society
ARPA	Advanced Research Projects Agency
BAFGOPI	Boston Area Faculty Group on Public Issues
Caltech	California Institute of Technology
CBW	chemical and biological weapons
CDTC	Combat Development and Test Center
CIA	Central Intelligence Agency
CN	chloroacetophenone (tear gas)
CS	o-chlorobenzylidene malononitrile (super tear gas)
CSRE	Committee for Social Responsibility in Engineering
DCPG	Defense Communications Planning Group
DEW	directed energy weapons
DM	diphenylaminechloroarsine (adamsite)
DOD	Department of Defense
ESSG	Army Corps of Engineers' Strategic Study Group
FAS	Federation of Atomic Scientists; later Federation of American Scientists
ICBM	intercontinental ballistic missile

IDA	Institute for Defense Analyses
I-Lab	Instrumentation Lab (later Draper Laboratory)
JRATA	Joint Research and Test Activity
MIRV	multiple independently-targetable reentry vehicle
MIT	Massachusetts Institute of Technology
MRI	Midwest Research Institute
NAS	National Academy of Sciences
NASA	National Aeronautics and Space Administration
NSF	National Science Foundation
OSRD	Office of Scientific Research and Development
OST	Office of Science and Technology
PSAC	President's Science Advisory Committee
Rad Lab	Radiation Laboratory (MIT)
R&D	research and development
RLE	Research Laboratory of Electronics (MIT)
SAC	Strategic Air Command
SACC	Science Action Coordinating Committee
SAGE	Semi-Automatic Ground Environment
SANE	Committee for a Sane Nuclear Policy
SDI	Strategic Defense Initiative
SDIO	Strategic Defense Initiative Organization
SDS	Students for a Democratic Society
SESPA	Scientists and Engineers for Social and Political Action
SIOP	Single Integrated Operational Plan
SIPRI	Stockholm International Peace Research Institute
SRI	Stanford Research Institute
SST	supersonic transport
TIBA	2,3,5-triiodobenzoic acid (defoliant)
TNW	tactical nuclear weapons
UCS	Union of Concerned Scientists
USDA	United States Department of Agriculture

SCIENTISTS AT WAR

Prologue

The Conscience of a Physicist

IN THE EARLY 1970s, a young physicist from Sandia National Laboratories arrived for a visit to Lawrence Livermore Lab, the epicenter of nuclear weapons research in the United States. The physicist passed through electrified fences and security checkpoints, duly impressed by his surroundings. It was, he later recalled, "a laboratory where . . . you really know you are in a weapons laboratory." Finally, he arrived at a small auditorium where an expectant audience awaited. At the designated moment, lab leaders strode onstage and began their presentation, which culminated in a dramatic revelation: the unveiling of a model of a nuclear warhead, its side panel removed to expose the ingenious organization of the components within. The visiting researchers were invited to take a closer look. As they solemnly marched by the "cut-away weapon," the young scientist suddenly experienced a sickening vision: "I had this feeling that I was at my mother's funeral, looking into the coffin." It was a queasy moment of epiphany. "I had not suddenly become a raging pacifist or something," he told an interviewer years later, "but I realized I didn't want to be part of it."[1]

That young physicist was Robert Park, who shortly after his visit to Livermore would leave his government-funded position at Sandia for an illustrious academic career at the University of Maryland and decades of service as director of public information for the American Physical Society (APS). Like many Cold War physicists, his career choice—to become a scientist in a field with obvious implications for national security—eventually forced him to confront a deep and troubling array of questions regarding the ethics of weapons research and the obligations of its practitioners.

Park's anguish was a poignant chapter in a story that had begun in the 1940s. Two years after the bombings of Hiroshima and Nagasaki, physicist Robert Oppenheimer famously observed that physicists had "known sin." In his reflections on the creation of nuclear weapons, and in the many hundreds of essays, memoirs, interviews, testimonies, and oral histories concerning the Manhattan Project scientists, the language of morality and ethics is ubiquitous. Physicists invoked notions of guilt, evil, obligation, patriotism, regret, and disillusionment. Many, like the refugee physicist Victor Weisskopf, had joined the project out of a sense of patriotism for the United States and a fear of Nazi development of atomic weapons. As Weisskopf recalled in a 1991 memoir, "There was never any thought of turning this down. How could I have refused an offer to join the best people in the country in a project of such enormous importance? How could I have refused to participate in the war effort of a country that had accepted and supported us so generously?" But as he worked on the project, Weisskopf's views evolved in confusing directions. He began to realize the bomb's full destructive potential, which frightened him, even as the intellectual challenges and satisfactions of his research grew more compelling. He recalled that: "Some of us, including myself, secretly wished that the difficulties would be insurmountable. We were all aware that the bomb we were trying to develop would be such a terrible means of destruction that the world might be better off without it. . . . Then, imperceptibly, a change of attitude came over us. As we became more deeply involved in the day-to-day work of our collective task, any misgivings that we had at the start began to fade, and slowly the great aim became the overriding driving force: We had to achieve what we set out to do."[2]

Even after Germany's surrender, Weisskopf and his colleagues, with one notable exception, continued with their work. Weisskopf later wrote, "The reaction most of us had was both interesting and somewhat depressing. It showed how deeply we scientists got attached to the task that was set before us and to the solution of the remaining technical problems. . . . in retrospect, I have often been disappointed that, at the time, the thought of quitting did not even cross my mind." Only Joseph Rotblat, the British physicist who would go on to found the international Pugwash arms control movement, resigned from the project.[3]

Some Manhattan Project scientists later recalled the German surrender and the Trinity test as turning points in their views of their own work; for others, it was the actual use of the bomb in Japan. Weisskopf was shocked at what he considered the unnecessary, and criminal, decision to drop a second atomic bomb on Nagasaki. For Weisskopf and the other scientists feeling the ethical burden of their wartime contributions, one pathway to redemption lay in political action. During the war years a small, prestigious, and influential group of scientists sought three goals: to avoid actual use of the bomb in favor of a "demonstration" in an uninhabited location; to share bomb information with the Allies; and, after the war, to establish international civilian control of nuclear energy. To pursue these goals, scientists drafted petitions, lobbied politicians, and founded the Federation of Atomic Scientists (later the Federation of American Scientists) and the *Bulletin of the Atomic Scientists*.[4]

Only in their last goal—civilian control of nuclear power—did the scientists achieve some success, with the creation of the Atomic Energy Commission. The failure to prevent the use of the bomb—and the contradiction between exquisite control over laboratory experiments and impotence in the face of military applications of new technologies—proved devastating to some scientists. Norbert Wiener, who had declined to work on the Manhattan Project but whose early cybernetic studies on aircraft technology enabled more efficient aerial bombardment, wrote in despair after the bombing of Hiroshima and Nagasaki: "Ever since the atomic bomb fell I have been recovering from an acute attack of conscience as one of the scientists who has been doing war work and who has seen his war work a[s] part of a larger body which is being used in a way of which I do not approve and over which I have absolutely no control. . . . I do not know how to publish work without making it available for the strongest hands and I do not like the strongest hands of the present time."[5]

Wiener's reflections raised a host of questions: To what extent could scientists—particularly physicists and chemists, whose work might yield new weapons technologies—dictate the terms and goals of their own research? To what extent should scientists connect their own research to its potential military applications, and to what extent, and by what means, could they exert influence over these applications? Wiener, along with Leo Szilard and a handful of other distraught scientists, responded to the horrors of the atomic bomb by switching fields entirely, retreating into biology, medicine, and other less clearly weapons-related areas. But most physicists and chemists rejected such action. All confronted the fundamental questions—whether particular developments in weapons technologies are "inevitable," whether work on scientific research with weapons applications is morally justifiable, and whether scientists have the power and obligation to deal with "the strongest hands" that control key military decisions—that would form the basis of a half century of ethical and political debates among scientists.

This book argues for the centrality of the Vietnam War in the evolution of scientists' ideas about ethical responsibility, from the individualized perspective prevalent during the post–Manhattan Project years to the radical structural critiques of the New Left that emerged in the late 1960s. In the aftermath of the atomic bomb, many elite scientists felt personally obligated to work within government advisory channels to influence the development and uses of Cold War science. The war in Vietnam challenged many assumptions embedded in this strategy, including the notion that individual experts could ethically and effectively work toward desirable ends within the powerful military industrial complex. Through the 1970s and 1980s, a growing contingent of elite scientists began to question their personal and institutional ties to the defense establishment. This shift in thinking was neither uniform nor seamless, but understanding its dimensions and the political context that produced it helps explain a number of key developments: the changing structure of government science advising; the character of scientists' activism throughout the Cold War, including the breadth and inclusiveness of the movement against the Strategic Defense Initiative; the shifts in academic policies concerning defense contracting; and the current challenges posed by attacks on scientific consensus.[6]

The horrors and moral anguish of the Vietnam War thus exposed generational and ideological divides among scientists. The war years remade the terms of the ethical debate itself, demanding answers not just in the realm of individual conscience, but on an institutional scale. At heart, this book attempts to elucidate and contextualize this wide range of ethical views and chart the ways that scientists themselves approached the scientific and moral challenges of the Cold War. It is the story of how academic scientists clamored for the expansion of government-funded research in the 1950s and how just a decade later, their campuses exploded in protests over these very research contracts. It is the story of Manhattan Project elders, New Left radicals, and their reconciliation in the cross-generational campaign against Star Wars.

✳ ✳ ✳

The heart of *Scientists at War* concerns the stretch of the Cold War from the launch of Sputnik in 1957 to the Reagan defense boom of the 1980s, but the debates of those years occurred in a world in which the status of scientists had been irrevocably altered by the development of the atomic bomb. World War II thrust the Manhattan Project generation of elite physicists and chemists into the political spotlight. Their understanding of the workings of an atomic nucleus brought them unparalleled access to policymakers and military leaders. Making the most of their own prestige and influence, many parlayed their concerns over atomic weapons into agitating for arms control and international cooperation. To accomplish these goals, they acted both as individuals and through new organizations, lobbying politicians and making public appeals. Although many disagreed with the decision to drop the atomic bomb in Japan and the campaign to strip Robert Oppenheimer of his security clearance in 1954, the Manhattan Project scientists did not harbor deep resentments against their government. By and large, they did not challenge existing political, military, or academic structures; rather, they worked within available channels and trusted in the power of persuasion to promote the goal of arms control. As their reflections attest, they tended to consider the ethics of weapons research in individualist terms; that is, they worried about their own contributions and felt the ethical burdens of their war work upon their own shoulders. It was the special responsibility of scientists to lobby for nuclear arms control, because scientists had created nuclear arms.[7]

Soon after the war, the same community of elite scientists who had unleashed the power of the atom was wracked by the anguishing debate over whether to develop the hydrogen bomb. On one side, physicist Edward Teller argued in favor of its development, in part because he believed that the advancement of scientific knowledge and technology was inevitable; better for the United States to develop the hydrogen bomb than someone else. As physicist Eugene Wigner explained, "Great weapons will always be developed soon after it is clear that they can be. And the hydrogen bomb was clearly possible in 1946; it would have been invented by 1960, even if Edward Teller had never been born." Unlike Wiener, who had worried about "the strongest hands" grabbing control of his research, Teller's mathematical collaborator Stanislaw Ulam felt plainly that it was not immoral "to try to calculate physical phenomena." After all, "even the simplest calculation in the purest mathematics can have terrible consequences," he wrote, affirming that theoretical research ought to be fundamentally separate from the moral dilemmas of its application. Of course, Ulam had been tasked specifically with determining, mathematically, whether a hydrogen bomb was possible; there was no mystery as to the applications of his research. Teller distilled the ethical problem more precisely, observing that scientists had an obligation to pursue an understanding of the physical world, including knowledge relevant to weapons production. "I am afraid of ignorance," he explained. On the other side were the many Manhattan Project physicists who deemed the "superbomb" an unnecessary and horrific weapon, best left undeveloped. As Hans Bethe recalled in a 1996 oral history, he disagreed deeply with Teller's attitude, explaining, "I would have been happy if we had remained ignorant."[8]

Bethe lost this debate, but the frustration he and many other arms control physicists experienced did not diminish their willingness to continue to work for government and military patrons. The Manhattan Project generation largely maintained the patriotic views of the war era, evidenced most dramatically in the enthusiasm with which they welcomed expanded opportunities for government service following the launch of Sputnik in 1957. In response to that dramatic world event, President Eisenhower called for a resurgence of American science and opened the doors of the White House to the top scientific minds of the nation by creating the President's Science Advisory Committee (PSAC) and a host of other new advisory mechanisms. Men like Bethe and I. I. Rabi used their new positions to push Eisenhower toward a nuclear test ban and a more moderate nuclear strategy. These efforts would reach fruition during the short-lived Kennedy administration, when PSAC members found an arms control sympathizer in Secretary of Defense Robert McNamara, and top science advisor Jerome Wiesner played a key role in the negotiation and ratification of the Partial Test Ban in 1963.

The affinity between McNamara and the PSAC scientists stemmed largely from McNamara's endorsement of a no-first-strike nuclear stance and a military approach rooted in "flexible response" rather than massive retaliation. But the same strategic outlook that appealed to scientists on arms control grounds would soon find its outlet in the jungles of Vietnam. Scientists working at the White House and in the Pentagon were drawn into the conflict, asked to provide scientific perspectives on how to use new and old technologies to wage war more effectively. Young physicists employed by the Institute for Defense Analyses' Jason group found themselves focusing less on techniques to enforce a comprehensive test ban and more on techniques to subdue guerrilla fighting forces in jungle conditions. Meanwhile, the use of nonnuclear tools of war such as tear gases, napalm, and chemical defoliants brought biologists, ecologists, and botanists into the ethical whirlwind as well, as these scientists debated whether such technologies themselves could be considered immoral, independent of the controversial conditions of their applications.

If the Manhattan Project had constituted problematic research in the service of a noble end, scientists' participation in the Vietnam War was perceived by many as problematic research in the service of a problematic end. George Kistiakowsky, a seasoned science advisor who opposed the war, nevertheless felt obliged to lead an effort to design an "electronic barrier" across North Vietnamese supply routes that he hoped would de-escalate the war and hasten its end. As had been true of the Los Alamos physicists urging a demonstration of the bomb, Kistiakowsky watched, devastated, as his research was instead used by military leaders to promote the very outcomes he had hoped to prevent. He, along with a small group of other disillusioned science advisors, resigned quietly from government service, maintaining detailed political correspondence with other scientists but refraining from the kind of outspoken activism increasingly on the rise on university campuses.

It was on these very campuses that a second generation of scientists was coming of age—the students and young faculty members whose beloved dissertation advisors had been wartime heroes at Los Alamos or the Radiation Laboratory at the Massachusetts Institute of Technology. In the context of Vietnam, however, the esteemed professors serving on President Johnson's PSAC suddenly seemed cautious, naïve, and complicit. Across the country, campuses erupted in antiwar protests, many of which targeted universities' lucrative defense contracts, classified research operations, and close relations between faculty and government. At the Massachusetts Institute of Technology, massive demonstrations led to the creation of the Union of Concerned Scientists and the university's eventual decision to jettison institutional ties to the controversial Draper Lab. At Princeton, where little weapons research was actually taking place, students and faculty nevertheless engaged in endless debates about the proper role of universities and scientists, questioning the concept of academic "neutrality." Similar arguments would wrack a host of professional societies, including the APS, which faced radical internal protests led by Charles Schwartz, the Berkeley physicist who had once been an adoring student of Victor Weisskopf.

Attacked on the left by antiwar protesters and on the right by the hostile Nixon administration, the Manhattan Project cohort and a wide circle of elite scientists struggled to find appropriate political and ethical ground. Nixon's initial support for an antiballistic missile system and the development of a supersonic transport proved the final straw for presidential science advisors. When a handful of advisors dared to oppose Nixon's desired policies in very public and persuasive ways, he dismantled the entire advising system set up in the aftermath of Sputnik. In the weakened advisory channels that replaced it, government science advisors were increasingly drawn from a pool of talented professionals with backgrounds in military and industrial work rather than the outspoken iconoclasts of Los Alamos. Lacking the relative independence of the Eisenhower and Kennedy years, federal science advisors were frequently called upon to rubber-stamp administrative decisions and pet projects.

Meanwhile, many of the antiwar groups of the late 1960s found new life in the resurgent antinuclear movement of the late 1970s and early 1980s. The stage was thus set for the dramatic political showdown prompted by President Reagan's announcement of the Strategic Defense Initiative in 1983, during which older and younger generations of scientists finally found common cause, attacking the program on both technical and moral grounds and employing tactics that ranged from personal persuasion to boycotts of the funding structures and institutions that facilitated its development.

* * *

In 1981, the APS conducted a survey of nuclear physicists, requesting both demographic information and reflections on the history and state of the field. When asked to assess how attitudes toward science had changed over the past several decades, respondents noted with near unanimity that the prestige of physicists had suffered greatly during the Vietnam years. They described the growing "mistrust of science and technology," how "the war in Vietnam led to suspicion of science and scientists," how physicists had been targeted by "vocal activist groups," and how "since about 1967, support for science had decreased and much of science itself [came] under fire from a variety of segments of society." As one Livermore physicist reflected: "When I first decided to major in physics, the public had very little understanding of what physics was, and especially why it might have any application to daily life. Shortly thereafter, public awareness of physics, and nuclear physics in particular, blossomed and the field soon became one of glamour in the public image. I have seen this glamour wear thin to the point that now science, particularly physics, and more especially nuclear physics, is suspect."[9]

As the development of the atomic bomb had raised deep moral and philosophical questions for physicists about the applications of their research, so too did the participation of scientists in the prosecution of the war in Vietnam and their indirect responsibility for the recycled and modified technologies employed in the service of controversial ends. In the face of deep and often vitriolic public scrutiny, many scientists were forced into self-reflection and self-defense. In 1967 Louis Fieser, the Harvard chemist who had invented napalm during World War II, observed defensively that back then, he "couldn't foresee that this stuff was [going] to be used against babies and Buddhists." But he argued that "the person who makes a rifle . . . he isn't responsible if it is used to shoot the President." At the other extreme, mathematical physicist William Davidon argued in the early 1970s that scientists and engineers had an obligation to halt their profession's complicity in the waging of the war in Vietnam—through political activity and, if necessary, industrial sabotage. The bombing of an applied mathematics laboratory at the University of Wisconsin epitomized one lethal outcome of such a view. In between these two extremes lay a myriad of other responses and approaches, including that of Arthur Galston, the botanist whose dissertation research had unintentionally sparked the development of Agent Orange, and who subsequently devoted decades of his life to trying to curb the use of what he considered to be a chemical weapon. The Vietnam War changed not just outsiders' attitudes about science, but scientists' attitudes about themselves.[10]

Within the dynamic field of Cold War science history, this book offers a number of key contributions.[11] Early chapters show how the Manhattan Project cohort of physicists promoted a powerful arms control agenda from within and without the halls of government, employing a variety of both technical and sweeping moral arguments. In particular, the close study of the debate over a nuclear test ban during the Eisenhower and Kennedy administrations challenges previous claims that high-level advisors relied on primarily technical assessments to persuade policymakers.[12] Subsequent chapters show clearly how this commitment to arms control—combined with the advisory opportunities expanded during the Sputnik boom years—facilitated a preference for the nonnuclear tools of limited war that would be employed in Vietnam, with disastrous consequences not only for the physicist advisors, but also for the biologists and chemists whose research suddenly became morally suspect. By revealing the limits of science advisors' influence as well as their wrenching internal debates about war and ethics, the analysis in these chapters offers a personal portrait of how Cold War science advising operated.[13] It also connects the activities of high-level science advisors with the developing ground-level scientists' movements of the Vietnam years, revealing how scientists themselves began to understand the "military-industrial-academic complex" and their agency within it in complicated and contradictory ways.[14]

The intellectual tumult of the Vietnam years also provided the context for much deeper debates concerning the epistemology of science itself, including the development and spread of provocative claims about the social construction of science and the implications of these new ideas for science policy and the political activities of scientists. This is not a work of philosophy but of history, and the intention here is to illuminate how and why aspects of New Left critiques of science gained credence during the Vietnam War and to explain some of the long-term consequences of that shift. Scientists themselves, particularly those elevated to administrative positions within academia during the Vietnam years, grappled with deep questions about how Cold War defense contracting was transforming the progression of scientific research.

Some of these arguments have been replicated by historians of science, most notably in the influential debate between Paul Forman and Daniel Kevles. In the late 1980s, Forman famously argued that scientists were in a sense duped by the postwar push for greater research and development funding. Heavily influenced by the work of technology historian David Noble, he contended that while scientists had gained immense funding, they had lost control. He asked, "What direction of the advance of science, and thus what kind of science, resulted from military sponsorship?" and determined that in the 1960s, certain fields, such as solid state physics, had flourished at the expense of less militarily relevant areas. He asserted that Cold War scientists suffered from "false consciousness," and were "focused so narrowly on immediate cognitive goals of their work as to miss its instrumental significance . . . to their military patrons."[15]

Forman's view was quickly contested. Kevles, for example, complained that Forman portrayed defense funding as having "seduced American physicists from, so to speak, a 'true basic physics,' encouraging them to the self-delusion that they were engaged in basic research of intrinsic interest while in reality they were merely doing the military's bidding." The truth, Kevles argued, was that there existed no single guiding federal agency setting research goals—rather, the funding system was varied and not monolithically organized, and scientists themselves often helped set "military technology policy."[16]

Though Forman hadn't posited the existence of a centralized conspiracy of military planners out to co-opt scientists, he worried about the effects of universities' growing dependence on defense dollars. Forman argued that as almost all physicists' labs and equipment began to depend on government and military funding for their existence, scientists' freedom to research could not help but be subtly influenced and unconsciously redirected. The broader phenomenon described here—the influence of funding decisions, experimental successes, and socially constructed reward systems on the paths of scientific research—had been documented and debated by many other historians of science, perhaps most notably beginning in the late 1970s and early 1980s in works by Bruno Latour, Donald Mackenzie, Peter Galison, and Ian Hacking, among others. All of these scholars were influenced, at least indirectly, by Thomas Kuhn's demonstration of the nonlinear progression of science and by the "Strong Program" of Edinburgh sociologists.[17] As Hacking wrote eloquently in 1986 during the Reagan defense boom, "When so much knowledge is created by and for weaponry, it is not only our actual facts, the content of knowledge, that are affected. The possible facts, the nature of the (ideal) world in which we live becomes determined. Weapons are making our world, even if they are never exploded."[18]

Kevles criticized Forman's failure to acknowledge the gray areas between basic and applied research, pointing out that something as fundamental and intellectually "pure" as particle physics simply could not have been pursued without government support; particle colliders are expensive, after all. While some topics might follow naturally from one another, and others were clearly valued for their potential application value, a third kind of research defied these categories or seemed to exist simultaneously in both camps—for example, fluid dynamics. As a result, Cold War military funding actually helped physics diversify. Kevles observed that by the mid-1950s, "What the diversification fundamentally signified was not the seduction of American physics from some true path but its increased integration as both a research and advisory enterprise into the national security system. . . . The closer civilian scientists came to the center of executive power, the better positioned they were to influence overall defense policy."[19]

Scientists at War examines the limits of that influence and the consequences for scientists. Although the emphasis is on ethical debate, this
book does not attempt to provide easy resolutions or a universal moral code
to follow; rather, it offers contextualized portraits of the myriad ways that
scientists themselves have answered powerful and important moral questions about their work and their participation in American life. These portraits illuminate how scientists have shaped and been shaped by Cold War
policy. They provide insights into moral and philosophical problems relevant
not only to the Cold War but to our own times, of importance not only to
today's practicing scientists and engineers but to all concerned citizens.[20]

ONE

The Sputnik Opportunity

THE MANHATTAN PROJECT veterans emerged from the postwar years with tremendous national and international stature. Though chastened by their failure to prevent an international arms race after the Soviet nuclear test of 1949, and scarred by bitter disputes over the development of the hydrogen bomb, the community of elite physicists was, at least outwardly, still uniformly committed to political action on behalf of an arms-control agenda. Even the hawkish Edward Teller, the physicist popularly dubbed "the father of the hydrogen bomb," served on the Board of Sponsors of the *Bulletin of the Atomic Scientists,* warning of the dangers of nuclear warfare. Still holding their monopoly on the understanding of precisely how the power of the atom could be harnessed in weaponry, these scientists found their expertise and advice in high demand in Washington, DC.

But nuclear policy in the 1950s was a messy business. Despite authorizing the use of atomic weapons in Japan and the development of the hydrogen bomb, President Truman had left little in the way of a coherent nuclear strategy for his successor, Dwight Eisenhower. Truman had presided over the crucial postwar years during which the Cold War deepened into irreconcilable global conflict: he had pledged support for anticommunist movements across the globe, implemented the Marshall Plan's aid to western Europe, oversaw the creation of the North Atlantic Treaty Organization, and waged the Korean War amid fears that the Soviet Union would seek expansion through proxy actions throughout the world.[1]

Initially, Truman's rhetoric of containment had emphasized nuclear "symmetry": U.S. military responses calibrated to the level of an enemy's provoking action. But the allure of the stockpile and pressure from the military services, particularly the air force, were too great. The state of nuclear weapons technology in the early 1950s required bomb delivery via aircraft, and the air force stood to gain enormously in stature and influence with a shift in security policy that emphasized nuclear capabilities. Pressed by the Joint Chiefs of Staff and the air force's Curtis LeMay, who headed the Strategic Air Command (SAC), Truman eventually approved a dramatic expansion of the nation's nuclear stockpile.[2]

Eisenhower's election in 1952 was thus a cause for optimism for the ascendant air force. Historian David Rosenberg writes in his chronicle of nuclear weapons policy during this period: "Where Harry Truman viewed the atomic bomb as an instrument of terror and a weapon of last resort, Dwight Eisenhower viewed it as an integral part of American defense, and, in effect, a weapon of first resort." Whatever Eisenhower's private views of nuclear weapons, his nuclear strategy, dubbed the New Look, relied on enormous first-strike capabilities that Secretary of State John Foster Dulles referred to as a "deterrent of massive retaliatory power." The new posture of asymmetry rested upon Eisenhower's publicly stated willingness to respond to minor threats with overwhelming, potentially nuclear, force.[3]

Curiously, the shift in policy arose from Eisenhower's desire to avoid the kind of "garrison state" he feared would be created with enormous military budgets. The New Look, with its emphasis on nuclear weapons rather than conventional forces, was *cheap,* and military expenditures, calculated as a percentage of GDP, actually declined during the Eisenhower years. Moreover, by promoting deterrence through the fear of nuclear attack and relying on allies to provide lesser conventional forces, Eisenhower hoped to prevent American involvement in protracted limited wars. He wanted to avoid another Korea.[4]

All this meant that in the mid-1950s, the nation's nuclear scientists enjoyed the benefits of government funding and political power. At the urging of Teller, the Lawrence Livermore Laboratory, a massive nuclear weapons research facility meant to rival Los Alamos, was constructed in Southern California in 1952. From 1953 to 1958, the lab's staff expanded from just under seven hundred to over three thousand employees, with a budget increase of over 1,500 percent. Meanwhile, air force leaders engaged in a process critics dubbed "bootstrapping": generating lengthy target lists that justified increased weapons stockpiles and overall funding. In 1953 the nuclear stockpile contained roughly one thousand weapons; by 1960 it had expanded eighteenfold. All the while, Curtis LeMay pushed for massive strike capabilities and expanded target lists of Soviet cities and industrial centers.[5]

* * *

The prominence of the SAC and the New Look did not go unchallenged. Arrayed against the SAC and its supporters were the Joint Chiefs of Staff, much of the army and navy leadership, and even consultants from the air force's spin-off RAND Corporation, all of whom rejected the SAC focus on urban areas as key targets. Within the army, Maxwell Taylor led the push for a reorientation toward limited war and conventional forces. He debated military priorities with the president on multiple occasions, urging "mutual deterrence" rather than "massive retaliation," but Eisenhower resisted.[6]

Scientists, caught in this infighting, were increasingly asked to weigh in on the value and feasibility of various weapons systems, even while some harbored hopes for a world without nuclear weaponry and a shift away from Eisenhower's New Look. Princeton mathematician John von Neumann had chaired a committee that recommended the development of intercontinental ballistic missiles, a technology that would strip the air force of its nuclear dominance by allowing nuclear weapons to be delivered via missiles rather than bombers. In March 1954 the Office of Defense Mobilization's Science Advisory Committee met with Eisenhower to discuss the state of the country's "technological capabilities" in the context of nuclear war. At the meeting, Eisenhower warned the scientists of the nation's vulnerability to a surprise attack and requested that they evaluate potential technical solutions. They established a special task force headed by a steering committee of James Killian from the Massachusetts Institute of Technology (MIT), Lee DuBridge from the California Institute of Technology (Caltech), Edwin Land from Polaroid, and a handful of other advisors, assisted by forty additional scientists and engineers serving as a "professional staff."[7]

The 1954 report of the Technological Capabilities Panel, as it came to be called, offered a prescient view of the future of the arms race. The panel urged numerous improvements: enhanced intelligence, including a program of high-altitude U-2 surveillance; expanded communications capabilities; greater support for basic science; and better preparedness in the case of a surprise attack. On this last point, the scientists pinpointed existing SAC weaknesses and urged that bases be "hardened" and nuclear resources be dispersed or airborne to prevent easy targeting. Most crucially, the panel predicted that by the end of the decade, the age of the bomber would wane and the age of intercontinental and intermediate range missiles would begin. As von Neumann had earlier recommended, the panel called for the development of both land-based and submarine-based missiles and for early "theoretical and experimental" investigation into antimissile defenses.[8]

Eisenhower was smitten by both the advisors and their advice; three years before Sputnik and the creation of the President's Science Advisory Committee (PSAC), the panel had established an important precedent regarding the value of science advising. He viewed the assembled scientific experts as honest, independent, and keenly intelligent, immune from the tawdry pressures of politics. But though the president expressed enthusiasm for the work of the Technological Capabilities Panel, he and the panel scientists parted ways on the fundamental strategy of the New Look. Killian and his colleagues had pushed for a renewed emphasis on technologies appropriate for nonnuclear "limited wars" to little avail. Killian himself later acknowledged that "in making these and other recommendations, the panel clearly was dissatisfied with the 'new look' defense policy and the concept of 'massive retaliation.'" But Eisenhower was not immediately persuaded.[9]

Nevertheless, Eisenhower's appreciation of the government advisors' work served an important, additional function. He was reaching out at the very moment that many in the scientific community were learning to fear the chilling power of Cold War anticommunism. In the spring of 1954, Robert Oppenheimer, accused of Communist sympathies, had been stripped of his security clearance after lengthy hearings, during which many of the leading Manhattan Project physicists testified on his behalf. Teller, the hawkish Hungarian physicist who had already clashed with his fellow Manhattan Project veterans over his work on the hydrogen bomb, had taken the opposing side, a position that would permanently alienate him from much of physics' academic elite. The case came to symbolize the excesses of McCarthyist America. Years later, Killian would credit the Technological Capabilities Panel with helping to re-knit the frayed relations between scientists and the federal government. The scientists on the panel were "citizens who felt an obligation to their country that overrode their dismay about a single administration," he wrote, and Eisenhower's respect and request for assistance paved the way for renewed cooperation. Killian's retroactive gloss likely overstated the panel's impact, but within three years, in the aftermath of Sputnik, the bonds between the White House and elite physicists would strengthen enormously.[10]

* * *

If the Manhattan Project catapulted elite scientists into the political sphere and the Oppenheimer case alienated them, the launch of Sputnik in October 1957 brought them into the government fold in even greater numbers. James Killian, the MIT president soon to be tapped as Eisenhower's special assistant for science and technology, later reflected that "while scientists possessed immense prestige in Washington during the years following World War II and historic actions were taken during the Truman administration to institutionalize science and technology in government, science had a uniquely close relationship to the presidency during Eisenhower's second term and extending into the Kennedy administration." Sputnik was the pivotal event. The Soviet satellite, at 184 pounds and just under two feet in diameter, orbited the earth every 96.2 minutes, emitting a steady stream of gentle radio beeps, a repeating audio reminder of Soviet technical superiority. The *New York Times* reported that the satellite launch was "an achievement of profound scientific significance for all mankind" but one "inevitably . . . sobering" for the United States. The public demonstration of Soviet achievements in space and rocket technology signified a potential "science gap" afflicting the United States. The event launched the creation of an extensive federal science advisory apparatus in the United States, a splurge in funding for research and development, and a short-lived "golden age" for the political influence of science advisors.[11]

Eisenhower had initially downplayed the significance of the satellite, but intense media coverage quickly required further presidential action. In response, Eisenhower announced the creation of a new cabinet position: the special assistant for science and technology, who would also head up the new PSAC. His choice for this influential position was James Killian, a non-scientist who had amassed an impressive track record in science administration, first with the World War II-era National Defense Research Committee and, since 1948, as president of MIT. The PSAC was not an entirely novel creation; Eisenhower had simply reconstituted and enlarged the Office of Defense Mobilization's advisory committee. Killian, a prominent member of the earlier group, had witnessed firsthand the respect accorded to its Technological Capabilities Panel in 1954. The panel had operated largely behind the scenes, however, and Killian was honored to chair the more exalted PSAC, which, as he put it, "was to be positioned at the very summit of government."[12]

In his letter to Killian explaining his new duties as the special assistant, Eisenhower directed him to keep abreast of "the use of science and technology in relation to national security" and to advise the president on all related matters. The president granted Killian "full access to all plans, programs, and activities involving science and technology in the Government, including the Department of Defense, Atomic Energy Commission, and CIA [Central Intelligence Agency]," and invited him to National Security Council and other classified meetings. He also tasked Killian with staffing and organizing the PSAC, which subsequently included an extraordinary collection of the nation's top scientists: Hans Bethe, the Cornell physicist and future Nobel Prize winner; Jerome Wiesner, the future president of MIT; George Kistiakowsky of Harvard; Edward Purcell and I. I. Rabi, both Nobel Prize winners in physics; Livermore director Herbert York (who considered himself the token representative from the "nuclear weapons establishment"); and others with similarly prestigious pedigrees.[13]

These scientists, in many ways the architects of the Sputnik boom, were largely academic physicists. Veterans of the Manhattan Project or radar research during World War II, they were patriotic, anticommunist, and idealistic, happy to offer part-time or full-time government service while maintaining their academic positions. Whatever disillusionment had been spawned by the Oppenheimer case, the PSAC scientists were enthusiastic about their new, expanded roles as government advisors; they considered national service and national security to be part of their obligation as scientists. As Killian recalled, perhaps a bit rosily in hindsight, "The group was held together in close harmony not only by the challenge of the scientific and technical work they were asked to undertake but by their abiding sense of the opportunity they had to serve a president they admired and the country they loved."[14]

Not surprisingly, much of the PSAC's early work concerned science and technology related to defense and the space program. Although some PSAC materials remain classified and no formal minutes of their meetings were kept, "in order to promote full and uninhibited discussion," internal summaries of their work are now available. During the first years of its existence, the PSAC participated in defense budget reviews, submitting "specific recommendations . . . concerning strategic delivery systems, air and ballistic missile defense, limited war, anti-submarine warfare, communications, and intelligence." The scientists' concerns were discussed at the top levels of government: by the president, the secretary of defense, and the Joint Chiefs of Staff. With its memberships organized into technology-specific panels, the

PSAC also provided "progress reports" for all the major "missile and satellite programs." George Kistiakowsky, who would succeed Killian as Eisenhower's second special assistant for science and technology, chaired the enormously influential PSAC Ballistic Missiles Panel. In March 1958 Kistiakowsky and Killian met with Eisenhower to discuss the ballistic missile program, after the panel had prepared a technical report building on the conclusions of the old Technological Capabilities Panel and offering a "national program for ballistic missile development over the coming years."[15]

The PSAC quickly revealed itself as unafraid to call for the cancellation of major weapons programs. In nearly every case, members opted to elevate technologies necessary for minimal deterrence over those required for massive first- and second-strike capabilities. Kistiakowsky's panel called for the cancellation of the liquid-fueled Jupiter missile, the continuation of the more easily transportable Thor intermediate-range missiles, and the accelerated development of solid, storable propellants. (These solid propellants would soon get their own ad hoc panel, led by Kistiakowsky and including the crème de la crème of academic and industrial chemists.) The panel offered qualified support for the land-based Minuteman Intercontinental Ballistic Missile Program, but reserved its greatest enthusiasm for the submarine-based Polaris missile system, which it deemed "less vulnerable (and of comparable cost to) the several land based systems proposed."[16]

From its earliest years of existence and despite changing panel membership, the PSAC was skeptical of the development of a large-scale antiballistic missile (ABM) system and checked its expansion. In the very first report by Jerome Wiesner's panel on ballistic missile defense, PSAC members noted the "extremely difficult nature and the great uncertainties involved" in interception, and as early as the spring of 1958, Wiesner was warning about "the decoy problem"—the challenge of identifying and destroying an incoming missile if it were surrounded by decoy projectiles. Though the panel supported an "experimental" ABM system, it opposed any "large-scale" development of the proposed Zeus system, partly on the grounds that it was vulnerable to "tactics of confusion and decoy." Instead, the panel recommended that at least through 1965, a better option to interception was "passive defense, i.e., dispersal, hardening, concealment, and quick reaction." In 1960 Stanford physicist Wolfgang Panofsky took over an expanded version of the panel, but despite the leadership change, "the Panel saw no reason to change the major PSAC conclusions previously drawn to the effect that large-scale production of Zeus was unwarranted, and that the presently configured system could not provide an effective defense against a determined enemy."[17]

Meanwhile, other PSAC panels explored the problems of antisubmarine warfare, nonlethal chemical and biological weapons, surprise attack, and arms control and the enforcement of a nuclear test ban. Outside of these explicitly defense-related topics, the PSAC issued reports on science education and basic research. The PSAC's influence extended beyond the panels and reports, however. PSAC members met formally and informally with a host of key government officials, and, particularly during the Eisenhower and Kennedy administrations, enjoyed extensive access to the president and top cabinet members. As Killian recalled, "PSAC was usually able to go directly to the president with its advice," and Eisenhower was known in later years to continue to refer to the PSAC as "my scientists." On his deathbed, Eisenhower reportedly told Killian that the PSAC had been "one of the few groups that I encountered in Washington who seemed to be there to help the country and not help themselves."[18]

✳ ✳ ✳

The influence of scientists expanded beyond the White House as well, through the reorganization and invigoration of advisory mechanisms in the Pentagon and the military services. The Air Force Scientific Advisory Board (SAB) expanded its membership from fifty-one in 1958 to eighty-eight in 1962, with the number of outside consultants more than doubling. The Department of Defense Reorganization Act of 1958 established a new Pentagon position, the director of defense research and engineering, reporting directly to the secretary of defense and tasked with supervising Pentagon research activity. Herbert York was the first to hold the new title. The Advanced Research Projects Agency (ARPA), created in February 1958 "as a separate operating agency with the Department of Defense," was dedicated to identifying and supporting innovative, cutting-edge science and technology projects. (After 1959, ARPA's chief scientist reported to the new director of defense research and engineering.) As the new agency alerted potential contractors, ARPA projects might have obvious military applications or not; the goal was to "'leap frog' the present state of the art in assigned fields to attain a dynamic, forward approach to defense of the United States." Flexibility and collaboration were key; for any given problem, "all possible approaches are considered and the selected approach is accomplished by utilizing the best capabilities of both government and industry." Early areas of research included ABM defense, materials research, toxicology, alternative energies (including nuclear and solar power), pre-NASA space technology, solid propellant chemistry for rocket development, and the VELA nuclear test detection program.[19]

Another key development of the post-Sputnik years was the Jason program's inauguration under the auspices of the Institute of Defense Analyses (IDA). IDA had been established in 1956 as a federally funded research and development center and served as a mechanism to provide civilian experts from academia to the Pentagon and military services, particularly the Weapons Systems Evaluation Group, itself a postwar creation linking military and civilian intellectual resources. James Killian served as IDA's first chairman of the board of trustees, and its founding member universities included Caltech, the Case Institute of Technology, MIT, Stanford, and Tulane. In the years after Sputnik, IDA's collaborations expanded beyond the Weapons Systems Evaluation Group to include work for ARPA and the launch of a special facility dedicated to mathematics and communications research located on Princeton University's campus. Columbia, Penn State, and Stanford joined as institutional members. In 1959 IDA created the Jason Division, envisioned as an "opportunity for outstanding academic physicists to devote their consulting time to scientific problems with defense implications while remaining in the academic community." Participants would devote one day a week to Jason work during the academic year and work several weeks during the summer, with pay at roughly $200 a day. Herbert York and John Wheeler personally sent recruiting letters to young academic superstars like Caltech's Murray Gell-Mann, describing the search for "15 or 20 of our best scientists with a strong interest in the defense of our country." The assembled scholars, who included multiple future Nobel Prize winners, would go on to provide significant advice to the Pentagon on weapons systems and arms-control measures.[20]

Thus, the new research and advisory mechanisms set up by Eisenhower in the aftermath of Sputnik enabled scientists to exercise their greatest political influence for arms control and disarmament during the Cold War. They were the same institutional ties that would later bind government scientists to the horrors of the Vietnam War.

✳ ✳ ✳

Beyond his own desire to expand scientific expertise in the government, Eisenhower also faced considerable public pressure to increase federal funding for science and engineering research and education. This was quickly translated into policy: The National Defense Education Act of 1958 provided federal money for scholarships and instruction in math and science. Federal funding for scientific research and development expanded dramatically, particularly to NASA, the Atomic Energy Commission (AEC), and the Department of Defense and the military branches. Five billion dollars was added to the defense budget of 1958. NASA's funding would double every year between 1958 and 1964. The budget for the National Science Foundation increased by 300 percent for 1959.[21]

These agencies, in turn, expanded their contracts with universities, where much of the actual research would take place. Around the country, the new funding fueled the creation or the expansion of a spate of special university labs, including MIT's Lincoln Lab and Instrumentation Lab (also known as the Draper Lab), Berkeley's Livermore Lab, and Stanford's Applied Electronics Lab. To paraphrase Randolph Bourne's observation that "war is the health of the state," during the late 1950s and early 1960s, the Cold War was the health of academic science.[22]

＊ ＊ ＊

As had been true during World War II, this expansion and coordination was encouraged and organized in large part by scientists, particularly academic scientists. Although a handful of prestigious advisors, including James Conant of MIT and Alan Waterman of the National Science Foundation, initially considered the National Defense Education Act unnecessary, a host of equally prominent experts strongly recommended expanded funding policies. The physicist I. I. Rabi, whom Eisenhower knew well from his years at Columbia University, was particularly enthusiastic. He encouraged the president to cite Sputnik as a justification for a more muscular science policy. Despite the cautious attitudes of a few high-level advisors, overall, scientists favored expanded science education and research support. The trend extended far beyond the moderate scientists who had Eisenhower's ear. At the hawkish end of the political spectrum, Teller promoted "more applied research" at universities, government labs, and industry. At the dovish end, Eugene Rabinowitch, founder of the *Bulletin of the Atomic Scientists,* encouraged the expansion of American science and applied research "unhampered by budgetary considerations."[23]

From almost the moment of its creation, the PSAC powerfully advocated for expanded support for science education and research. In December 1958 the PSAC Panel on Research Policy, chaired by physicist Emanuel Piore, released a report, "Strengthening American Science," which urged better national coordination of key research programs, resulting in March 1959 in the creation of the Federal Council for Science and Technology. Caltech president Lee DuBridge headed a Science and Engineering Education panel, which predictably urged improved standards and "public support to meet these goals." Berkeley chemist and Nobel laureate Glenn Seaborg oversaw the production of "Scientific Progress, the Universities, and the Federal Government," a report emphasizing research and training. Other panels tackled government research and development policies, including support for national laboratories and general contracting mechanisms.[24]

In the summer of 1960, the PSAC issued an urgent call for increased federal support for basic science, "in order for the United States to maintain its pre-eminent position in basic science." Because basic research is "high risk," the scientists argued, many "profit-motivated individuals and corporations" were unwilling to offer sufficient investment and support: "Only the Federal Government can afford to support basic science on the scale necessary to insure continued national progress." National progress included national security, living standards, and public health. In this vein a host of PSAC panels focused on specific areas of scientific research that merited greater support, the "hitherto little supported areas of science" now deemed to be "of critical importance to the national welfare." These included "materials science, oceanography, and atmospheric science" and emerging "interdisciplinary" fields in which research was conducted in "both universities and government-owned establishments." In conjunction with the Federal Council for Science and Technology, the PSAC panels recommended the establishment of "interdisciplinary laboratories on university campuses" and long-term support for research facilities and training.[25]

* * *

Scientists themselves helped create the pivotal "Sputnik moment." "Sputnik had no automatic political valence," writes historian David Kaiser. "Techno-political events rarely do." Rather, "determined lobbying" by key Manhattan Project veterans transformed the launch of a satellite into "a political event requiring a specific political response." Kaiser suggests obliquely that self-interest may have motivated the lobbying efforts of scientists in the aftermath of Sputnik. Certainly, they—and their research institutions—benefited financially from the new celebration of science and technology expertise.[26]

But for many future PSAC members, Sputnik raised genuine concerns about Cold War security and the state of American science. Perhaps even most importantly, the elevation of elite physicists to positions of political and military influence also offered an opportunity more precious than the millions of dollars of research funds. It was a chance for nuclear redemption in the form of arms control. Yet within a decade, nearly every aspect of the Sputnik advising boom would be challenged by a new generation of students and science activists: the role of academic scientists as government and military advisors, the Pentagon sponsorship of scientific research, the expansion of government-funded interdisciplinary labs, and the rise of what would come to be called the "military-industrial-academic complex."

✳ ✳ ✳

Even as Eisenhower promoted the nuclear-heavy New Look, his administration also oversaw a parallel, contradictory effort to negotiate a test-ban treaty, nonproliferation measures, and disarmament with the Soviet Union. In the winter of 1954, the United States conducted a series of nuclear tests from facilities at Bikini Atoll, part of the Marshall Islands in the Pacific, including the "BRAVO test" of a fifteen-megaton hydrogen bomb on March 1. Newspapers initially reported that the March 1 blast had "obliterated its test island" and "unleashed violence so tremendous that even its designers were amazed." But within days the press was reporting something else: the presence of a Japanese fishing boat that had been "showered with radioactive ash" during the testing. The ship, the *Fukuryu Maru,* had been logged outside the location of the testing area, yet all twenty-three of the fishermen aboard suffered symptoms of radiation sickness and were immediately hospitalized. Pressed for an explanation, the AEC's Admiral Lewis Strauss finally attributed the disaster to "an unexpected shift in the direction of the prevailing winds in the higher altitudes," suggesting that officials had been surprised only by unpredicted weather, not unpredicted aspects of thermonuclear detonation. Later reports estimated that over three hundred—and perhaps thousands—of observers and island residents had been exposed to varying levels of radioactive fallout.[27]

Almost immediately, the international community registered its disapproval; a spate of countries, including Japan and India, called for an immediate testing halt, as did antinuclear activists in the United States. Much of the early controversy focused on the dangers of fallout, the radioactive particles produced by atmospheric testing that could contaminate air, water, and land, endangering human health. As the Duke University physicist L. W. Nordheim put it, "Dangers due to radioactivity seem to stir human emotions to a much greater extent than other man-produced hazards; understandably, since radioactivity can be neither seen nor felt . . . [it] can travel in the air over large distances, and its very nature and action are unknown and unfamiliar to most people."[28] In 1955 the *Bulletin of the Atomic Scientists* published an influential series of related articles by physicist Ralph E. Lapp. A Manhattan Project veteran and former head of the nuclear physics division for the Office of Naval Research, Lapp had been one of the signers of the 1945 University of Chicago petition urging President Truman not to use the bomb on Japanese civilians. Now, shocked by the BRAVO test, he turned his efforts to research and writing, producing exposés on the dangers and the extent of the BRAVO fallout, government secrecy on the matter, and the inadequacies of current civil-defense preparations. The articles presaged a full-length 1957 book, *The Voyage of the Lucky Dragon,* a best-selling account of the ill-fated Japanese fishing ship.[29]

More formally, the Federation of American Scientists (FAS) called on the United Nations to conduct a study of the "potential dangers in atomic and thermonuclear bomb tests" on the grounds that the BRAVO test had shown that "effects cannot be restricted within national boundaries, and that the lives and health of people in other countries are endangered." The group warned of possible radioactive contamination of the earth's atmosphere, worsening international relations, and the risk of genetic mutations. The time had come, they wrote, to bring "one facet of the atomic armaments race and the threat of war into the spotlight of human morality."[30]

As warnings about the fallout risks of aboveground nuclear explosions proliferated, arms- control advocates saw a new potential path to disarmament. In November 1954 David Inglis, a physicist at the Argonne National Laboratory, founding FAS member, and Manhattan Project veteran, wrote an impassioned appeal for a testing ban in the *Bulletin.* His goal was not simply to reduce the risks of fallout but to alter the course of the arms race itself. He acknowledged that an internationally enforced test ban "provides no disarmament and provides arms limitation only indirectly by limiting the development of new types of arms." But it could nevertheless achieve something important: *"It would slow down the rate of development of new techniques of offense and allow the techniques of defense to come closer to catching up."* He imagined a world in which the two superpowers would maintain only aging stockpiles of "old-fashioned" H-bombs but would pursue cutting-edge new technologies of detection and interception. Meanwhile, in the absence of testing, no other nations could develop thermonuclear weapons. Inglis was part of a small but influential group of scientists urging support for this "small but significant and practical measure of arms limitation." Inspired, the FAS, after conducting a poll of its membership, added a nuclear test ban to its political goals.[31]

Even on the pages of the *Bulletin,* some disagreement about the necessity of a test ban persisted, with scientists weighing in on both sides of the issue. (Tellingly, Inglis's initial statement had been accompanied by a sidebar advertising employment at Lockheed's Missile Systems Division.) Such dissension existed at the highest level of government as well. The BRAVO disaster had prompted the first of a series of Soviet proposals for disarmament measures and a testing moratorium. State Department officials endorsed a two-year testing halt but faced opposition from the Pentagon, the Joint Chiefs, and the AEC. Eisenhower himself seemed uninterested. In the aftermath of the BRAVO test, the president was advised by his Special Committee on Disarmament Problems, led by Harold Stassen, the special assistant for disarmament, with representatives from the State Department, the Pentagon, the CIA, and the AEC (whose contingent included Princeton's John von Neumann). Teller and Strauss argued strenuously against a testing halt, and their position proved ascendant throughout Eisenhower's first term.[32]

The remaking of federal science advising after Sputnik changed the dynamics of the debate. The influx of new PSAC scientists, many of whom had sharp political and personal disagreements with Teller, slowly helped shift attitudes about testing. Though Eisenhower's various advisory groups had failed to reach a consensus on the issue, in the spring of 1958, Soviet Premier Nikita Khrushchev and Eisenhower both agreed to study the possibility of a monitored test ban. A PSAC panel on Arms Limitation and Control was established to research the "military and technical aspects of possible arms limitation agreements" and ultimately recommended the establishment of an executive office devoted to arms control, the U.S. Disarmament Administration. Hans Bethe chaired an ad hoc PSAC working group dedicated to the test ban. The year 1958 also saw the origins of the VELA nuclear test detection program, a collaborative effort by the Pentagon and the AEC to explore techniques to monitor nuclear testing remotely. Much of the research and design was conducted under the auspices of Los Alamos and ARPA, with support from the PSAC and significant advisory input from the young members of IDA's Jason group. By 1960 VELA would have three component programs underway, corresponding to the major testing environments. VELA-Uniform covered seismic and other forms of monitoring to detect and distinguish underground nuclear tests from earthquakes or nonnuclear explosions; VELA-Sierra explored earth-based means of detecting high-altitude tests; and VELA-Hotel was concerned with space-based monitoring technologies. Without adequate monitoring to prevent cheating, the United States and the Soviet Union were unlikely to agree to a test ban. (Not surprisingly, VELA was an enormously popular project for arms-control-minded scientists and other intellectuals. In April 1961 Grayson Kirk, the president of Columbia University, wrote to PSAC member Lee DuBridge to request a meeting to discuss "how best university scientists can contribute to the success of the nation's efforts in seismology," specifically, ARPA's VELA program.)[33]

Outside the government, a small but influential circle of biologists began raising powerful concerns about the health risks of fallout from atmospheric testing. Barry Commoner, a World War II veteran and Washington University professor whose research included the biological effects of radiation, helped launch the Greater St. Louis Citizens' Committee for Nuclear Information and a widely publicized study of strontium-90 levels in children's baby teeth. He was joined by Caltech Nobelist Linus Pauling, who in the spring of 1957 began circulating a petition calling for an end to nuclear testing. In a matter of days, over two thousand scientists had signed on, largely biologists, chemists, and medical researchers. The FAS endorsed Pauling's efforts, promoting a test ban both as a means to reduce fallout and as a step toward general disarmament. That fall, the *New York Times* printed a full-page advertisement by the newly created National Committee for a Sane Nuclear Policy calling for a test ban. Within a year, the organization's membership had swelled to over twenty-five thousand, and Dulles and Eisenhower felt, in the words of a later internal history, a "growing pressure on the United States Government to make some move toward the cessation of nuclear testing."[34]

In response to the substantial domestic and international clamor for a testing ban, on July 30, 1958, a new compromise proposal came from an unexpected source: Livermore's Edward Teller and the AEC's Willard F. Libby. Teller, fearing that Eisenhower would be pushed to embrace a sweeping, comprehensive ban, called instead for a "limited test moratorium." To debate the details of this plan, in August the State Department convened a new entity, the "Committee of Principals," initially "an ad hoc group of high ranking officials meeting . . . on an issue too urgent and specialized to be put into the machinery of the National Security Council." Attendees included Allan Dulles of the CIA; John McCone and Gen. Alfred Starbird of the AEC; Killian; and Jerome Wiesner, a PSAC member and professor of electrical engineering at MIT. Though little documentation of this first meeting has survived, the Committee of Principals surely influenced Eisenhower's announcement later that year of a testing moratorium and the commencement of test-ban negotiations with the Soviet Union.[35]

In the absence of comprehensive polling or survey data, it is difficult to gauge scientists' levels of support or opposition to the ban. But scientists' political activity on behalf of the test ban was far more expansive than the public movement against it. Thousands of prominent scientists signed petitions in favor of the test ban, but no organized movement of scientists existed to oppose it. Reporters who covered the debate sought opposing voices and, apart from Teller, found only a handful of scientists willing to explain on record why they hadn't signed. For the most part, these men cited concerns about the proper political role of scientists. For example, in 1957 J. H. Hildebrand, a chemist at the University of California, told *Science* that he opposed entering "the realm of international diplomacy where a scientist possesses no peculiar knowledge or wisdom." As Herbert York reported confidently to Congress, "The majority [of scientists] agrees with me rather than Dr. Teller . . . they are in favor of proceeding with the test ban."[36]

The United States conducted its last aboveground test before the moratorium on October 30, 1958, at Yucca Flats in Nevada, and negotiations began the following day at the Geneva Conference on the Discontinuance of Nuclear Weapons Tests. The Committee of Principals met throughout that autumn and more than twenty-five times in 1959 and 1960. During the same period, Eisenhower attempted to end the infighting between the air force and the other military branches by calling for the development of a "Single Integrated Operational Plan" (SIOP) to govern nuclear weapons policy. The resulting plan reflected the priorities of LeMay's SAC, to the consternation of army and navy leaders. Eisenhower quickly dispatched his science advisor, George Kistiakowsky, to evaluate the plan. Kistiakowsky sided squarely with the SAC's critics, opposing both the specifics of the plan and the general principles of massive retaliation. The SIOP would "lead to unnecessary and undesirable overkill," he explained. It was not necessary "to kill 4 and 5 times over somebody who is already dead." Unlike similar criticisms by

Killian set forth in 1954, in 1960 Kistiakowsky's cautionary words carried weight with Eisenhower.[37]

Eisenhower was convinced, but 1960 was too late for a major policy change. And on May 1, 1960, an American U-2 surveillance plane was shot down over the Soviet Union, and relations between the two superpowers quickly deteriorated. Test ban negotiations collapsed. The breakdown coincided with the presidential transition in the United States. The incoming Kennedy administration, elected in part on a dubious "missile gap" platform, would inherit the SIOP and its accompanying factional and strategic tensions.

∗ ∗ ∗

In his parting speech, Eisenhower famously warned of the deep structural changes wrought by the onset of the Cold War, particularly "the acquisition of unwarranted influence, whether sought or unsought, by the military industrial complex." He had a special message for scientists as well. He worried that the massive new funding streams in place could overshadow the natural curiosity of scientists, steering research toward militarized ends. And despite his admiration for his own science advisors, he imagined a future in which "public policy could itself become the captive of a scientific technological elite." The outgoing president had few specific prescriptions for how to avoid these outcomes, and his own policies had increased their likelihood. As PSAC member Herbert York later reflected, "Eisenhower believed both in the necessity of having a military-industrial complex and in the problems and dangers it brought with it."[38]

With this ambivalent legacy, and with the majority of his science advisors staying on through the next administration, Eisenhower left office with the New Look still in place and with prospects for a permanent test ban dim at best.

TWO

The Moral Case for a Test Ban

BY THE END of the Eisenhower administration, the United States—and, publicly, the Soviet Union—had professed support for a testing ban and for radical disarmament measures. As a 1960 State Department press release explained, the "ultimate goal" of negotiations was "general and complete disarmament under effective international control," including the "cessation of production" of nuclear weapons and "their complete elimination from national arsenals," except for any deemed appropriate for "an international peace force."[1] But such language did not reflect the actual priorities of U.S. policymakers; it clearly served other Cold War purposes. As with much high-level political decision making, in reviewing the Kennedy test-ban negotiation process it can be difficult to determine which exchanges were examples of honest disagreements and which were conducted in the interests of political or propagandistic advantage. Speechwriters carefully crafted the public rhetoric of the Cold War, which, on both sides, tended toward bombast, hostility, and sweeping idealism.

Inside the Kennedy administration, ideological and strategic differences shaped policy debates, as did partisan concerns and interdepartmental rivalries. The following account draws only slightly on the speeches of Kennedy and Khrushchev; rather, much of the evidence is culled from internal memos, reports, and the minutes of what appear to be fairly candid meetings of the Committee of Principals assembled to work on problems of testing and arms control. Some of the technical discussions relevant to the test ban remain classified, particularly on topics such as tactical nuclear weapons, the neutron bomb, and aspects of Soviet surveillance. Nevertheless, enough documentation exists to provide a partial glimpse into the Kennedy admin-istration's approach to arms-control negotiations, the contributions of sci-entists, and how both moral and technical concerns shaped the character of the debate.[2]

* * *

Newly installed in office, President Kennedy quickly established arms con-trol as a key goal of his administration. He oversaw the elevation of the U.S. Disarmament Administration from its position within the Department of State to full-fledged agency status in the form of the U.S. Arms Control and Disarmament Agency (ACDA), headed by Director William C. Foster. To ensure that the necessary ACDA legislation was passed, Kennedy was aided by the Federation of American Scientists (FAS), who held a "series of briefing breakfasts" for key members of Congress and their staffers on the subject of disarmament. Kennedy also continued to pursue a test ban, the Eisenhower-initiated moratorium, and the practice of convening the Committee of Principals charged with addressing concerns about testing.[3]

Perhaps most importantly, Kennedy's selection of Jerome Wiesner as his special assistant for science and technology ensured that the topic would receive substantial attention. Born in Detroit in 1915, Wiesner received his PhD in electrical engineering from the University of Michigan and worked on acoustics research, including an unlikely stint traveling through the rural United States with folklorist Alan Lomax, collecting music samples for the Library of Congress. During World War II, he joined the radar research team at the Massachusetts Institute of Technology's Radiation Laboratory (MIT Rad Lab) and served a short term at Los Alamos before returning to MIT. By 1961 his research credentials primed him for a powerful parallel career as an arms-control advocate, whose selection as Kennedy's science advisor won the unalloyed admiration of the FAS, who noted with pride Wiesner's involvement in the international Pugwash conferences on disarmament.[4]

During the presidential campaign of 1960, Wiesner had become a blunt policy confidante of Kennedy, providing the candidate with detailed back-ground on major science- and technology-related concerns, including the case for arms control. This included a frank discussion of nuclear strategy and an analysis of the competing views within the government and mili-tary. As Wiesner summarized it, the major debate was whether to maintain "only an adequate deterrent capability" or to pursue "a massive first-strike capability." The issue "has divided the Pentagon for fifteen years," he wrote, "with the Army and the Navy on one side and the Air Force on the other; an issue so basic that it essentially controls almost every other military decision." The army and navy wanted to ensure that the nation could retaliate after a Soviet attack—that is, they wanted to ensure some measure of nuclear survivability—while the air force, led by LeMay, preferred to focus on the U.S. ability to cripple the Soviet Union in a single strike.[5]

Wiesner patiently explained the tenets of the air force approach to Kennedy, including its self-serving reliance on a large bomber force. His skepticism was obvious. Although he acknowledged that "our forces were awesome and the posture taken by LeMay and Dulles sufficiently belligerent to make the Soviet leaders very cautious," he did not believe that sufficient intelligence and bombing capabilities existed to execute the air force vision of both knowing and destroying all Soviet weapons locations at once. Moreover, the Strategic Air Command (SAC)'s influence in military planning had led to a monopolization of resources that starved other worthy endeavors.[6]

This misallocation was especially problematic in light of the necessary transition away from air force bombers carrying nuclear weapons to "a major dependence upon missiles," beginning with the Atlas ICBM (intercontinental ballistic missile). In explaining this shift to Kennedy, however, Wiesner emphasized the difficulties inherent in the new programs, particularly the challenge of estimating Soviet capabilities and resources. The Polaris and Minuteman ICBM programs would be improvements over the SAC bomber system, but not without their own dangers and deficiencies.[7]

Wiesner further cautioned Kennedy against the use of tactical nuclear weapons. When the United States had a nuclear monopoly, he wrote, tactical nuclear weapons had been a means to compensate for the Soviet Union's "numerical advantages" in conventional forces and to "deter any Soviet ground movements." But this deterrence grew suspect as Soviet nuclear stockpiles increased and "more and more people, especially in Europe, began to doubt that we would carry out our retaliatory blow against the Soviet cities if the Soviet armies moved into Western Europe." Deterrence based on tactical nuclear weapons yielded a particularly high-risk kind of stability.[8]

In addition to these technical and logistical arguments, Wiesner urged Kennedy to reject the air force nuclear doctrine for a single, overarching reason: to de-escalate the arms race. The emphasis on massive attacks unnecessarily antagonized the Soviet Union. "In spite of what we said," wrote Wiesner, "it must have been impossible for most of the Soviet military men to accept our claims that SAC was primarily retaliatory in intent. At a minimum, the Soviet military leaders were provided with a convincing argument for building up their forces." A smaller deterrent force of missiles and bases could help reduce "political controversy and embarrassment." Wiesner's stance was clear. "For the next few years it is probably wise to maintain some air defense," he wrote, but "the new administration should examine critically the wisdom of maintaining the present level of air defense effort indefinitely."[9]

Kennedy soon tapped Wiesner as his science advisor and chose another nuclear non-hawk, forty-four-year-old Ford Motor Company president Robert McNamara, as his secretary of defense. Though McNamara had worked with Curtis LeMay in analyzing air force efficiency during World War II, he had not followed in his footsteps in support of developing massive first-strike capabilities. McNamara did not call for total disarmament or an end to research on nuclear weapons, but he was deeply skeptical of their actual use. (He later reported that he had advised both Kennedy and Johnson never to use nuclear weapons.) Almost immediately upon appointing him, Kennedy tasked McNamara with a thorough review of Eisenhower's proposed military programs and defense budgets for the current and subsequent fiscal years. The result was an early articulation of McNamara's approach to national security: a dramatic turn away from the New Look and its priorities in favor of deterrence and the flexibility to respond to small-scale foreign conflicts through nonnuclear means.[10]

To begin, McNamara laid out his central security objectives. The United States had obligations to protect its allies and the "free world" in general, but how did these translate into strategy and policy? McNamara explicitly rejected the SAC doctrine of massive first-strike capability: it was not necessary to maintain a "pre-attack" massive stockpile of nuclear weapons as the SAC claimed; rather, the nation needed only "survivable retaliatory power," which would be sufficient to ensure deterrence. To this end, McNamara recommended improved protection of military forces and weapons through expanded air defense, civil defense, accelerated work on the Polaris program and the hardened bases necessary for Minuteman missiles, and improvements in the "highly vulnerable" systems of command and control. (McNamara's promise to develop "systems with greater endurance and flexibility under conditions of thermonuclear attack" would fit well with the creation of the decentralized ARPANET, the precursor to the modern internet.)[11]

But McNamara went further than just endorsing this alternate view of deterrence. He criticized the excessive dependence on nuclear weapons more generally, complaining that focusing on missiles at the expense of other military options limited U.S. decision making. "We have been forced into a single strategy for retaliation," McNamara explained. "At the present we have little ability to make decisions in the event of an attack." Current war planning relied heavily on a nuclear response, which was limited and dangerous in its own right, but the Pentagon suffered as well from an "over-emphasis on general war" in the first place. What about smaller, localized conflicts—what McNamara referred to as "limited war"? Eisenhower may have hoped that the New Look would deter U.S. involvement in limited wars by dint of its nuclear emphasis, but McNamara saw only the danger—and likelihood—of nuclear overkill. "Our forces designed to fight overseas, those we would call on to fight in limited conflict, are, in fact, strongly oriented in their war plans, current capabilities, material procurement, and research and development, towards general nuclear war," he warned. "This is at the expense of their ability to wage limited and especially non-nuclear war." With an eye toward potential interventions in Southeast Asia, Latin America, the Middle East, and Africa, McNamara called for an expanded "ability to deal with guerrilla forces, insurrections, subversion . . . sub-limited war capabilities."[12]

McNamara's budget statement was a blueprint for the two key components of his strategic legacy: the emphasis on choice and options, which came to be known as "flexible response," and the commitment to developing the capability for military intervention with varying and increasing degrees of nonnuclear force, or "graduated escalation." Both would soon guide American involvement in Vietnam. But for arms-control scientists in 1961, Mc-Namara's declared interest in a host of postcolonial disputes was less important than the underlying nuclear message. McNamara wanted to "raise the threshold of our local non-nuclear defense capability, and reduce our dependence on nuclear war." He not only adamantly rejected the New Look but offered financial resources to back his words. His budget review noted pointedly that "we are doing too little research and development on nonnuclear weapons" and called for a "substantial increase" for nonnuclear armaments. Arms-control scientists, elevated to unprecedented political influence after Sputnik, suddenly had an ally at the very top of the Pentagon.[13]

✳ ✳ ✳

Less than two months after the inauguration, the Committee of Principals, held over from the Eisenhower years, was hard at work reviewing definitions of terms and inspections requirements for various test-ban proposals. The new secretary of state, Dean Rusk, had expressed qualified support for arms control, including a test ban, and early meetings featured numerous debates on the technical aspects of an enforceable ban. At issue was the problem of underground testing. While atmospheric tests were easier to detect and therefore monitor, underground tests were trickier and in some circumstances difficult to distinguish from earthquakes or other seismic activity. During its meetings in early March 1961, the Committee of Principals reviewed the state of detection technology and the requirements for an underground ban. They discussed the number of likely annual earthquakes in Soviet territories and how many on-site inspections would be necessary to ensure that no clandestine testing was taking place. On-site inspections were a particular sticking point with Soviet negotiators.[14]

Even in these technical discussions, political objectives, rather than scientific evidence, shaped the debate. Glenn Seaborg of the Atomic Energy Commission (AEC) wanted the number of inspections linked to the number of earthquakes with no upper limit, a standard unlikely to meet Soviet approval; Wiesner offered a far more flexible view. He observed that 'one-for-one' inspections were not necessary, since any inspections would be a powerful deterrent against cheating, and, realistically, 'one clandestine nuclear test would not be significant.' Neither Wiesner nor Seaborg disputed the seismological data or the state of U.S. detection techniques, but their interpretations and recommendations plainly revealed their differing political priorities.[15]

Two influential reports from this period further reveal how technical information was marshaled in arguments both for and against a test ban. An "Ad Hoc Panel on the Technological Capabilities and Implications of the Geneva System," led by Bell Labs physicist James B. Fisk, had been convened in the winter of 1961 to provide technical expertise to John McCloy, Kennedy's disarmament advisor. The panel was almost entirely composed of scientists, with a handful of military representatives, and included the physicists Hans Bethe, Harold Brown, Richard Latter, J. Carson Mark, Wolfgang Panofsky, and Herbert York; the scientists Frank Press and Herbert Scoville; and Gen. Alfred Starbird, Gen. Austin Betts, and retired army general Herbert Loper. Although much of the panel's March 1961 report remains classified, a redacted and "sanitized" version released in 2008 reveals lengthy explanations of U.S. detection capabilities and limitations and analyses of several predicted outcomes for test-ban scenarios.[16]

The Fisk panel wrote that the problem of the test ban was not one "where positions should be controlled by the technical issues." Political and military considerations remained paramount. Nevertheless, the most crucial technical aspect of the test ban was "verifying violations." On this topic, the report noted that although the United States possessed a long-range detection system with "acoustic, seismic, electromagnetic, and radioactivity components . . . deployed around the USSR and its satellites," it could not definitively distinguish underground nuclear explosions from natural seismic activity. "There is no known way to identify an underground nuclear explosion by its seismic signals alone," the panelists observed. To resolve this problem, the panel urged restored and expanded funding for the VELA detection program.[17]

The report also addressed other technical concerns: the reliability of the U.S. stockpile (for which the panelists did not find testing to be urgent), the ability of nonnuclear countries to develop nuclear weapons (the panels doubted a ban would prevent the development of "a simple, heavy" fission bomb), the construction of an antiballistic missile (ABM) system (even with testing, the panel deemed the prospects for ABM development "not encouraging"), and U.S. advances in new classes of weaponry, including lightweight warheads and the "enhanced radiation" neutron bomb. On this last point, the panel observed that progress "depends in various degrees on testing." But the larger question was one of security: what would be the overall consequences for U.S. security if there were no test ban, a comprehensive and enforced test ban, or a test ban under which the Soviets engaged in clandestine testing?[18]

Answering these questions required an evaluation of the country's strategic posture. The panel determined that for massive first- and second-strike capabilities, the particular refinement of warheads via testing mattered little; quantity, not quality, was what mattered. For deterrence, with its emphasis on survivability and mobility, the development of lighter warheads might be more important. In either case, the panelists observed, existing resources were vast. The warheads currently available in both systems could "completely over-kill the population and over-destroy the floor space of urban area targets by blast and fire." The panel thus predicted that with no test ban, additional testing would improve first-strike survivability for both sides.

With an enforced test ban, both sides could still potentially "maintain a very strong deterrent strategy" simply on the basis of weapons that could be stockpiled without further testing. Even with an unenforced test ban with maximal Soviet cheating, U.S. deterrence capabilities might be "degraded" but still effective.[19]

The report was not an uncritical push for a comprehensive test ban, but its structure—opening with an acknowledgment of the limitations of detection and enforcement but closing with a relatively mild assessment of the worst-case scenario of Soviet cheating—provided valuable fodder for the administration's test-ban advocates. John McCloy, Kennedy's chief disarmament advisor, summarized the panel's work in a memo to the president: "This report, in my judgment, from a technical standpoint buttresses the conclusion that it is in the overall interest of the national security of the United States to make a renewed and vigorous attempt to negotiate a test ban agreement." McCloy referred to "the consensus of scientific thinking and analysis contained in this report," thus preempting any claims that technical aspects were in dispute. The risks of clandestine Soviet testing could be ameliorated with further VELA research, he noted, but in any case, they were risks that were "on balance, worth taking." He closed with a final appeal for a test ban that went beyond fears of fallout or hopes to freeze a U.S. stockpile advantage. A test ban "would be a significant step in the field of arms control," he wrote. Like the FAS scientists, McCloy viewed a test ban as a stepping-stone to disarmament.[20]

Naturally, the report had its detractors as well. One member of the Fisk panel, the retired army general Herbert Loper, attached a dissenting view to the final report. Many of his objections remain classified, but he closed with an argument reminiscent of the hydrogen bomb debates. "Any action on the part of the United States which denies its scientific and engineering community the opportunity to apply its maximum capabilities to its nation's defenses cannot result in a military advantage to the country," he wrote pointedly.[21]

More influentially, the Fisk panel's implications were explicitly refuted in a similarly themed RAND report issued a week later, one of whose authors, Richard Latter, had been a member of the Fisk panel. From its first sentence, the RAND report set out to challenge the Fisk view:

There is near-universal agreement among scientists that certain nuclear tests cannot be detected. Despite this, there is strong pressure for a test ban based on the supposition that nuclear weapons technology has reached that point of diminishing returns where no new discoveries can upset the balance of military power. In a word, that it does not matter whether or not we continue to test or whether or not the Soviets cheat on a test ban agreement.

This RAND special report identifies five new aspects of the nuclear problem that say, in sum: it does matter that we test. The report has been reviewed by Dr. Edward Teller and John S. Foster of Livermore Radiation Laboratory who verify its technical authenticity.[22]

The authors argued that should a test ban be imposed and Soviet testing proceed anyway, the United States would suffer numerous disadvantages. Soviet researchers would likely discover unforeseen vulnerabilities in U.S. forces. The United States would miss out on key ABM defense advances. If anything, a test ban might lead to a greater arms buildup, since the United States would be forced to diversify and expand its deterrent forces.

Like the Fisk report, the RAND work also emphasized the lack of technical disagreement among scientists, thus characterizing the nature of the debate as strategic and political. The authors offered a single recommendation: "adhere to a principle of adequately controlled tests." Since adequate control did not yet exist for undergrounding testing, any test ban should be partial, applying only to "atmospheric tests, some space tests, and underground tests above a threshold." Thus, even with its implicit criticism of the Fisk report, the RAND scientists also endorsed a test ban, only modified with an underground threshold high enough to ensure that cheating could be effectively monitored.[23]

The difference between the technical aspects and the political implications of the comprehensive test ban merit further clarification. Whereas the Fisk committee members and the RAND experts agreed that remote monitoring of low-yield underground nuclear tests was, with current detection techniques, difficult if not impossible, they differed on the political, moral, and strategic implications of this fact. How threatening was clandestine Soviet testing? Did the risks outweigh the potential benefits for arms control and peace? While these questions were debated at the highest levels of government, many arms-control scientists took up the study of detection techniques as a means to contribute resources and expertise in the service of arms control. If detection techniques could be improved, then the political scales might tip toward support for a comprehensive test ban. As it turned out, even when detection technology improved, the moral and political arguments on both sides remained unchanged.

✳ ✳ ✳

The greatest opposition to a test ban came not from the scientists of the President's Science Advisory Committee (PSAC) or RAND, but from nonscientists at the Pentagon and in the arms services who were anxious to resume testing. In the spring of 1961, National Security Council staffer Robert Komer complained that the Soviets had "impaled" the United States on "the hook of a self-imposed test ban." He worried that the Soviets were planning to test in secret.[24] Meanwhile, the Joint Chiefs were preparing their own attack on arms control, noting in a May report that "no practicable arms control agreement and inspection systems can be envisaged that would eliminate the danger of surprise attack altogether." In other words, arms control would make the country less secure.[25]

Through the spring of 1961, Kennedy and his top advisors met to sort through the conflicting recommendations and determine the U.S. negotiating strategy. The problem of inspections loomed large. McCloy reported on Soviet unwillingness to allow on-site investigations, while Herbert Scoville of the Central Intelligence Agency (CIA), generally a strong advocate of arms-control measures, objected to the prospect of Soviet inspectors arriving in Nevada to examine test areas.[26]

In the meantime, Komer, the Joint Chiefs, and Harold Brown, the new director of defense research and engineering at the Pentagon, were all refining their arguments in support of a resumption of testing, on the grounds that new advances could be made with applications to limited nuclear war and an ABM system. In a memo to the president, McNamara endorsed these views, noting that testing could contribute to "the whole spectrum of defensive and offensive weapons systems." Kennedy, however, worried that resumption could trigger a backlash in public opinion.[27]

At least in part, the resurgent call for new tests resulted from the failures of negotiation. At the meeting of the Committee of Principals on May 23, 1961, Ambassador Arthur Dean reported that prospects for a test-ban treaty seemed 'pretty dim.' Without Khrushchev's acceptance of on-site inspections, even Wiesner worried about the consequences of an unenforced ban. Throughout the spring and summer of 1961, then, discussions drifted away from hopes for a test ban to debates over if, when, and how the United States should resume nuclear testing. Wiesner hoped that at the very least, new tests would not produce fallout. Rusk and McNamara seemed to agree. Marc Raskin, an advisor to McGeorge Bundy, argued that testing would hurt the

United States in the eyes of the world, would "quicken the pace of the arms race, not slow it down" and, should it lead to expanded production of tactical nuclear weapons, raise the risk that limited U.S. interventions abroad would lead to nuclear war. The economist John Kenneth Galbraith wrote to Kennedy opposing a resumption of testing.[28]

Even the nation's top nuclear scientists were not anxious to resume testing. Norris Bradbury, head of Los Alamos, confirmed to Wiesner that renewed testing was not imperative; resumption would not lead to "radical change in national strength" or "new dimensions in warfare." Plenty of bomb-related work could be done in a laboratory setting, and crucial delivery systems could be tested regardless of a nuclear moratorium. But Bradbury's relaxed attitude was not simply a matter of minimal scientific need. He, like Wiesner, was committed to arms control and the important step that a test ban represented. He wrote, poignantly: "The current test ban negotiations, although disappointing, represent the first real attempt to alter the course of history with respect to the nuclear arms race. It will be a grave disappointment to have to admit failure of even this poor attempt . . . So serious does the eventual world nuclear weapon situation seem to be that, without clear evidence that the Russians were testing, I would personally prefer that the United States not be the first to resume this activity."[29]

Bradbury attached a statement from his Los Alamos colleague Carson Mark, who viewed the weapons situation with a similar lack of urgency. In his view the country already possessed a vastly destructive stockpile of effective weapons. "No advances by testing can alter the fact that with systems available both to us and to potential opponents each can inflict physically and psychically insupportable damage on the other," he wrote. "Improved designs may make it easier and cheaper (in some sense) but not more fearful; while less advanced designs may make it more costly and cumbersome but not less certain."[30]

In late July, PSAC's Ad Hoc Panel on Nuclear Testing, led by Stanford physicist Wolfgang Panofsky, issued an influential report on the resumption of testing that largely echoed the conclusions of the earlier Fisk study. Overlapping members of the panel included Bethe, Panofsky, Press, and Fisk himself, joined by Los Alamos's Norris Bradbury, Livermore's John Foster, and Harvard's George Kistiakowsky. The new document opened by emphasizing that technical concerns were not the most important factors at hand: "the final decision on whether or not to resume testing also involves very important non-technical or military issues." Nevertheless, from a technical perspective, resumed testing would likely speed the development of lighter war-heads, "enhanced neutron radiation weapons," ICBMs, and an ABM system. If only atmospheric testing were banned, but undergrounds tests allowed to proceed, this research would not be "seriously impaired," only "more dif-ficult and costly." From a larger strategic standpoint, however, a complete, observed test ban would probably freeze the U.S. advantage in place, while unlimited testing by both sides would likely allow the United States and the Soviet Union to "approach the same level of warhead technology." Like the Fisk report, the Panofsky panel examined possible outcomes of cheating and observance and noted that in all cases, whether counterforce or deterrence was the goal, the United States already possessed "over-kill" capabilities.[31]

At the top levels of government, the report was received in predictable ways. Arms-control advocates touted it as a strong case against resumption. Undersecretary of State George Ball agreed, urging Kennedy to hold off on testing in the hopes of successfully negotiating a test-ban treaty. He sum-marized the Panofsky report for the president as a confirmation that testing decisions "can be governed primarily by non-technical considerations." AEC head Glenn Seaborg declared himself "in general agreement" with the Pan-ofsky report, though he added his own request for expedited preparations in the event of testing resumption and expanded yield limits for laboratory testing in the meantime.[32]

On the other side, high-level Pentagon and military officials registered their dissatisfaction. Livermore's John S. Foster, a member of the panel, wrote separately to Wiesner to explain that while the Panofsky group had disagreed about "the question of urgency," they still had concluded that the United States should eventually resume testing. Harold Brown, director of defense research and engineering, complained to Wiesner that the Panofsky report had underestimated the risks of Soviet cheating and ignored the potential of testing to contribute to the development of an ABM system. He hoped that underground testing would resume "as soon as it is politically expedient." McNamara, suspecting that the Soviets might already be testing clandes-tinely, reiterated his support for a test ban but recommended that testing preparations commence nonetheless.[33]

Maxwell Taylor agreed, arguing that testing was critical for the develop-ment of lightweight warheads and tactical nuclear weapons. Taylor, a deco-rated veteran and former army chief of staff during the Eisenhower years, had publicly repudiated the rigidity of the New Look in his 1960 book, *The Uncertain Trumpet*. Though Kennedy would later appoint him as chairman of the Joint Chiefs of Staff, in 1961 he was serving as a military advisor to the president. He now wrote to Kennedy, "Unless the most compelling of political arguments can be adduced against it," testing should resume.[34]

The Joint Chiefs offered the harshest assessment of the Panofsky report. Rather than interpreting or tweaking its findings to support resumption, they directly challenged the findings themselves. They complained of inaccuracies, "unconfirmed intelligence estimates," and conclusions that were "conjectural and subject to gross error." The Chiefs pointedly disagreed with the implication that "there is little urgency connected with US resumption of testing," asserting instead that new tests were both urgent and necessary. Moreover, they staunchly opposed a continued ban on atmospheric testing, citing a report from the chief of the Pentagon's Defense Atomic Support Agency arguing that "world-wide fallout from past tests has not produced a biologic hazard."[35]

The criticism of the Joint Chiefs disturbed Kennedy. Were technical and intelligence estimates actually in dispute? The president wrote to Taylor in confusion: "The Joint Chiefs took a very strong position against the Panofsky report on testing. I wonder who prepared their analysis. Was it done by one, or two, or three men? Was it done outside of the Defense Department by a group of scientists, or what?" He seemed surprised that the Joint Chiefs could take the position they had when "the Chairman of the AEC seems to find himself 'in general agreement in the findings and conclusions of the report.'"[36]

The following day, Kennedy raised the issue at a meeting of the National Security Council, during which Panofsky personally presented his panel's report. McNamara, Allan Dulles of the CIA, and Wiesner were all in attendance, as were the heads of the two nuclear labs, John Foster of Livermore and Norris Bradbury of Los Alamos. The ensuing discussion, at turns bitter and fractious, revealed deep differences in both understanding and interpretation of the issues at hand. As Panofsky pled his case, Gen. Lyman Lemnitzer, chairman of the Joint Chiefs, argued with CIA head Allan Dulles over the quality of U.S. intelligence regarding the Soviet stockpile. Kennedy eventually ended the bickering by requesting that Taylor, Lemnitzer, Dulles, and Panofsky meet separately to 'define the disagreements and narrow them if sensible.' In the meantime, the president seemed swayed by the need for eventual resumption of testing. He worried about the negative political effects of new tests given the upcoming meeting of the General Assembly of the United Nations, but he promised to reach a decision soon.[37]

That decision came less than two weeks later, in favor of resumption. But before any official U.S. statements could be made, the Soviet Union rendered the political anxiety moot by announcing their own resumption of nuclear tests. Avoiding any mention of the parallel decision making at home, White House speechwriters demonized the Soviet move in the harshest language possible. "The Soviet government's decision to resume nuclear weapons testing is in utter disregard of the desire of mankind for a decrease in the arms race," one press release read. Robert Komer, grateful that the United States was still several weeks away from test readiness, now urged Bundy to play up the delay, making the most of "looking peaceful while they look warlike." Edward R. Murrow, then serving at the U.S. Information Agency, recommended holding off on testing and using the time to advance U.S. propaganda: "This can be done not only by the exposure of Soviet duplicity, but also by playing heavily upon the fears of hazards to health and future generations." Moreover, he observed, "This time, if properly employed, can be used to isolate the Communist Bloc, frighten the satellites and the uncommitted, pretty well destroy the Ban the Bomb movement in Britain, and might even induce sanity into the SANE [Committee for a Sane Nuclear Policy] nuclear policy group in the country." Murrow's advice revealed with painful clarity that while the words of Wiesner, Bradbury, and other insider arms- control advocates might be respected by the administration, the activism of men like Barry Commoner and Linus Pauling was not.[38]

* * *

As U.S. testing preparations began in earnest in the autumn of 1961, the next set of high-level debates addressed the nature and extent of the tests. Wiesner, dogged in his arms-control efforts, hoped at least to prevent atmospheric tests. Late in September, he warned Bundy that there were no "critical requirements for nuclear tests in the atmosphere" and the minimal advantages ought to be weighed against "the political problems from fallout." Wiesner was not present at the subsequent meeting of the Committee of Principals, but his viewpoint was expressed ably by Bundy and Rusk. Although both McNamara and Seaborg professed strong support for atmospheric testing, Bundy and Rusk warned of the widespread international and domestic opposition to testing in general, and to fallout in particular.[39]

Rusk also inserted the prospects of a test ban back into the discussion, reiterating that despite the Soviet actions, a testing agreement was still in the national interest, and 'If the USSR says it is now ready to sign a test ban treaty, we would presumably agree.' Rusk suggested that the United States offer the test-ban treaty again to the Soviet Union, and if they refused, only then commence testing. Over the next few weeks, consensus settled around Rusk's proposal; appearing on "Meet the Press" at the end of the month, Seaborg stated publicly that should the Soviet Union accept the terms of the U.S. comprehensive test-ban proposal, the United States would forego atmospheric and underground testing. But with continued Soviet opposition to on-site inspections, no agreement would be reached.[40]

Throughout the winter and early spring of 1962, Wiesner swam against the tide of sentiment favoring a resumption of atmospheric testing. As details of the Soviets' massive series of tests emerged, some members of the Committee of Principals grew skittish about even pursuing a test ban at all. Whereas Rusk held fast that a treaty was still in the national interest, William C. Foster worried that Soviet achievements had shifted the advantage away from the United States. Wiesner, however, was staunch in his position. In November 1961 he told the committee 'that the key issue was not whether the United States was equivalent to the Soviet Union in every aspect of nuclear weapon technology but whether the United States is missing any of the things it should have for its security.' Soviet gains and possible imbalances were not the only factors relevant to testing decisions, he pleaded.[41]

Wiesner's prestige and influence were challenged by the reemergence of Livermore's Edward Teller, who had played little role in the Kennedy-era test-ban discussions to that point, other than to review the earlier RAND report on the subject. Now, at the president's request, he provided detailed advice on a proposed course of atmospheric tests. Teller paid lip service to the cause of arms control, noting that he "had hoped" that the new tests would not be necessary but "in this hope I have been wrong." Without atmospheric resumption, he warned, "a dangerous situation will arise in the mid-1960s," when the Soviet Union succeeded in applying their newfound expertise into "a hard-hitting first strike force." Many of Teller's specific assessments remain classified, but his support for testing was clearly linked to his desire for improvements in lighter, cheaper warheads and an ABM system. If anything, he endorsed a more expansive series of tests, with additional testing in "deep space."[42]

Teller also delivered a decidedly nontechnical message, an appeal for Kennedy to restore the reputation of nuclear weapons researchers. "The men working on the development of nuclear explosives," Teller wrote, have been "subjected to considerable strain" because "public opinion continued to frown upon [their] activities." This had created a manpower problem, in which it had become "increasingly hard to induce excellent young people" to pursue weapons research. To ameliorate the situation, Teller invited Kennedy to visit the two weapons laboratories and perhaps "make a public statement directed to the scientists of our Country," explaining that "the development of nuclear explosives can be used to provide us with the strength that insures peace."[43]

Teller's rhetoric—invoking the dire specter of Soviet superiority and describing nuclear power as "the strength that insures peace"—could not have been further from that of Wiesner. In December 1961, Wiesner once again urged Kennedy to reject atmospheric testing. There was no technical or military need, he wrote, and even if some military advantages were to be gained, the tests were still "not critical or even very important to our overall posture." Without testing, the nation could still "maintain an extremely effective deterrent." (It was deterrence, not counterforce, that concerned Wiesner.) His final plea reiterated that the question of atmospheric testing was a matter of politics, not technology. Wiesner wrote encouragingly to the president, "You have the flexibility to make whatever decision on this matter best supports your broader foreign policy and national security objectives."[44]

Wiesner's allies repeated this argument. Raskin implored Bundy not to allow atmospheric testing on the grounds that "The world today is searching for some kind of moral and political leadership . . . For the first time in many years the United States can reclaim such moral and political leadership." The historian and presidential advisor Arthur Schlesinger Jr. described the problem in similar terms. "Technical evidence will not yield a clearcut answer" to the question of atmospheric testing, he wrote to Kennedy. "The decision, in short, is back in the political field." He cited a Gallup poll showing that popular opinion was evenly divided on the issue.[45]

Among the president's top advisors, however, opinion was overwhelmingly in favor of atmospheric testing. As Bundy acknowledged in a memo to the president, he, Rusk, Seaborg, McCone, Brown, and Lyndon Johnson all supported atmospheric testing. But Bundy was candid: the final decision belonged to the president, not his advisors. "I believe that if you personally care enough, and want to make the argument strongly enough, you can carry a decision against atmospheric testing with the Congress and the country," he wrote in a memo. Despite his own view that, "on balance," the military advantages were real and the political risks and advantages "even," Bundy acknowledged to the president that an atmospheric ban was a plausible, "safe" option. If Kennedy chose to pursue it, his advisors would support him. But Bundy also cautioned the president that the decision should be "yours alone—not yours with support from politically vulnerable disbelievers like Wiesner." Though Wiesner's advice was sought and treated with respect at high-level meetings, Bundy clearly considered him a potential political liability.[46]

Indeed, Wiesner was growing more agitated in his position by the day. As support coalesced around the resumption of atmospheric testing, he urged Kennedy even more vigorously to pursue not just an atmospheric ban, but a comprehensive test ban. In January 1962, the air force's Twining Commission delivered an influential report favoring a massively expanded testing program to Curtis LeMay, and Bundy forwarded a summary to the president, alerting him that it was "probably of high political importance." That same week, Wiesner urged Kennedy to consider "two more comprehensive proposals which I believe would have greater political appeal than an atmospheric test ban." To achieve the comprehensive ban, the United States should drop its requirements for on-site inspections down to the three that the Soviets had proposed. The ban could then be used as the "first stage" of an even more sweeping disarmament proposal. If the Soviets agreed, wrote Wiesner, "We will have made a great gain for world peace."[47]

Less than a month later, Wiesner offered the president an even more radical proposal, a test ban and disarmament plan with three components: "1) a complete ban on nuclear weapons tests in all environments; 2) the cessation of all research and development on nuclear weapons; and 3) a complete cut-off of the production of fissionable material except for agreed quantities to be used for peaceful purposes." Rather than quibbling over "elaborate" inspections requirements in the service of more moderate measures, the drastic terms of the complete testing and research ban would, obviously, "justify a higher level control." Each side would have twenty annual inspection opportunities, and the ban on weapons research would be enforced "by placing permanent inspectors in all weapons laboratories and by maintaining a check on the activities of all scientific personnel previously engaged in weapons work." Wiesner's reaction to the prospect of resumed atmospheric testing was to support a sweeping ban on all nuclear weapons research.[48]

At the very least, Wiesner's proposals expanded the spectrum of possibilities open to the president, shifting any compromise position closer to the side of arms control. And, should Kennedy consider an atmospheric test series as a final, intensive research period before the imposition of a ban, Wiesner had provided support for the strongest arms-control outcome yet. Kennedy, in fact, had evinced interest for just such an outcome. In late February 1962, even as the decision was being made to resume atmospheric tests, Kennedy met with Wiesner and top representatives from ACDA. As described in the meeting minutes, the president "indicated the great importance he attaches to being prepared to offer a test ban treaty immediately," along with "across-the-board cuts in armaments." Wiesner's dogged insistence had kept a comprehensive ban in the picture even in the face of unified opposition among Kennedy's top cabinet members and the impending onset of atmospheric testing.[49]

* * *

Outside of the halls of government, of course, far greater support for a comprehensive ban and opposition to atmospheric testing existed, particularly among elite scientists. In mid-February, 147 Cornell staffers urged Kennedy via telegram not to resume tests. Less than two weeks later, 72 Cornell faculty members, led by mathematician Jacob Wolfowitz, sent an alternate message: a general expression of "confidence" in the president to make a wise decision and faith that should Kennedy opt to resume testing, it would be because it was necessary, not because he was swayed by reckless or irresponsible forces on either side of the debate. (Thus, even the opposition message did not explicitly support new tests.) The FAS reiterated their call for an atmospheric ban, noting that resumed testing implied "that our security can in the long run be maintained solely by military strength" rather than by working politically for peace. In addition to this political argument, the FAS warned of the danger of global fallout, which, though it would affect "only a very small fraction of the world's population," could nonetheless harm those who "have no voice in the decision to test." Since both sides already possessed the technical power to destroy each other, the FAS argued, "the social and political repercussions [of testing decisions] are quite as important as, and perhaps even more important than, the technical and military factors."[50]

When Kennedy did finally announce the new series of tests, which included atmospheric detonations, his national speech emphasized reluctance and thoroughness. "Every alternative was examined," he explained. "No single decision of this Administration has been more thoroughly or more thoughtfully weighed." Kennedy recounted the Soviet violations of the voluntary agreements of 1958. He assured his audience that top advisors, including "the most competent scientists in the country," had reviewed testing policy. With the "unanimous recommendations of the pertinent department and agency heads," Kennedy announced that he had authorized the AEC and the Defense Department to conduct a new series of tests, beginning in April, both underground and in the atmosphere over the Pacific.[51]

Kennedy discussed at length the problem of radioactive fallout and his hopes for an eventual test ban. Anticipating an international outcry at the testing resumption, Kennedy reminded potential critics "that this country long refrained from testing, and sought to ban all tests, while the Soviets were secretly preparing new explosions." He emphasized efforts to minimize the inevitable but small amounts of additional radiation anticipated, regretting that he was forced to "balance" the hazards of radiation against "the hazards to hundreds of millions of lives" at stake in the arms race. The recent Soviet tests—including tests of smaller and more explosive weapons—required a U.S. response. In this context, Kennedy observed, "If we are to maintain our scientific momentum and leadership—then our weapons progress must be not limited to theory or to the confines of laboratories and caves." The United States had tried, but "the basic lesson of some three years and 353 negotiating sessions in Geneva" was that the Soviet Union had refused an enforced ban, preferring an "uninspected moratorium" under which they could test in secret. Blame for the failure of arms-control efforts lay with the Soviet leadership.[52]

* * *

Although the *New York Times* editorial page offered strong support for the president's decision, in other quarters the prospect of resumption met with condemnation. Linus Pauling responded with a scathing telegram to Kennedy, highlighting the fallout risks to unborn children and referring to atmospheric testing as a "monstrous immorality" that could render the president as "one of the most immoral men of all times and one of the greatest enemies of the human race." The *Los Angeles Times* interviewed four West Coast scientists for an article titled "Sky Tests Win Support of Scientists," but two of the subjects couched their support in bittersweet terms. "I was sad about it but there was no other course," observed the Manhattan Project pioneer and Nobel laureate Harold Urey. California Institute of Technology geneticist E. B. Lewis offered similar sentiments: "It's very discouraging, very depressing."[53]

The FAS took a more muted position than that of Pauling, officially opposing atmospheric testing but expressing appreciation for Kennedy's overall commitment to disarmament. The following month, however, a soul-searching essay on "The Future of FAS" demanded a reassessment of the organization's goals and tactics, given the dramatic shifts in nuclear politics since its World War II-era founding. Michael Amrine, a former editor of the *Bulletin of the Atomic Scientists,* observed that FAS scientists no longer held a monopoly on nuclear expertise, nor was their political influence as powerful as it may once have been. He noted, "When our Council voted to ask the President not to resume nuclear tests, we did so quite openly, knowing that he could and probably did have information—as well as councils of scientific discussion—not open to us. Just what is it that we, the FAS brains, knew that the President might not? What special virtues do our councils possess?"[54]

One new approach came from Leo Szilard, the nuclear pioneer turned Pugwash member who had met with Khrushchev personally in the fall of 1960 to urge him toward arms control. Through the winter of 1962, he crisscrossed the country on behalf of what would become the Council for a Livable World, recruiting over two thousand volunteers to commit 2 percent of their income to elect pro-peace candidates. It was not enough for him—or other elite scientists—to attempt to persuade politicians personally, he explained to his university audiences, for "these distinguished scholars and scientists would be heard" but not necessarily "listened to." Votes were what mattered. In pursuing his radical dream of disarmament and the abolition of war, Szilard urged that most traditional of American activities: electioneering.[55]

✳ ✳ ✳

While Szilard, the FAS, and other activist groups criticized the "Dominic" series of atmospheric tests that were underway through the spring and summer of 1962, inside the administration, Wiesner kept up the pressure for strong disarmament measures and for a comprehensive test ban, now urging that the threshold for underground explosions be dropped entirely. In a detailed memo to the president, Wiesner fleshed out his reasoning in both the required "technical-military" terms and in his own moral and political language. In a prescient analysis, Wiesner noted that the military supporters of the threshold saw it as a loophole, fueling "hope for future tests" that could be conducted within the limits of the treaty.[56]

Technically, Wiesner argued, the Soviets were correct to complain that a treaty with a threshold was not a true ban, since low-yield tests—anything less than the equivalent of a 4.75 magnitude earthquake—could continue underground according to current proposals. In Wiesner's estimation, removing the threshold would make Soviet cheating *more* difficult, not less so, since any seismic event would then be open to investigation. In the meantime, with no major technical advances on the horizon, "the threshold proposal is fundamentally a scientifically indefensible position." Rather than get bogged down in exactly what could or could not be detected, the United States was better off dropping the threshold entirely.[57]

There was, in fact, a breakthrough of sorts on the horizon. Beginning in 1958, the air force scientists who ran the detection system for nuclear explosions had begun to suspect that their estimates of naturally occurring Soviet seismic activity might by incorrect. By 1961 VELA researchers working with more accurate instrumentation systems expressed similar doubts, and further investigation confirmed that the United States had drastically overestimated the likely number of earthquakes that could potentially be confused with underground nuclear tests. (The revised number, in fact, largely agreed with the estimates the Soviets had offered in 1958.) At the same time, detection technologies themselves had improved, allowing longer-range monitoring from fewer observation stations.[58]

The technical revisions did not substantially shift the existing support and opposition to a comprehensive, no-threshold ban, however. At the Principals' meeting in July 1962, ACDA proposed just such a ban, with 'no right to test at all' and a monitoring network of international and national resources. Due to the new seismic information, the required number of detection stations and on-site inspections could be reduced to numbers more amenable to the Soviets. Should that plan fail, a ban on atmospheric testing alone could be a backup. While Wiesner endorsed the first option, representatives from the AEC preferred the second. Seaborg told his colleagues that 'AEC's chief concern was the effect of stopping underground testing on the vitality of our laboratories,' and Leland Haworth, the AEC commissioner, added that 'an atmospheric ban, with continued underground testing, would permit us to maintain a posture of readiness.' Suddenly the sticking point was no longer the problem of Soviet cheating, but the effect of a test ban on U.S. nuclear weapons development.[59]

Other Principals quickly rallied around the atmospheric ban. Rusk, noting that Soviet negotiators might not accept even a reduced number of on-site inspections, preferred to start with an offer of an atmospheric treaty and then try "making it comprehensive as soon as possible." In terms of world opinion, atmospheric tests were all that mattered. Murrow agreed. But Wiesner persisted in his call for more sweeping measures. 'Why not try to get a total ban,' he asked his colleagues. Why start with the limited, narrowest option? In a memo written that same day, Bundy summarized the new data and its possible consequences for the test ban to the president. The ultimate decision rested solely with Kennedy, and Bundy reminded him of the political stakes. The test ban was critical for "preventing proliferation," but reducing treaty demands—even if supported by technical advances—might be perceived as caving to Soviet wishes, and could "hand the opposition a juicy issue for November." Bundy warned Kennedy that "many of your advisors will argue . . . that this is no time for concessions of any sort," but he credited Wiesner's tireless efforts to keep the full range of treaty options open. He counseled the president: "What you can do—and here, by a good deal of in-fighting, Jerry [Wiesner] and I have fully preserved your freedom of choice—is to decide."[60]

✳ ✳ ✳

The political pressure regarding the appearance of concession was perhaps more powerful than Bundy anticipated. Even William C. Foster of ACDA sided with Rusk in suggesting that an atmospheric ban be presented first, so as not to appear to be "scaling down" the "safeguards . . . necessary for national security." In Congress, Rep. Chet Holifield of the Joint Committee on Atomic Energy warned the president not to make any hasty negotiating decisions based on the new data. California Republican Craig Hosmer expressed skepticism about the "supposed advances in the science of seismology" and called the media coverage of the Advanced Research Projects Agency's Project VELA nothing more than "propaganda to try to drum up support for the test ban treaty highlights." Edward R. Murrow "wouldn't recognize a seismograph if he saw one," Hosmer complained, and VELA didn't justify "accommodation" to the Soviet position unless one read the evidence after "putting on rose colored glasses and chewing tranquilizers." More broadly, the test ban itself was an empty "symbol of peace," whose value had "become artificially inflated by the hypnotic effects of constant misleading propaganda." Amid stiff Congressional opposition and the constant reminders from the AEC that underground testing would "keep our laboratories alive and vigorous," by the end of July, Kennedy had been convinced that political will was allied against any kind of treaty. Better chances at Senate approval for any agreement would have to wait until the fall.[61]

Unfortunately for Kennedy, the autumn of 1962 was no time for a massive public campaign on behalf of a test ban. Though September and early October saw continued discussions of various disarmament measures and additional monitoring techniques (including the Pugwash-recommended system of "black boxes"—automatic devices to detect and record seismic activity within a host country), the mid-October discovery that the Soviet Union was constructing ballistic missile sites in Cuba swept attention elsewhere. As the Cuban Missile Crisis unfolded, Kennedy's top advisors proposed diplomatic and military responses that could take the United States to the brink of nuclear war, and the American public faced the terrifying nuclear stakes of the Cold War with fear and uncertainty.[62]

Historians have debated the effect of the Cuban Missile Crisis on Kennedy, particularly whether the president emerged in the aftermath with a deeper awareness of the full risks of nuclear war and therefore more firmly committed to an arms-control agenda. At the very least, the history of 1961–1962 shows that Kennedy's starting point was already a position relatively hospitable to arms-control advocates, as evidenced in the test-ban treaty meetings that were already underway before the missile crisis began. And the infamous "Thirteen Days" of the crisis only temporarily diverted personnel and resources away from test-ban treaty planning. Lower-level deputies stood in for their absent bosses at the November 1 meeting of the Committee of Principals, but top advisors returned ten days later, pessimistic but present and ready for discussion. William C. Foster of ACDA observed mournfully that the missile crisis had revealed the perils of the arms race and the rationale for seeking 'a more stable world.' Rusk reported that prospects for disarmament negotiations with the Soviets looked "gloomy" but noted that the president still hoped to make progress on a test ban.[63]

By February 1963, Kennedy and the Principals were back to work vigorously debating threshold levels and numbers of on-site inspections (the Soviets wanted no more than three; McNamara was prepared to offer six). Maxwell Taylor, now the chairman of the Joint Chiefs, had joined the Committee of Principals and represented a strong new voice opposed to a comprehensive test ban. To bolster the Joint Chiefs' case, the 1962 Twining report prepared for Curtis LeMay had now been updated to take into account the recent atmospheric tests by both the United States and the Soviet Union. Signed by Livermore's John Foster, Simon Ramo of Thompson Ramo-Wooldridge, Los Alamos mathematician Stanislaw Ulam, Princeton physicist John Wheeler, UCLA chemist William McMillan, and Edward Teller, the report asserted that "a test ban would involve greater risks to the national security than perhaps have been realized." The administration's claims that a ban would protect the U.S. nuclear advantage and minimize the risks of a surprise attack were, "from a scientific and military viewpoint . . . not valid."[64]

In a separate appendix, William McMillan, who in addition to his academic work served as a RAND advisor and chairman of the Defense Research and Engineering Ad Hoc Weapons Effects Group, laid out a detailed list of potential tests, ordered by priority, that would be eliminated with a test ban. These included tests of reentry vehicles, communications systems, and the survivability of hardened missile sites, as well as tests relevant to ABM system capabilities, implying that gains in all these areas would be lost with testing restrictions. Another appendix quoted a report from the Air Force Scientific Advisory Board on potential new advances in tactical nuclear weapons. It was not true, the authors argued, that no new technological leaps were on the horizon. "If we do not keep the scientific leadership," they warned ominously, "others will take it." Distaste for the administration's position was clear: "If the United States is to renounce a revolutionary device which others can then secure without our knowledge, no portion of this responsibility should attach to the Air Force."[65]

In a dramatic closing section, the report reiterated Teller's concerns about the effects of the test ban on morale and recruitment at the weapons laboratories. The authors then repeated the old hydrogen bomb arguments in stark terms: "The technology of nuclear energy has its own laws and its own internal structure. It can be stopped by the hand of man as little as the advance of weaving machinery could be stopped by the Luddite mobs. But we can be stopped by our own actions. Out of a developing nuclear technology, we see emerging better defense and peaceful uses of nuclear energy. But the most important peacetime use of nuclear energy is the preservation of the peace." The implications were clear: scientific knowledge and technical development were inevitable. Voluntary avoidance of research was naïve, misguided, and dangerous; it would not prevent the development of weapons by others, and it would endanger U.S. nuclear superiority and there-fore, national security.[66]

* * *

Resistance in Congress was also still strong, particularly after a closed ses-sion in which representatives from Los Alamos and Livermore described what kinds of clandestine testing might take place and what potential "sig-nificant advances" the Soviets might achieve. Meanwhile, Wiesner tried to combat the dire warnings of the updated Twining report by reassuring the president that the additional tests proposed by the AEC were "unimportant." By the time of the April 1963 meeting of the Committee of Principals, ten-sions were high, but a fragile foundation of support for a comprehensive treaty seemed to exist. Maxwell Taylor reported that the Joint Chiefs still consid-ered a treaty with no threshold to be "unsatisfactory," but meeting minutes confirmed a consensus "that the text was adequate and that a test ban treaty was still in the national interest of the United States." William C. Foster re-ported that the only unresolved issues were conditions for peaceful nuclear research under the Plowshares program. On this front, Seaborg hoped to allow for additional testing, while Wiesner warned that Plowshares tests, no matter how innocently described, 'almost certainly would contribute to weapons development.'[67]

Outside of government, arms-control advocates rallied to the cause. In May 1963 a prominent contingent of scientists began circulating a public statement supporting a test ban. They were led by David Inglis of the Ar-gonne National Laboratory, who had first called for a test ban in the pages of the *Bulletin of the Atomic Scientists* in 1954. Of the nine major signa-tories, four—Freeman Dyson, Donald Glaser, Hans Bethe, and Francis Low—were members or advisors of the Institute for Defense Analyses' Jason program, created during the Sputnik boom. (Other Jasons added their names as cosigners, along with Harvard's Salvador Luria, Matthew Meselson, and a dozen others.) The scientists emphasized the advances in detection techniques from 1958 to 1963 and reiterated that both sides al-ready possessed "over-kill capabilities" that would maintain deterrence even with the most stringent testing ban. A test ban, in their view, would be a powerful tool in reducing the risks of the arms race and nuclear war. *New York Times* coverage quoted the statement's moral message—that the ban stood in "the best interests of the United States and world peace"—and cited by name the three Nobel Prize–winning signers: Donald Glaser, James Watson, and Albert Szent Gyorgyi. Within weeks, a wide range of intellectuals and cultural icons had signed on to similar and expanded statements of support, as prominent scientists reached out to the leading lights of arts and literature, including Leonard Bernstein, John Steinbeck, Aaron Copland, John Huston, and Elia Kazan.[68]

The primary obstacles to a successful treaty, of course, were not Plowshares tests or a lack of high-profile support, but Soviet recalcitrance and Senate opposition. The comprehensive treaty proposal sent by the Kennedy administration after the April Principals' meeting had elicited a frosty Soviet response. A new round of negotiations with Khrushchev was announced in early June. With the Principals' tentative consensus on a comprehensive ban and the stirrings of a movement of scientists and intellectuals prepared to lobby Congress and public opinion, the prospects for success seemed promising. The *New York Times* called the planned talks "a glimmer of hope." But three days later, the newspaper reported that Khrushchev had once again rejected U.S. inspection terms, indicating that he might even withdraw entirely the previous Soviet acceptance of three on-site inspections. At the Committee of Principals' meeting on June 14, Rusk quickly laid the main issues on the table: he had little hopes for the current talks, given Khrushchev's claim "that inspection is a form of espionage" and U.S. unwillingness to sign a treaty without any on-site monitoring whatsoever. But, Rusk noted, the president's commitment to arms control ran deep. He relayed a message from Kennedy, that "this may be our last chance to avoid a larger and more difficult arms race . . . In 10 or 20 years it will be important that the US made as great an effort as possible to achieve a test ban."[69]

Perhaps Rusk hoped that these stirring words would rally the Principals; instead, the meeting rapidly degenerated into an angry, fractious argument over U.S. security, the worth of a test ban, and the Joint Chiefs' threat to testify against the treaty in Congress. As before, Maxwell Taylor complained that the current treaty draft would permit Soviet cheating, but now he implied that the military leadership was prepared to state their opposition openly in Senate hearings. Rusk seemed shocked that the Joint Chiefs would dare to oppose the official administration line, noting pointedly that he "would not feel free to take a foreign policy position that disagreed from that of the President." The room erupted in argument. Wiesner asserted that the matter was political, not technical, and that whatever their theories about Soviet cheating, "the laboratory directors are not in a position to judge the overall policy considerations." Rusk noted bitterly that he "didn't think Edward Teller was talking as a technical man when he talked about the test ban." Put on the defensive, Taylor retorted that the Chiefs 'were conscientious men who were sincerely concerned about our national security,' and invited ACDA representatives to convince the military leadership that their concerns were unfounded. When pressed, he conceded that the Joint Chiefs might be able to support an atmospheric-only test ban. As McNamara weakly called for more studies and meetings, Rusk, incredulous, replied that "members of the Committee had all agreed, he thought, that the risks to national security from an unlimited arms race outweighed the risks inherent in a test ban treaty." But the Joint Chiefs had not, it seemed, and with their political threat to undercut the proposed ban in Congress, the prospects for a comprehensive treaty looked as dim from the American side as they did from that of the Soviets'.[70]

* * *

The stage was set, then, for the success of the compromise partial test-ban treaty, which forbade testing in all environments—in the atmosphere, outer space, and underwater—except underground. The treaty satisfied the moral arguments about the dangers of fallout and offered a tentative step in the direction of arms control, with hopes for future advances. More importantly, it was acceptable to Khrushchev, and prospects for ratification seemed promising in the U.S. Senate.

For the most part, arms-control scientists supported the treaty. Biologist and antinuclear activist Barry Commoner later called it a triumph for environmentalism. The FAS sent physicist and Jason member Freeman Dyson to testify before the Foreign Relations Committee in support of the ban, and after ratification, the FAS hailed the treaty as "a first and significant step to slow the pace of the arms race and reduce the danger of nuclear war." Kennedy's PSAC offered a strong statement of support in mid-August, timed to encourage Senate passage. Although the scientists acknowledged that the treaty raised "many important questions other than those of a technical nature," they focused on the technical reasonableness of its terms. It could be adequately enforced, and it would not prevent the development of defensive techniques for hardening missile sites or exploring an ABM system. In reality, with the elimination of the underground component of the treaty, little technical debate existed within the government leadership, but the patient explanations of scientists seemed a useful tool for rallying public opinion in support of the ban. In a press conference held at Los Alamos, lab director Norris Bradbury told a pool of reporters that the treaty would cause little or no change in the staffing and research routines of the lab. Bradbury tried to appeal to both sides of the debate. The test ban would not halt nuclear research or, realistically, the development of new types of weapons, he noted reassuringly. And fundamentally, the importance of any bomb research—and the ethical justification for the existence of Los Alamos—was deterrence. "Los Alamos," he reflected, "has no fondness for atomic weapons per se; people in the Laboratory don't work on atomic bombs because they like to kill people, think of them killing people. They have only worked in this over the last 20 years because we thought in some way this provided a strength for the country to avoid war, to bring about, ultimately, as there seems today to be a start, a step toward the abolition of war."[71]

The words of Bradbury and the PSAC were carefully planned, as was the administration's public-relations campaign in support of the treaty, which drew heavily on the efforts of prestigious scientists. In a note to Ted Sorensen from Kennedy staffer Fred Dutton, Dutton summarized the work of pro-test-ban efforts by business and agricultural groups, churches, the American Federation of Labor-Congress of Industrial Organizations, and other "voluntary constituent organizations." Of the latter category, Dutton wrote that he wished to "contain" groups like "the Friends, SANE and others who have had a major interest in a test ban" because "I personally do not think that we pick up any support, but only suspicion, if they lobby the Hill or get out in front publicly on the treaty." Instead, Dutton hoped to elevate the actions of elite scientists, such as the Nobel Prize winners who had already registered their support, and a small group of "particularly effective" men, including Kistiakowsky, Killian, Rabi, and York, who were already planning a series of private meetings with key Senators. Another contingent of life scientists had been tapped to "dramatize the fall-out problem." Though many of these men maintained moral and ethical commitments to arms control not entirely unlike the convictions of SANE members and other activists, the administration preferred their ability to emphasize technical and scientific language when offering their political endorsement of the treaty. In his history of the early years of the PSAC, Zuoyue Wang has described this kind of political maneuvering in another way. The PSAC, in his analysis, acted according to "their recognition of the necessity to view scientific and technological solutions within a social and political context," which "underlined their insistence on examining not only the means, but also the ends of technological programs of the government." In the case of the test ban, the PSAC, led by Wiesner, was committed fully to both the technical means and the moral and political ends.[72]

* * *

Historians have vigorously debated the roles of Eisenhower, Kennedy, and scientist-activists in the passage of the Partial Test Ban.[73] Recent scholarship has credited the importance of the Sputnik-inspired expansion of science advising with exposing top government leaders to powerful arms-control arguments and diluting the influence of Edward Teller. The new coterie of PSAC advisors entering government in 1957 presented a near-unified front in favor of the test ban and arms control more generally, revealing Teller as a mouthpiece, albeit an influential one, for only a tiny scientific minority. Under the influence of esteemed PSAC members such as Bethe and Rabi, Eisenhower agreed to a testing moratorium in 1958 and entered negotiations for a comprehensive test ban, which, despite periodic setbacks, would culminate in the success of the partial ban during the Kennedy administration. This view is corroborated by Herbert York, the former Livermore director and Advanced Research Projects Agency chief scientist, who later wrote that the approach to arms control "underwent a sea change at the White House level" in the years after Sputnik; a change that included his own growing political commitment to securing a comprehensive test ban.[74]

But could science advisors and activists have achieved their goal of a comprehensive test ban had they taken stronger arms-control positions or chosen different tactics? One recent assessment has suggested that Kennedy's advisors were paralyzed by McCarthyism and therefore relied on narrow technical arguments instead of the kinds of moral claims about fallout made by "outsider" science activists like Linus Pauling. But PSAC members, many of whom were also members of the FAS, had long warned of the dangers of fallout as well. Fallout was a risk relevant only to atmospheric testing, however, and not underground testing, where the technical problems of monitoring were at issue. (Atmospheric tests were easy to detect; underground tests were not.) A partial test ban, which would eliminate atmospheric testing but not underground testing, sufficiently addressed the moral argument concerning fallout risks so heavily promoted by Pauling but also exposed the limits of such arguments for the promotion of a comprehensive ban.[75]

The more sweeping moral case against underground testing was one rooted in eventual disarmament, a far more radical position than the antifallout position. Moreover, to make a purely moral case, that is, to ignore the technical concerns over whether the Soviet Union could effectively cheat the system, meant committing to a form of disarmament that was potentially unilateral. Although some PSAC scientists could offer only qualified support for such a position, Wiesner and York both argued on multiple occasions that moral and political concerns outweighed the minor technical hurdles to a comprehensive ban. In 1963 York told a Senate committee that "this dilemma of steadily increasing military power and steadily decreasing national security has no technical solution." The only answer was peaceful political arms-control negotiation.[76]

More realistically, implicating science advisors in Eisenhower's failure to secure a test ban (and Kennedy's ultimate achievement of only a partial ban) likely overstates the role of both scientists and domestic politics in the outcome. In 1960 neither a moral case against fallout nor a disarmament rationale for a comprehensive ban could repair the damage of the U-2 incident. A broader, international view of the Cold War suggests the importance of the U-2 incident in ending test-ban hopes at the conclusion of the Eisenhower administration (since Khrushchev angrily withdrew from negotiations in response) and the changing postures of the Soviet Union—rooted in political developments in Germany, China, Romania, and Cuba, among other places—for the eventual passage of a limited ban in 1963. Historian Vojtech Mastny writes that for all the parties involved, "higher priorities" of Cold War politics in 1958–1962 left test-ban negotiations "an exercise in exasperating futility" accompanied by "similarly sluggish nonproliferation talks." Not until most of these other priorities were resolved could any kind of test-ban treaty succeed. During the Kennedy administration, U.S. negotiators proposed both a comprehensive test ban and a partial test ban to the Soviet leadership, and it was only the latter that proved acceptable to Khrushchev.[77]

The participation of scientists in the treaty process nevertheless deserves attention, particularly as an indication of the political and ethical strategies of prominent scientists during this critical Cold War moment. The scientists' movement in support of the test ban included the work of outsiders like Pauling, international organizations like Pugwash, and the efforts of insiders such as Kistiakowsky, Wiesner, and York. All of these scientists operated according to ethical impulses that pushed them toward reducing the risk of nuclear war, and all of them, even Pauling, assumed that appropriate tactics included trying to influence the policymakers in Washington through traditional means, whether via the circulation of petitions and public statements, providing analyses of technical problems, or personal advice and persuasion.

∗ ∗ ∗

In the end the treaty passed the Senate by a vote of 80–19, though Senators added stipulations ensuring that weapons production would continue despite the test ban. Nuclear stockpiles increased during the Kennedy administration, such that by 1964 there was "an increase of 150 percent in the number of nuclear weapons available, a 200 percent boost in deliverable megatonnage, the construction of ten additional Polaris submarines (for a total of 29) and of 400 additional Minuteman missiles (for a total of 800) above what the previous administration had scheduled." Though the Committee of Principals continued to meet up until the week before Kennedy's assassination, no new progress would be made on a comprehensive test ban for another three decades.[78]

Outside the halls of government, research on nuclear weapons and an ABM system would proceed apace at the nation's weapons labs, but, as Teller had predicted, by the mid-1960s the morale of lab staffers and the prestige of their work had indeed begun to decline. A 1965 *Washington Post* article about weapons scientists observed that "few young scientists and engineers regard nuclear weapons work as the cutting edge of science as it was when Fermi and Oppenheimer, Rabi and Teller were at Los Alamos." Los Alamos in the 1960s, the authors noted, was "suffering from the same doubts that plague a middle-aged man who wonders whether he is as secure as he thought." Worried about national security, their jobs, and their declining social status, many nuclear scientists threw their political support behind the arms buildup and against restrictions on testing or other arms-control measures. This shift in thinking about nuclear weapons research—from its 1945 image as simultaneously glamorous and destructive work demanding personal atonement to its 1960s incarnation as regular employment requiring political protection—offers a glimpse at some of the ways in which ideas about weapons work changed as the Cold War progressed.[79]

Although some arms-control activists were dismayed by the compromised outcome of their efforts, the consequences of the test ban for the Manhattan Project generation of physicists were largely positive. Elevated to high government service by their expertise and their desire to contribute to the expansion of American science, PSAC members during the Eisenhower and Kennedy administrations had seized the opportunity to promote arms-control ends, with notable success. In 1963 their public prestige and influence were as great as ever, as evidenced in the prominent role Kistiakowsky and others played in securing Senate passage of the test ban and in the weight afforded by the popular press to any scientist with a "Nobel laureate" appellation and an opinion on weapons research. In the late 1970s, Killian would look back on the 1950s and recall that "those were memorable and exciting times when government, industry, and the universities felt themselves in a symbiotic relationship and achieved a powerful creative collaboration." Though many PSAC advisors of this period welcomed Eisenhower's famous words about the dangers of a military-industrial complex and a "scientific technological elite," events of the late 1950s and early 1960s suggested ways in which prestigious scientists could work within the system toward salutary ends.[80]

Wiesner embodied this prestige and access. In the most intimate, high-level meetings of the Committee of Principals, he consistently offered an impassioned, uncompromising voice in favor of the most radical arms-control positions of anyone in the room. He had the president's ear and was respected, listened to, and consulted. Years later, one of Wiesner's colleagues recalled that "the Test Ban for Jerry was a great achievement, and a great failure (because it was only partial)." When Wiesner's successor, Donald Hornig, was announced shortly after the successful passage of the test ban, Wiesner prepared to leave government with every reason for pride in his contributions to government service.[81]

Fittingly, Wiesner exited his advisory role, where he had pushed successfully for expanded federal funding for academic science, to resume his academic and administrative work at MIT. Even before Sputnik, MIT had been a major recipient of defense research contracts, but the post-Sputnik boom saw the dramatic expansion of research money and military-related work. The 'military-industrial-academic' complex, a step beyond the expansive system about which Eisenhower had warned in 1960, was in full swing in Cambridge, Massachusetts. When Wiesner returned to his academic home in 1963, this work was still considered prestigious and worthwhile. Less than ten years later, the campus would erupt in protests over the very defense contracts and government affiliation Wiesner had welcomed as a member of the PSAC. The shift in attitudes toward the relations among government, military, academia, and industry was dramatic, but not surprising; between 1963 and 1973 lay the tragedy, devastation, and polarization of the Vietnam War.

THREE

The Science of Nonnuclear War

SHORTLY BEFORE THE 1960 ELECTION, Jerome Wiesner wrote a lengthy memo to John F. Kennedy discussing key military, technical, and political problems facing the country. Although Wiesner wrote mainly about nuclear weapons and the arms race, he also warned the future president about the prospect of "limited war"—nonnuclear military conflicts that might include "jungle fighting in the far east," and could require "specialized weapons for the different areas where we might have to fight and specialized for the military situation involved." Wiesner's memo reflected much of the work and many of the concerns of Eisenhower's President's Science Advisory Committee (PSAC). Earlier that year, Harvard chemist George Kistiakowsky had sent Wiesner a PSAC budget review containing similar warnings. "We believe there are serious deficiencies in our limited war capabilities," wrote Kistiakowsky. He urged the renewed development of smaller conventional weapons to correct what he perceived as a dangerous imbalance favoring the nuclear arsenal. He blamed military leaders for developing "the big weapons systems required for general war to the neglect of the specialized weapons and systems needed to deal with limited war." Kistiakowsky wanted improved surveillance and reconnaissance and expanded research on nonnuclear technology, including chemical and biological weapons (CBW). On the latter topic, Kistiakowsky criticized the navy and air force for failing even to request funds for "delivery and dissemination of these agents." In these discussions, Kistiakowsky summarized the conclusions of two key PSAC panels—the eight-member Panel on Biological and Chemical Warfare, which in 1959 had recommended expanded CBW research programs, and H. P. Robertson's Limited War Panel, which criticized both the lack of conventional weapons research and the "organizational structure of the military establishment" responsible.[1]

Kennedy's selection of Robert McNamara as secretary of defense thus pleased science advisors who supported arms control and for whom "flexible response" was a welcome alternative to Eisenhower's defense posture based on the threat of massive nuclear strikes. Within the first two months of Kennedy's presidency, the Pentagon was echoing Wiesner's warnings. A Defense Department report released in January 1961 noted that "Of all the areas around the periphery, U.S. ability to conduct a limited war in mainland Southeast Asia is the most questionable. . . . The United States and its allies presently do not have an adequate capability for counter-guerrilla type limited military operations." As one staff memo put it, "Rather than major limwar a la Korea we must be most prepared to fight on the order of Laos, Vietnam, Congo, or Lebanon." McNamara himself reiterated this point in his February 1961 budget review, shortly after he took office as secretary of defense. He warned Kennedy that "We have too little ability to deal with guerrilla forces, insurrections, [and] subversion" and recommended a "substantial increase" in relevant research and development.[2]

McNamara and the government scientists in the Kennedy administration defined limited war in terms of its nonnuclear character.[3] Seymour Deitchman, a Defense Department and Institute for Defense Analyses (IDA) physicist who wrote a 1964 book on limited war dedicated to Kennedy, located the origins of U.S. limited war policy in the Korean War, which, in his view, exemplified a conflict with limited scope and objectives. Although Korea had initially prompted a brief reactionary turn toward "massive retaliation," Cold War realities and the election of John F. Kennedy had quickly corrected the course of U.S. policy, in his view. Deitchman described the Kennedy administration as representing "the complete renewal of leadership in the areas of foreign and defense policy," lauding especially its "flexibility" in its actions to "resist nonnuclear aggression."[4]

From the moment Kennedy took office, his science advisors and his Pentagon leadership, united in their skepticism of the previous massive nuclear buildup, pushed for renewed emphasis on research and development for conventional weapons appropriate for limited warfare and counterinsurgency. And they found receptive audiences among Kennedy and his top military brass. Faced with cutbacks in other areas, the Joint Chiefs "unanimously favored" the Defense Department's proposed increase of $100 million for "limwar" research and development. They welcomed additional funding packets to the three service branches for work on "weapons for guerrilla warfare" (army); "new assault helicopters" and "improvements to existing ordnance, biological and chemical weapons, and fuzes" (navy); and the "development of sensors for use against small tactical targets under all-weather conditions, low level attack capabilities, [and] improved anti-personnel weapons and improved fuzes" (air force). In March 1961 the army's chief of research and development ordered the rapid intensification and expansion of research related to special warfare, including guerrilla combat. The army added a limited-war laboratory at its Aberdeen Proving Ground in Maryland, staffed largely with civilian engineers, with approval for biological and chemical research, including defoliant systems. The navy similarly set up a "Sub-Limited Warfare Research Project" at its testing facilities in China Lake, California, and the air force's Special Air Warfare Center at the Eglin AFB in Florida led the service's counterinsurgency-related research into new aircraft design, munitions including napalm and white phosphorus, and chemical defoliants.[5]

Kennedy, in turn, continued to encourage the work of his PSAC on limited-war problems. As tensions with the Soviet Union escalated over crises in Cuba, Laos, and Vietnam, Kennedy created a "Special Group" devoted to counterinsurgency (which itself oversaw a special research and development committee) and by 1963 had established a Committee of Principals on Chemical and Biological Weapons. Members of Kennedy's cabinet encouraged this shift through their own enthusiasm for clever new technological possibilities. Deputy National Security Advisor Walt Rostow, for example, cheerfully referred to the antiguerrilla "special gadgets" being developed by the army as "fun and games." Perhaps most importantly, in June 1961, the Defense Department's Advanced Research Projects Agency (ARPA) launched Project AGILE, an R&D program devoted to nonnuclear "remote area conflict" in coordination with defense contractors like Raytheon and Sperry. The Pentagon had quintupled ARPA's budget for research into counterintelligence, surveillance, and psychological warfare. Research topics now included improved communications for use by "friendly indigenous forces," new "tactical helicopter" technology, "incendiary weapons," and lightweight rifles. Under ARPA auspices, some of the Vietnam War's most controversial weaponry, including defoliants and napalm, would be refined and prepared.[6]

* * *

Where did scientists fit in this new world of limited-war preparation and shifting military research priorities? At the top advisory levels, they stood exactly where they had already been. Much of the roster of Eisenhower's original PSAC remained intact; Jerome Wiesner, Herbert York, George Kistiakowsky, James Killian, I. I. Rabi, and many other key committee members had all been serving continuously since 1957. By and large, they were patriotic men anxious to continue their government service, pleased by McNamara's defense budget review, with its lack of emphasis on nuclear weaponry, and sympathetic to Kennedy's anticommunism.

The same was true of scientists at large. In the armed services and at the Pentagon, top brass could reach out to the scientists already employed within the existing network of universities, private industry, and nonprofit organizations created by Eisenhower-era defense contracts. In a letter to Kennedy in April 1961, Wiesner extolled the virtues of these networks, lauding the contributions of government-sponsored university labs at the Massachusetts Institute of Technology, Columbia, and the University of California, which contributed to defense research and development, as well as nonprofit groups like IDA and RAND, which offered operations analysis. Wiesner assured Kennedy that this type of contracting was the best "means of getting highly skilled and critically needed assistance which could not otherwise be obtained."[7]

ARPA took the lead in recruiting for limited-war research. To work on the problem of guerrilla warfare, ARPA created several new fulltime positions for "generalist" scientists and organized a twenty-member advisory team, headed by physicists Lloyd Berkner and Luis Alvarez, to report to ARPA and the PSAC. In August 1961 State Department staffer Robert Johnson described the ease with which ARPA could reach out to the corporate-academic-military networks—now "captive" resources—that had been expanded during the Eisenhower administration. Johnson wrote to Rostow: "Letters have been sent to all of the 'captive corporations' and to the universities that might help . . . in this problem area [guerrilla warfare]. As specific research problems then arise, an appropriate institution will be asked to make available a particular scientist to work on it. The scientist would be sent to the field to study the problem and then to come back and work out a solution." Hypothetical planning for guerrilla warfare lacked the intense intellectual excitement that had drawn scientists to the Manhattan

Project, but Johnson hoped that the participation of distinguished PSAC members would help "to convince physicists and others that these problems were as important as the problem of developing nuclear weapons on which they had worked during the war."[8]

* * *

Within the next five years, the networks described by Wiesner and Johnson would be tapped in pursuit of strategies and technologies related not just to limited war in general, but to specific military needs in Southeast Asia. The expansion of science advising during the Eisenhower administration and the appeal of Secretary of Defense McNamara's reduced emphasis on nuclear weapons thus created a cadre of loyal, committed scientists ensconced at all levels in the Kennedy administration—just in time for the rapid expansion of U.S. involvement in Vietnam.

During the Kennedy and Johnson years, presidential science advisors, Pentagon consultants, and military researchers would be called upon to contribute to the development of weapons technology and strategic planning necessary for counterinsurgency operations in Southeast Asia. Scientists pioneered and then evaluated the efficacy of defoliant operations, assessed the merits of bombing campaigns, proposed new applications for sensors and communications technology in order to design "electronic barriers," and offered countless other forms of input and criticism. Scientists both within and without government reacted to the expansion of the war with a range of responses, from disillusionment and protest to fervent support and voluntary assistance. Within the ranks of government and military advisors, some scientists found their consciences challenged by the work they were asked to perform. Others felt a responsibility to mitigate a bad situation by using technology to prevent escalation. Yet a third group, proud of their contributions to the war effort, stepped into the public spotlight in order to counter what they perceived as uninformed outside criticism. This outside criticism came not just from student protesters but from academic scientists as well, many of whom conducted their own independent evaluations of the effects of new weapons technologies, particularly the controversial uses of defoliants and tear gases, which they considered forms of chemical warfare.

As a few contemporary observers pointed out, scientists' contributions to the war in Vietnam consisted largely of "applying engineering skills to produce weapons and equipment from items which were already available," including napalm, defoliants, and sensor technologies. There was no mobilization on the scale of the Manhattan Project or the Office of Scientific Research and Development.[9] Some top scientists also participated in high-level decisions concerning bombing escalation and anti-infiltration techniques. This small group perhaps earned a share of the stigma accompanying explicit weapons and war work, although they later suffered criticism disproportionate to their actual contributions. But the stigma of Vietnam extended far beyond just this handful. The deep anger and opposition to the war subjected every new wartime technology, and in most cases, recycled wartime technology, to scrutiny and outrage. Some of this criticism came from scientists themselves, who occasionally succeeded in shaping and reforming policy. At the same time, antiwar sentiment fueled a broader attack linking American science to the military-industrial complex. Critics condemned scientists' funding ties to the Pentagon, their advisory roles (even if their advice was often far less hawkish than actual enacted policy), and, in a deeply critical analysis, their perceived complicity in nearly every dehumanizing aspect of U.S. foreign policy, consumer culture, and capitalism itself. That the key architects of the war—from McNamara to Westmoreland—regularly invoked the language of science and experimentation only magnified the opposition.

The war in Vietnam drew on the entire range of scientists who had offered their services in the aftermath of Sputnik: PSAC members, lab personnel, industrial researchers, and academics working as part-time consultants with IDA, the Jasons, RAND, ARPA, and the military services themselves. For many of these scientists, what began as support for non-nuclear weapons alternatives and peaceful space research soon meant implication in a war that was long, bloody, and, by the 1970s, massively unpopular.

∗ ∗ ∗

Research and development went hand-in-hand with U.S. military needs. Vietnam was hardly a foreign policy priority during the early Cuba-oriented years of the Kennedy administration, but the conflict between the anticommunist government of Ngo Dinh Diem in South Vietnam and the nationalist insurgency of the National Liberation Front troubled Kennedy as it had Eisenhower. The continued presence of U.S. advisors and encouragement from the U.S.-backed Diem regime positioned the region as a key site for weapons experimentation. As early as April 1961, Rostow encouraged Kennedy to send "a military hardware research and development team to Vietnam." A month later, the Department of Defense (DOD) prepared a "Concept of Action," offering contingency planning for the inevitable moment Kennedy decided "conditions in Vietnam are critical." The document provided administrative suggestions for the formation of a Vietnam task force, stated general U.S. policy ("pacification," "stabilization," and "unification") and recommended that the United States loosen its strict interpretation of the Geneva agreements. The authors also pushed expanded weapons-related research, urging leaders to "concentrate U.S. military research and development to develop better military equipment for use in resolving insurgency problems in Vietnam. The area should be treated as a laboratory and proving ground, as far as this is politically feasible."[10]

Implementation of this final recommendation took shape almost immediately. Plans for testing the army's Mohawk aircraft in the "actual combat environment" of South Vietnam were put in place in the spring of 1961, while a "Limited War" R&D "Task Group" organized a visit to South Vietnam in July. By the spring of 1963, the Joint Chiefs had ordered the creation of a Joint Operations Evaluation Group "to test tactical concepts and doctrine" on the ground in South Vietnam.[11]

Most notably, ARPA's Project AGILE included the establishment of ARPA-run "Combat Development and Test Centers" (CDTC) in South Vietnam and Thailand, tasked with taking on both the technological as well as the ideological challenges of counterinsurgency. By the fall of 1961, the Saigon CDTC, housed initially in the Joint Command headquarters, employed ten full-time civilian and military personnel, as well as fifteen other professional and clerical staffers. As one Pentagon report explained, the centers would "provide a mechanism through which the special talents of the U.S. scientific laboratories and industry may be brought into physical contact with the problems of South East Asia on a continuing basis." The center therefore hosted extended ARPA-funded visits from civilian scientists and other technical experts on loan from industry, to conduct research on the problems of guerrilla warfare and "infiltration in remote areas." Projects included night-vision technology, lie detectors, and the testing of new lightweight weaponry, including the Armalite AR-15 rifle, a model deemed more appropriate for the "smaller stature and body configuration" of South Vietnamese soldiers. With the goal of preventing North Vietnamese soldiers' passage into

South Vietnam, the lab workers developed chemical markers, scents detectable to dogs, and other "labeling agents" designed to identify interlopers, as well as acoustic and magnetic sensors triggered by clandestine movement. ARPA also arranged, with encouragement from the Diem regime, preliminary defoliant research for the purpose of clearing the jungle canopy that shrouded routes into South Vietnam. A military status report from 1961 predicted the CDTC would become "one of our most important agencies for determining and field testing materiel and doctrine peculiar to anti-guerrilla warfare." (Eventually, test-center control would largely be turned over to the South Vietnamese military. By 1965, U.S. officials were describing the lab as South Vietnam's "counterpart agency" to ARPA.)[12]

In 1961 Pentagon visitors to the lab acknowledged that "no sure-fire, absolute, and very few 'secret weapons' are on the immediate horizon," but praised the lab for its imaginative research nonetheless. Indeed, much of the new equipment actually generated by this expanded funding push was rather mundane. A 1963 military listing of completed and distributed counterinsurgency-related items included a "small fire bomb" but also new lightweight poplin uniforms, hammocks, ponchos, jungle boots, tropical hats, nets, audiovisual equipment, Styrofoam boats, and signal flares. Nevertheless, the idea that Vietnam—its countryside and its inhabitants—constituted a lab bench for U.S. weapons scientists would have a dramatic effect on both the short- and long-term character of the war. The history of tear gas deployment and defoliant use for jungle-clearing and crop destruction illustrates some of the consequences of this attitude.[13]

While the use of external substances to promote or deter plant growth had existed for decades, the key research relevant to the defoliant use in Vietnam began in the mid-1930s, with the discovery by academic and industrial scientists that certain organic acids could mimic two key kinds of plant hormones, auxins and ethylenes. Building on this work, Camp Detrick researchers focused on two phenoxyacetic acids that worked through similar means: 2,4-dichlorophenoxyacetic acid (abbreviated 2,4-D) and 2,4,5-trichlorophenoxyacetic acid (2,4,5-T). The Department of the Army contracted with the University of Chicago for further research on the acids, and accompanying experimentation at Camp Detrick included aerial test spraying in the Everglades. As one botanist later recalled, the sudden abundance of research money to university botanists was "greeted like manna from heaven."[14]

The army rejected the use of defoliants explicitly for crop destruction during World War II, out of concern that it would be perceived as chemical warfare. (German crop destruction—via flooding—was later deemed a war crime at Nuremberg.) Nevertheless, the military branches continued to conduct in-house research on herbicides throughout the 1950s, including the development of the first "tactical herbicide," Agent Purple, during the Korean War. From 1954 to 1964, funding for CBW research would rise by 1000 percent. Camp Detrick would expand into the massive Fort Detrick, occupying over a thousand acres of land, complete with livestock farms and manufacturing facilities. Throughout the 1950s, the U.S. Army Chemical Corps and the Crops Division of the Biological Warfare Laboratories, based at Fort Detrick, undertook "the evaluation of thousands of compounds for herbicidal activity" including 2,4-D, 2,4,5-T, and cacodylic acid, an effective grass and rice killer with a high arsenic content. Other key experimentation sites included Eglin Air Force Base (AFB) in Florida, Fort Drum in New York, Fort Ritchie in Maryland, and Dugway, Utah.[15]

At the same time, American chemical companies were also hard at work producing commercial herbicides and pesticides. As early as 1948, a Monsanto plant in West Virginia had begun mass production of 2,4,5-T, later a key component of Agent Orange. A recent Pentagon report on the history of military herbicide research and use stressed that the development of commercial herbicides and weed killers has followed a separate path from the military development of "tactical herbicides," but the two undeniably share crucial chemical ingredients. When the time came for mass production of the tactical herbicides, chemical companies such as Dow and Monsanto already had the necessary equipment and expertise.[16]

✳ ✳ ✳

In Vietnam, the use of herbicides greatly appealed to military planners determined to halt guerrillas who used the cover of foliage to cross from North Vietnam into South Vietnam, or who planned ambushes from behind obscuring tree canopy and roadside vegetation. The Diem regime enthusiastically endorsed the idea, having already sought American technical assistance not only for defoliation but for crop-spraying activities as well. Since much of the available botanic research concerned the effects of chemicals on North American plant life, planners promoted additional defoliation research both at Fort Detrick and on the ground in South Vietnam, hoping to refine concentration calculations and delivery methods for maximal effect in the Vietnamese jungle. In the summer of 1961, the first shipments of defoliants and related supplies arrived in Saigon for testing, and a South Vietnamese helicopter, outfitted with American equipment, flew its first defoliation test in August. Meanwhile, scientists at the Plant Sciences Laboratories at Fort Detrick, mobilized through Project AGILE, undertook new research on jungle defoliation and began stocking large quantities of 2,4,5-T and cacodylic acid, the chemicals that would soon be adapted into Agent Orange and Agent Blue. In Saigon, CDTC planners, mulling the problem of crop destruction and awaiting additional chemical supplies, proposed using napalm to burn off mature rice crops.[17]

By November 1961, McNamara and Deputy Secretary of Defense Roswell Gilpatric were urging Kennedy to approve an expanded defoliation program with multiple long-term goals: to clear roadside foliage in order to lessen the risk of ambush; to remove obscuring vegetation in the vicinity of enemy bases and infiltration routes in order to allow better surveillance; and to destroy rice, manioc, corn, and other crops in order to starve the Viet Cong into submission. The Defense Department assured the president that the chemicals to be used were "commercially produced in [the United States] and have been used for years in industrial and agricultural plant growth clearing operations" with "no harmful effects on humans, livestock or soil." In fact, as a later Pentagon history reported, the exact composition of the herbicides to be tested had been developed by military scientists and technicians at Fort Detrick and elsewhere; they were not available commercially, nor were they regulated by the U.S. Department of Agriculture (USDA). (Although some of the *components* of the herbicides had been produced commercially, the particular "formulations and concentrations . . . greatly exceeded how the commercial components of these tactical herbicides . . . were formulated and used in the United States.") The companies contracted to manufacture them—Dow, Monsanto, DuPont, and others—worked according to military specifications.[18]

But Kennedy's advisors explained only that additional testing would be needed to tailor the existing defoliants to the Vietnamese terrain. U. Alexis Johnson, deputy undersecretary of state, urged that experimental defoliant operations be carried out on areas of the Vietnamese jungle to determine maximal effectiveness. McNamara later explained to Kennedy that herbicides and delivery methods would have to be adapted to the "great variety of vegetation" and weather conditions in Vietnam. The necessary tests would take place over operationally important areas, such as ammunition dumps and roadsides, so as to gauge any immediate military advantages. An internal document from the period called for the "experimental defoliation" of selected regions. Thus, researchers would "test" the effectiveness of the new defoliants on Vietnamese targets by using them on Vietnamese targets.[19]

✳ ✳ ✳

With presidential approval secured in January 1962, expanded testing and spraying began, carried out by the air force's newly created Operation Ranch Hand. The first chemicals tested were Agents Purple, Pink, Green, and Blue, named for the color-coded identifying stripes on their supply drums. As a later Pentagon history would describe it, the effort was "a test program for evaluating tactical herbicides for vegetation control in South Vietnam." It began with American C-123 planes spraying swaths of the roadside outside Bien Hoa with Agent Purple, a phenoxyacetic defoliant, and included constant evaluation and additional research by visiting scientists. James W. Brown, an army scientist who had served stints at Fort Detrick and Camp Drum, oversaw much of the testing at the CDTC. In Vietnam, Brown was frustrated by the lack of botanic expertise on the local flora. Whereas McNamara demanded limited experimentation devoted solely to obtaining operational results, Brown and his ARPA colleagues often submitted lengthy, technical reports, trying in to fill in gaps in botanical knowledge by detailing growth cycles of local fauna and other key observations.[20]

In the spring and summer of 1962, the Defense Department sent Brigadier Gen. Fred Delmore, head of the U.S. Army Chemical Corps' Research and Development Command, to lead a follow-up team of experts, including USDA scientists Warren Shaw and Donald Whittam, ARPA's Levi Burcham, and the Chemical Corps' Charles E. Minarik. This group carefully assessed different types of vegetation, growing seasons, defoliant viscosity, and the effects of droplet size. They devised quantitative ratings systems in keeping with McNamara's known affinity for statistics and recommended additional testing and experimentation with additives. While the team expressed doubts about the effectiveness of current spraying technologies, they endorsed crop destruction and emphasized how successfully the herbicides had cleared Vietnamese mangrove swamps. Top advisors seized on these findings:

Michael Forrestal, aide to McGeorge Bundy, praised the group's work to Kennedy and enthusiastically promoted extensive mangrove spraying with the assurance that "mangrove growth has no economic value." The experiences of the visiting scientists illustrate two important aspects of herbicide use in Vietnam: the lack of extensive background research on local ecology and the effects of defoliants before test spraying began, and the tendency of military officials to endorse scientific evidence that bolstered existing plans to increase defoliant use.[21]

While the USDA scientists conducted their research in Vietnam, other army and air force scientists participated in additional defoliant testing at multiple sites at home and abroad. These included over 3,400 acres in western Thailand; 150 acres of Canadian forest in New Brunswick; and areas throughout the Southern United States, Hawaii, and Puerto Rico. Fort Detrick researchers conducted tests of Agent Orange and Agent White in Georgia swampland and alongside mountainous stretches of Tennessee Valley Authority power lines and oversaw testing on the island of Kauai carried out by the University of Hawaii's Department of Agronomy and Soils. ARPA researchers conducted test sprayings of hundreds of acres leased from private landowners in Texas and oversaw the defoliation of tropical forest areas in Puerto Rico. Additional research on herbicide characteristics and handling procedures occurred at army and air force labs at Eglin AFB, McLelland AFB, Kelly AFB, Aberdeen, and, primarily, Fort Detrick. The work at these test sites and labs focused military attention on the chemicals that would come to dominate aerial spraying in Vietnam: Agents Blue, White, Purple, and Orange.[22]

✳ ✳ ✳

President Kennedy approved the initial testing program in January 1962; within the year he authorized additional mangrove spraying efforts, and, after much wariness, crop destruction programs. Early results were mixed, but the rest of the decade saw a massive campaign of chemical spraying, primarily to clear forest canopy and other foliage surrounding communication lines, transportation routes, and base perimeters but also including, according to the *New York Times,* "50,000 to 75,000 acres" of crop destruction in the spring of 1965. By the end of the war, Ranch Hand pilots had dropped over fifty million kilograms of herbicidal chemicals on over 4.2 million hectares, largely in South Vietnam, but with a small number of missions in Cambodia, Laos, the demilitarized zone, and North Vietnam. Twenty percent of South Vietnam's jungle area was sprayed and over a third of its mangroves. In 1965 Agent Orange became the dominant chemical used, and the peak years of spraying occurred between 1967 and 1969. The chemicals included phenoxyacetic acids and arsenical compounds that mimicked plant hormones to disrupt growth (Agents Orange, Green, Pink, Purple) and desiccants made from cacodylic acid (Agent Blue). The eleven million gallons of Agent Orange, accounting for roughly 60 percent of all the chemicals sprayed, were produced largely by a handful of Pentagon-contracted companies, including Dow Chemical, Monsanto, Hercules, and Diamond Shamrock.[23]

Cautious debate accompanied the decision to use defoliants in Vietnam. Public concern would run parallel to the physical increase in defoliation operations, particularly later in the decade as reports of ecological destruction in Vietnam coincided and contributed to the growing environmental movement in the United States. But in the early 1960s, conversations about the risks and merits of herbicide use occurred largely behind closed doors in Washington and mostly concerned evaluations of the efficacy of the chemicals involved and the potential propaganda risks associated with defoliation. Sensitive to world opinion, military planners acted with caution: the first shipments of chemicals arrived in unmarked drums, and early testing plans called for disguising U.S. planes with Vietnamese insignias.[24]

Perhaps of most concern was the risk that crop destruction would draw accusations of chemical warfare. In the early days of defoliant decision making, Kennedy's advisors warned him of the risks of using herbicides, particularly for crop destruction. As Gilpatric acknowledged to Kennedy in 1961, "The use of chemicals to destroy food supplies is perhaps the worst application in the eyes of the world." Rostow wrote a telling memo to Kennedy in November 1961 explaining that a presidential authorization was needed for chemical crop destruction because "this is a kind of chemical warfare." Others in the president's inner circle assured him that such action did not constitute chemical warfare, however, nor did it violate international law—the British, they argued, had used similar tactics in Malaya. Kennedy was skeptical.[25]

The definition of chemical warfare was a tricky thing. Language banning "asphyxiating" and "poisonous" gases had been inserted into the Treaty of Versailles in the aftermath of the deadly use of mustard and chlorine gas during World War I but applied only to the defeated country of Germany.

In 1925 the United States had signed but not ratified the Geneva Protocol banning the first use of chemical and biological weapons. The protocol prohibited "the use in war of asphyxiating, poisonous or other gases, and of all analogous liquids, materials or devices" as well as "bacteriological methods of warfare." This left a number of questions unanswered: Did "asphyxiating" gases include tear gas, as the British had argued? What was meant by "other gases"? Where did chemical defoliants and incendiaries fit?[26]

No answers were forthcoming. During World War II, the United States had refrained from defoliant use but had dropped napalm, an incendiary considered a chemical weapon by some critics, during firebombing campaigns in the Japanese theater.[27] Franklin Roosevelt had specifically enunciated a no-first-use policy for chemical and biological weapons but had also authorized CBW-research programs headquartered at Camp Detrick. No clear delineations regarding nonlethal chemicals had been set. As critics would later point out, postwar army manuals informed soldiers that "the United States is not a party to any treaty, now in force, that prohibits or restricts the use in warfare of toxic or nontoxic gases, of smoke or incendiary materials, or of bacteriological warfare." In the mid-1950s, the Army Chemical Corps was expanded and reorganized to accommodate expanded CBW-research programs. Although Roosevelt's no-first-use policy had never been publicly revised, a 1959 Congressional resolution reiterating the restriction was soundly defeated in the face of vehement opposition from the State Department and the DOD.[28]

Top decision makers in 1961 drew on this complicated history in their debates over crop destruction. While they argued, however, the practice was already underway through the policy of "testing" on targets and the provision of chemicals to the South Vietnamese military. A State Department memo from the spring of 1962, before approval for the full-scale crop destruction program had been secured, noted that "results of the few crop destruction experiments reported to ARPA not conclusive." The memo further instructed officials in Saigon that should the program be approved, chemicals were to be supplied to the South Vietnamese on a "covert basis" and efforts made to "disassociate US publicly with actual operations which would be conducted solely by [the Government of South Vietnam]." Some reports have suggested that the first crop destruction operations began even earlier, in the summer and fall of 1961 when the first defoliant shipments arrived in Vietnam, but these early efforts likely consisted of South Vietnamese troops hand-spraying crops and dropping the chemicals from their own helicopters, without the use of U.S.-provided aerial spraying technologies.[29]

In the meantime, early noncrop defoliation was attracting significant anti-American press, both in Vietnam and throughout the Soviet bloc. In a way, the attention was liberating. As McNamara wrote to JFK shortly after one such media barrage, "I am inclined to believe that the propaganda impact has now been made and that we can use herbicides without causing a serious new international incident." This rationale, combined with input from the ARPA scientists confirming the efficacy of the chemicals on mangroves, underlay official approval for widespread mangrove spraying. Kennedy's "hold-off order" on aerial crop destruction, however, remained in effect through the summer of 1962. Kennedy worried that enemy crops would be indistinguishable from friendly crops and that the negative propaganda regarding "food warfare" would be extreme. By the fall, State Department frustration was mounting. Frederick Nolting, the U.S. ambassador to South Vietnam, reported with annoyance that the optimal testing window for crop destruction during the growing season was closing and that "Washington should consider seriously giving us authority to work out with GVN [the government of South Vietnam] another target area for test and evaluation purposes where possibility effective results could be maximized. Without carrying out such test operation careful preparation for which will take some time, we will never, really be able determine whether crop destruction can be effective weapons against VC without involving serious disadvantages to our side. Feel constrained point out that matter of crop destruction has been under consideration for long time, with GVN still waiting for decision from us. However we will be able soon inform them at least of decision proceed with test operation of spraying from air." Nolting's words typified the language of testing and experimentation constantly invoked by U.S. advisors in Vietnam.[30]

✳ ✳ ✳

Eventually reassured by Vietnamese officials that propaganda risks could be minimized and targets chosen responsibly, Kennedy finally authorized chemical crop destruction in South Vietnam in October 1962, to be approved on a case-by-case basis based on initiating requests from Saigon. Spraying began the following month, with South Vietnamese helicopters dropping U.S.-supplied chemicals over 750 acres of rice, beans, and manioc in Phuoc

Long Province, followed soon after by targets in Thia Thien Province. American advisors were not allowed on board the helicopters during these initial spraying operations.[31]

In February 1963 the first American press reports of Ranch Hand crop destruction missions appeared. That spring, a United Press International reporter based in Saigon and a correspondent for the *Minneapolis Tribune* provided corroborating accounts describing the use of crop-killing defoliants in the central highlands. Kennedy began to rethink his decision. After a meeting with British Ambassador David Ormsby Gore, during which the British officials present doubted the effectiveness of the defoliants and warned that Asians had a historical aversion to unknown chemicals, the president requested a review of all defoliation and crop destruction programs.[32]

Meanwhile, Kennedy faced contradictory pressures resulting from the new publicity. What McNamara had initially considered liberating now had the effect of hardening the American commitment to chemical herbicides. As one official wrote in April 1963, even though "defoliation is at best only partially effective militarily," stopping the use of herbicides now "would tend to confirm Bloc charges and invite further such campaigns because of their proven effectiveness against us." The issue, as was the case in so many other instances of Vietnam decision making, was credibility. To Kennedy's advisors, a retreat in defoliation policy constituted a retreat in the war at large.[33]

But increasingly, the "charges" were coming not from the Soviet bloc, but from Western scientists and political activists. In Congress, Senate Majority Leader Mike Mansfield and Wisconsin representative Robert Kastenmeier worried publicly about the reported crop destruction. In the *New York Times,* the British philosopher Bertrand Russell characterized the use of chemicals in Vietnam as an "atrocity" of "chemical warfare." In March 1963 the *New Republic* published "One Man's Meat," an article damning American deployment of chemicals in Vietnam. The magazine noted that herbicides marketed in the United States contained detailed warning labels conveying their toxicity and the risks of human exposure, suggesting that their use in war violated the 1925 Geneva Protocol. The article also offered a new, influential analysis of the use of herbicides as weapons: it was both the ends and the means that mattered. Defending defoliants on the grounds that they were simply common chemicals publicly available in the United States, as the *New York Times* had in response to Bertrand Russell, was insufficient; it was the *manner* in which they were used, in heavy concentrations and sprayed over large swaths of land, and the end effects of that use—crop destruction, widespread human exposure—that elevated them to the status of chemical weapon.[34]

The *New Republic* article was also emblematic of a nascent public awareness of the ecological risks associated with chemical herbicides and pesticides. In 1962 Rachel Carson's bestseller *Silent Spring* offered a devastating account of the risks of DDT use. Carson's form of environmentalism, rooted in an awareness of the delicate balances required for the survival of complex ecosystems and drawing on analyses of new chemicals and their potential damage to plants and animals, found a receptive audience in a nation primed by civil-defense exercises, fears of nuclear fallout, and, increasingly, direct experience with smog and pollution.[35] In time the ecological costs of the war in Vietnam would be a powerful link bridging the environmental and antiwar movements of the late 1960s, but in 1963 the public outcry was small and easily ignored; the efforts of Bertrand Russell and the *New Republic* had little discernible effect on policy. Kennedy's requested task force conducted its review, and existing crop destruction policies were reaffirmed in October 1963. The following year, the air force's Ranch Hand program was officially made permanent, and what had been an experimental unit with "temporary duty status" became a fixture of the American presence in South Vietnam.[36]

* * *

Lyndon Johnson's accession to the presidency after Kennedy's assassination saw the continuity and expansion of existing commitments to counterinsurgency and nonnuclear weapons technology. As McNamara reminded the Democratic Platform Committee in August 1964, the new president had declared that "the United States is, and will remain, first in the use of science and technology for the protection of its people." McNamara emphasized that in the mid-1960s, overwhelming nuclear might was no longer "enough," and new weapons technologies were needed: "The effectiveness of the strategic nuclear deterrent we have assembled against our enemies has driven them to acts of political and military aggression at the lower end of the spectrum of conflict. The Communists now seek to test our capacity, our patience, and our will to resist at the lower end of this spectrum by crawling under the nuclear defenses of the free world. . . . in the twilight zone of guerrilla terrorism and subversion."[37]

McNamara's views continued to be popular with many scientists, and the presidential election of 1964 further cemented the support among scientists for the Johnson administration. Chapters of groups such as "Engineers and Physicians for Johnson and Humphrey" and "Scientists and Engineers for Johnson" proliferated, spurred by hopes for arms control and fears of the extreme hawkishness of Johnson's challenger, the Arizona conservative Barry Goldwater. George Kistiakowsky, Herbert York, and Jerome Wiesner were all members. Donald Hornig, the Los Alamos veteran and Princeton chemist chosen as Johnson's science advisor, spoke to one such group in South Carolina two weeks before the national election. "Although traditionally scientists and engineers don't take an active role in political campaigns," he observed, the high stakes of the election had brought greater political involvement. As Hornig put it pointedly, "Many [scientists and engineers] have been involved in the development of military power which could, if improperly used, destroy mankind. They want to assume that, if possible, it not be used and that the responsibility for its use be exercised at the highest level of the government—in fact, by the President and by a responsible President."[38]

As promised, after his election President Johnson followed through on the commitment to emphasize limited-war weapons and technologies. Pentagon cost-cutting efforts from the early 1960s, including efforts to shift from cost-plus contracting to competitive fixed-price contracts, were matched by an 800 percent increase in counterinsurgency programs. In Vietnam, Westmoreland oversaw the creation of the Joint Research and Test Activity to coordinate the work of the three services and ARPA. By 1966, its 150-person staff included thirty "civilian engineers and scientists."[39]

Crop destruction operations likewise expanded after 1964 into more populated VC-controlled areas in central Vietnam. Beginning in December 1965, aerial defoliation also extended across the border into Laos and included a small number of unpublicized crop destruction missions. In 1965 State Department officials estimated that 9,310 hectares (roughly 23,000 acres) of cropland had been destroyed, representing fifty-four million pounds of food. The results seemed to justify the high social costs; although planners privately dismissed a RAND study warning of the adverse effects of spraying on popular attitudes, they nevertheless continued their efforts to dissociate the U.S. publicly from crop destruction operations. Operation Farmgate employed mixed U.S.-Vietnamese crews piloting air force C-123 planes deliberately sporting only Vietnam Air Force markings. American officials reported in August that after a U.S.-led defoliation mission outside

Bien Hoa resulted in damage to the trees and vegetables of two neighboring villages, "intensive psywar is being conducted" and "prompt compensation" pursued in order to minimize backlash.[40]

Despite these efforts, public criticism, sporadic in the early 1960s, grew increasingly widespread by the middle of the decade. In reaction to reports that the strategic hamlet of Cha La had been accidentally defoliated, the *Washington Post* ran an op-ed that referred to defoliants as "chemical weapons" that were "totally unsuited" for the war in Vietnam. A pair of *New York Times* articles by Charles Mohr, published in December 1965, also proved influential. Mohr estimated that the U.S. military was dropping "enough chemicals to cover 20,000 acres a month," and described the progress of defoliation in evocative language: "Within three days after a single spraying with the non-poisonous weed killer, the effects are noticeable. Within a week there is an 'autumn' look. But three months must pass before the 'winter in Vermont' effect is achieved. Four months after that, however, the foliage begins to grow back." Mohr also reported that although "air force officers say they are forbidden to discuss it," herbicidal operations in Vietnam included deliberate crop destruction. Mohr described the spraying of "small areas of major military importance where the guerrillas grow their own food" with an effectiveness rate of 60 to 90 percent if the chemicals were applied during the growing season. He also cited attempts to destroy existing rice caches ("Even with thermite molten-metal grenades, it virtually will not burn.") or render them unpalatable through the addition of "yellow dye and shark repellant."[41]

* * *

The use of chemicals for defoliation and crop destruction were not the only military actions that attracted criticism. Increasingly, antiwar critics' arguments included damnations of another controversial military practice: the use of gas. The problems posed by "nonlethal gases"—specifically tear gas, super tear gas, and the nausea-inducing adamsite—elicited internal debate among Pentagon and administration insiders as well as the public at large.[42]

As early as 1961, military planners had discussed the experimental use of gas in Vietnam. In a report prepared after a visit to South Vietnam by Taylor, Rostow, and other Pentagon, Central Intelligence Agency, and State Department personnel, officials proposed using "anti-personnel" chemical warfare agents in areas where "the population may be essentially 100% Viet

Cong." For example, they explained, "a part of Zone D might be used as a proving ground with perhaps advance notice being given . . ." The chemicals, presumably provided by the United States to South Vietnamese soldiers, would not be subject to "the usual objections" since "this proposed application would be in one's own country." In other words, the United States would be exempt from blame if *South Vietnamese* military forces used gas against their guerrilla opponents. Of course, the report continued, "From the point of view of political acceptability, incapacitating agents are probably to be preferred to lethal agents, but on the basis of technical feasibility, only the latter may be possible. This remains to be determined."[43]

While there is no evidence of explicitly lethal gas use during the war, the attitude suggested by the document was one of enthusiasm for experimentation ("proving ground"), especially if it were the South Vietnamese who actually carried out the work. Indeed, the following year saw the first transfer of tear gases to the government of South Vietnam, to be used at their discretion with no particular U.S. approval process. The three major gases eventually provided to South Vietnamese military forces were tear gas (chloroacetophenone, abbreviated CN) and super tear gas (o-chlorobenzylidene malononitrile, abbreviated CS), first delivered in 1962, and adamsite (diphenylaminechloroarsine, abbreviated DM), a nausea-inducing chemical provided at least as early as 1964.[44]

Both CN and DM had originally been developed during World War I, while CS had been discovered in the United States in 1928 and refined as a weapon by British researchers in the 1950s. Although solid at room temperature, the chemicals could be packed into grenades and sprayed as aerosols to create gaslike clouds. Of the three, the flowery-smelling CN produced incapacitating conditions of the shortest duration: intense irritation to the eyes, skin, and respiratory passages lasting several minutes. CS's effects were similar but lasted five times longer and could include nausea as well. By far the most potent of the chemicals was DM, nicknamed "vomit gas" by the British soldiers who had used it in Bahrain. DM produced irritation to the eyes, nose, and throat, as well as intense nausea and vomiting for a period ranging from thirty minutes to several hours. Army instruction manuals from the early 1960s warned that DM should not be used in "any operations where deaths are not acceptable." In 1965, at least three American companies were manufacturing versions of these gases commercially: Federal Laboratories in Saltsburg, Pennsylvania; Lake Erie Company in Cleveland, Ohio; and Fisher Laboratory in New Jersey.[45]

On multiple occasions in the late fall and early winter of 1964–1965, some or all of these gases were used in the course of military operations in Vietnam. During riots in Saigon in November 1964, South Vietnamese military forces reportedly used CS against protesting Buddhist monks and students. Later that winter, a mission to rescue American prisoners believed to be held at Tam Giang in Xuyen Province employed CS and smoke grenades dropped from a helicopter, but "no contact was made." A similar effort two days later at Thanh Ham in Tay Ninh resulted in the use of 100 CS grenades, 550 grenades containing a mixture of CN-DM, and another three hundred pounds of CS dropped by helicopter on an area later determined to be unoccupied. In late January 1965, the South Vietnamese Air Force conducted a search-and-destroy mission on Phu Lac Peninsula in Phu Yen Province, employing 900 grenades containing either CS or CN-DM, as well as additional helicopter-dropped CS. Westmoreland reported to officials in Washington that tear gas had wafted into a nearby village, prompting a Radio Hanoi report of chemical weapons, but "the effect was very slight." A Defense Department cable reported that eighty-eight Viet Cong had been killed during the raid, but that the Vietnamese Army's use of "CS and CN/DM agents . . . left much to be desired." Nevertheless, "this was a first experience for the troops involved and lessons were learned."[46]

Two months after the Phu Yen mission, two young Associated Press reporters, Peter Arnett and Horst Faas, broke the story to a global audience, describing the experimental use of nonlethal gases in Vietnam. In the flurry of press conferences and newspaper coverage that followed, administration insiders emphasized the nonlethality of the gases and reiterated that any "experimentation" had been one of tactics, not new "secret types of military gas." The tactic in question, the dropping of tear gas on villages from helicopters, was explained as "a way of attacking guerrillas mixed in with civilian populations without killing the civilians or destroying their villages." Secretary of State Dean Rusk publicly asserted that the gases were not prohibited by the Geneva Convention of 1925 and were best described as "gases which have no lethal effect, which have a minimum disabling character." Meanwhile, inside the administration, McGeorge Bundy summarized Westmoreland's report of the December incidents of tear gas use for President Johnson: "It sounds to me like no one even cried." Johnson in turn professed to have had no prior knowledge about the gas use, but he nevertheless expressed frustration with critics who were more concerned that someone's eyes had watered than with "our soldiers who are dying."[47]

But public opinion balked at reports from the *Los Angeles Times* and other newspapers that the gas used had been a mixture of tear gas and adamsite, a combination whose reported vomit-inducing properties could last hours and which, in prolonged, heavy doses, could potentially cause death. Soviet bloc news outlets referred to the gas use as chemical warfare, and newspapers in London and Paris published critical editorials. In the *New York Times*, Max Frankel's front-page coverage placed the tear gas and nausea agents squarely in the context of World War I-era blister gases and quoted at length Senator Wayne Morse, who considered adamsite use a violation of international law.[48]

* * *

Public criticism of gas use forced an internal review and a standoff between Pentagon leadership and Kennedy's coterie of science advisors. On March 25 Wisconsin congressman Robert Kastenmeier, with fifteen congressional cosigners, wrote to President Johnson in reaction to the reports of gas use. In 1959 Kastenmeier had introduced a resolution reaffirming American policy against the first use of biological weapons or "poisonous or obnoxious gases." The measure, although largely a reiteration of the Roosevelt policy, had been defeated at the urging of the Defense Department. Now Kastenmeier requested an official statement of administration policy on the use of gas and demanded that decision-making authority be transferred from local commanders to the president himself. He quoted antigas statements by Eisenhower and Roosevelt and warned that "the first use of gas in warfare, however innocuous its variety or effective its results, subjects the using country to the censure of the civilized world."[49]

Government officials quickly assured Kastenmeier that the gases were humane alternatives to more lethal tactics and in no way constituted gas warfare. As such, it was appropriate for local commanders to authorize their use. Only true chemical weapons—not the "riot control agents" used in Vietnam—required presidential approval. Moreover, they explained, the chemicals had been used in cases in which "the Communist Viet Cong had taken refuge in villages, using innocent civilians as shields. Riot control agents were employed in an attempt to subdue the Viet Cong without exposing the South Vietnamese civilians and prisoners being held by the Viet Cong to injury from more lethal weapons such as rifles and machine guns." The unsuccessful outcomes of these actions were not mentioned.[50]

For the benefit of Kastenmeier and his congressional cohort, Pentagon officials defined CN as "a lacrimatory agent which is an irritant to the respiratory passages and sometimes to the skin, and which incapacitates for about three minutes"; CS as "a more recently developed lacrimatory agent with effects similar to CN but which, in concentration, sometimes leads to nausea and which incapacitates for 5–15 minutes"; and DM as "a pepper-like irritant to the eyes, throat and mucous membrane, which may also cause vomiting and which incapacitates for 30 minutes to two hours." But internal correspondence with the office of Donald Hornig, Johnson's beleaguered science advisor, suggested far greater risks, particularly from DM. Hornig's office described DM:

> Adamsite (DM) Medical effects—when inhaled, causes pain, malaise with aching in eyes, joints, and teeth. Similar to influenza in effect. Vomiting occurs because of pain and malaise. Action comes in two to four minutes and lasts for 2–4 hours. The safety factor is between 8 and 10-fold, i.e., a dose 8–10 times that causing incapacitation from pain and malaise may cause death. (This is a small safety factor particularly if children and old people are exposed). At the incapacitating dose, the mortality rate may be as high as 1 percent.

> Prior use—Used in combination with tear gas for riot control as in Koji-Do, Korea, during POW riots.

> Commercial Availability—Unknown Available to US Police Departments

According to Hornig, DM was potentially lethal, with an estimated period of effectiveness twice the duration reported to Kastenmeier. Hornig recommended an immediate halt of the use of both CS and DM.[51]

✳ ✳ ✳

McNamara responded to the outcry by ordering a moratorium on tear gas use in Vietnam, an unpublicized ban that lasted from January through the end of the summer of 1965. The same period saw a dramatic escalation of the war, including the onset of the massive bombing campaigns of Operation Rolling Thunder. In September, Leon Utter, a young marine colonel apparently unaware of the ban, ordered the firing of forty-eight CN grenades into a tunnel and bunker system in which over four hundred Vietnamese soldiers and civilians had been hiding. The action succeeded in forcing the occupants out of the underground complex and prompted West-moreland to request a lifting of the ban so that tear gases could be employed to clear tunnels and underground shelters as an alternative to other, more lethal, weaponry. (Alternate responses to occupied tunnels included the use of flamethrowers, grenades, sealing, and wholesale demolition.) Westmore-land's recommendation was supported by Chairman of the Joint Chiefs of Staff Earle Wheeler, who informed McNamara that the "Mighty Mite," a lightweight blower designed to fill tunnels and underground spaces with gas quickly, had already undergone testing in Vietnam. He assured McNamara that any policy shift would be kept quiet: there would be no public announce-ments, and responses to any questions would be deliberately "low key." Mc-Namara was persuaded; he informed President Johnson that he favored the shift and, following Wheeler's advice, passed along instructions to Westmo-reland to refer to the chemicals in public as "tear gas" rather than chemical agents and to emphasize their humanitarian aspects. With Bundy's assur-ance that "even the New York Times is resoundingly with us on this," the ban was lifted, and by early October tear gas—CN and CS—was once again in use on the ground in Vietnam.[52]

The decision making had not been unanimous, however. Arthur Gold-berg, the U.S. ambassador to the United Nations, objected to the resumed use of gases out of concern for bad publicity.[53] Donald Hornig, who had pre-viously urged a halt to the use of both CS and DM, reiterated his opposi-tion to McGeorge Bundy two weeks after reports surfaced of Col. Utter's use of tear gas. Hornig wrote to Bundy:

> The effects produced by CS are much more violent than those resulting from CN, including chest constrictions and bronchial symptoms; there is less knowl-edge of the toxicity of CS in closed spaces; and in high concentrations it may produce nausea. The different between the effects of CS and CN would cer-tainly be apparent, and this might appear to some critics as an escalation of normal tear gas.
>
> More important, however, is the fact that the use of CS may considerably complicate our public relations problem. Our case for using tear gas would be considerably strengthened by the fact that tear gas is widely accepted as a humane civilian riot control agent. I think, however, there is a real ques-tion as to whether CS has in fact been used in this country or abroad and, if so, in more than a few scattered cases, by civilian authorities in normal riot control circumstances. I have not been able to obtain clear answers to these

questions which we must certainly understand if authorization is granted for the use of this agent.

I am also not aware that the DOD has made the case that CN will not adequately accomplish the purpose of clearing Viet Cong and civilians from tunnels and enclosed areas. In the event that CN proves ineffective for this purpose, CS can always be subsequently authorized.[54]

Hornig's concerns evoked McNamara's trademark commitment to graduated escalation—if CN was sufficient to achieve the objective, why resort to the more dangerous CS unnecessarily? But when McGeorge Bundy forwarded Hornig's memo to McNamara, McNamara was unmoved. He responded by producing an alternate source of technical expertise to counter Hornig, noting pointedly that "our Army technical experts assure me that CS is not as lethal as CN when used in inclosed [sic] spaces." He also dismissed Hornig's complaint of the lack of historic examples of civilian use of CS as irrelevant, since most social disturbances were quelled by military—not civilian—forces. He attached a list of examples of such use, ostensibly to demonstrate CS's acceptability. The list noted that U.S. military forces had used CS against civil rights demonstrators in Cambridge, Maryland, and Oxford, Mississippi, and against rioters in Panama. CS had also been used by the West German border police, the British forces in Cyprus, the French in Algeria, and the South Vietnamese military against Buddhists and student protesters.[55]

McNamara's response confirmed that on the technical and ethical questions of tear gas use, the president's science advisor was not the last word. Despite a shared arms-control commitment, a small rift had opened between the PSAC and the Pentagon, and the influence of the old guard of advisors was on the wane. The military-industrial-academic networks forged during the Eisenhower era had borne new fruit: McNamara now had his own set of competing experts.

FOUR

Into the Ethical Hot Pot

IN THE EARLY 1940s, a young botany graduate student, Arthur Galston, was researching auxins and the physiology of soybean flowers at the University of Illinois. His work focused on the ways that a compound called 2,3,5-triiodobenzoic acid (TIBA) could help speed the flowering process. In the course of his research, however, Galston noticed that at high concentrations, TIBA could cause abscission: the weakening of cellulose at the juncture of leaf and stem, resulting in the shedding of leaves. Galston finished his degree in 1943 and devoted the following three years to unrelated war research on rubber production. Unbeknownst to him, researchers at Camp Detrick had noticed his graduate school discovery and had undertaken a new research program on TIBA with an eye toward the potential tactical uses of chemical defoliation in World War II's Pacific theater. Galston was shocked when, in 1946, two Camp Detrick senior scientists visited him at the California Institute of Technology (Caltech) to inform him that his thesis work had served as a model for current military research on defoliants.[1]

For Galston, who viewed his graduate work as essential to the development of Agent Orange, the reports of crop destruction and defoliation in Vietnam were shattering. Galston suddenly saw himself implicated in a potential ecological nightmare; years later, he described his graduate TIBA research as "the scientific and emotional link that compelled my involvement in opposition to the massive spraying of these compounds during the Vietnam War." The use of Agent Orange in Vietnam was deeply disillusioning—an ethical turning point that, in Galston's words, "violated my deepest feelings about the constructive role of science, and moved me into active opposition to official US policy."[2]

Galston immediately composed a critical resolution to present at the annual meeting of the American Society of Plant Physiologists. When the Executive Committee declined to present it, Galston began collecting signatures on a similar petition, written in the form of a letter to President Johnson. The document expressed the "serious misgivings" of about a dozen scientists, particularly regarding the unintended ecological consequences of spraying, the persistence of chemicals in the soil, and the possible food-denying effects on Vietnam's children. Johnson failed to reply, but an undersecretary of state wrote back to Galston, assuring him the chemicals used were harmless and that crop destruction only occurred in remote areas, with advance warning.[3]

John Edsall, the Harvard biologist, had a similar experience. In 1960 members of the Society for Social Responsibility in Science had hosted an open forum on chemical and biological weapons (CBW) at Harvard's Sanders Theater. Some of the scientists in attendance belonged to the Boston Area Faculty Group on Public Issues (BAFGOPI). BAFGOPI boasted over ninety members in 1962, mostly professors at New England universities, including Edsall, the radical Nobel-winning Massachusetts Institute of Technology (MIT) biologist Salvador Luria, Harvard biologist Matthew Meselson, linguist Noam Chomsky, physicist Bruno Rossi, and others. BAFGOPI had been created through the circulation of an open letter to President Kennedy on civil defense and saw itself as an "informal organization . . . limited to members of the academic community" who were protected by academic freedom and united by arms-control convictions. Members focused on foreign and nuclear policy, including nonproliferation, civil defense, and the budding conflict in Vietnam.[4]

Early in 1966 Edsall wrote to McNamara, expressing his opposition to crop destruction, and Maj. Gen. Michael Davison replied, asserting that the chemicals used were common weed killers that "harm neither humans nor animals, and do no harm to the soil or water supplies in the concentrations used." With letter-writing efforts at a seeming dead end, Galston and his colleagues turned their attention to circulating petitions and urging the American Association for the Advancement of Science (AAAS) to open new scientific inquiries into defoliation. It was a quest that would culminate, four years later, in the creation of an $80,000 Herbicide Assessment Commission and a Department of Defense review.[5]

In the meantime, the press coverage of defoliation had attracted nonscientist attention as well. In March 1966 a landscape architect, Robert Nichols, staged a hunger strike to protest crop destruction in Vietnam and the hypocrisy of U.S. policymakers who claimed that the United States was assisting Vietnamese food production. In response, State Department spokesman Robert McCloskey acknowledged publicly the U.S. role in crop destruction but argued that only "one-third of one per cent of the total area under cultivation in Vietnam" had been affected. McCloskey's attempt at reassurance foundered, however, as critics added the confirmed charge of intentional starvation to their allegations of chemical warfare.[6]

✳ ✳ ✳

As public outcry over gas use merged with antidefoliation activism, accusations of chemical warfare proliferated, bringing botanists and biologists into the political fray. In the fall of 1966, a small group of Harvard scientists discussed taking public action to express their discomfort with the U.S. use of chemicals—both defoliants and gases—in Vietnam. The group included biologist John Edsall, physicists Freeman Dyson and Bernard Feld, *Bulletin of the Atomic Scientists* cofounder Eugene Rabinowitch, and the young Harvard biologist Matthew Meselson, who at twenty-nine was an advisor to the Arms Control and Disarmament Agency, a former student of Linus Pauling at Caltech, and, along with Franklin Stahl, hero of the famous Meselson-Stahl experiments detailing the character of DNA reproduction. The result was a highly publicized letter signed by twenty-two prominent scientists ("including seven Nobel laureates," noted the *New York Times*), urging President Johnson to declare, as policy, that the United States would not initiate chemical weapons use. But the scientists went further, explaining why supposedly "nonlethal" gases and defoliants represented a far more serious threat than might initially appear. Gases that might not directly cause death could nonetheless be used in lethal ways, they argued. For example, "When, in Vietnam, we spread tear gas over large areas to make persons emerge from protective cover to face attack by fragmentation bombs or when we use tear gas so that a moving target cannot move so fast, we use gas to kill." (Two years later, Seymour Hersh would make similar allegations in a widely cited article in the *New York Review of Books*, documenting at least two major aerial bombing attacks in 1966 and 1968 that had been preceded by widespread tear gas drops.)[7]

Perhaps most importantly, the scientists argued that the weapons represented a dangerous world precedent: "The United States, along with other nations, recognizes that the use of even the smallest nuclear artillery shell in war would raise issues of extreme gravity. It would break down barriers to the use of more powerful nuclear weapons, and no one could tell where the escalation might end. The use of chemical or biological weapons, even relatively mild ones, involves similar dangers."[8]

The letter circulated through the top levels of the Johnson administration, and although Johnson himself advised officials not to answer questions regarding the letter, his staffers readied themselves for a media barrage. They drafted answers to expected questions and circulated the results among officials, urging an emphasis on the "nontoxic" nature of the substances used and the careful restrictions on gases and herbicides that would prevent any escalation to more dangerous chemicals. The *New York Times* could report only that "despite protests by 22 leading American scientists," the Pentagon had confirmed the continuation of chemical use. But in the face of the scientists' criticism and in the aftermath of a massive, failed attempt to defoliate and burn the Boi Loi woods in South Vietnam using napalm and white phosphorus, top decision makers quietly eased the number of crop destruction and defoliation missions in Vietnam.[9]

In a related attempt to allay public concerns, the United States also endorsed a Hungarian-introduced United Nations resolution affirming the Geneva Protocol, but with the explicit caveat that the protocol had been intended to curb poisonous gases, not the use of common riot-control agents. This exclusion did not satisfy Meselson and his colleagues, who embarked on an expanded petition drive that would ultimately collect thousands of new signatures, including 127 members of the National Academy of Sciences (NAS). Beyond Meselson, other scientists and experts at Harvard were also linking defoliants and gas under the common heading of chemical warfare, with all its horrifying implications. In January 1966 Victor Sidel and Robert Goldwyn, two doctors from Harvard Medical School writing in the *New England Journal of Medicine,* disputed the nonlethality of the gases, noting that "they can kill under certain circumstances: extremely high concentration of agent of highly susceptible victim, such as the very young, the very old, or the very sick." Defining chemical warfare as "the employment of chemicals toxic to men, animals or plants"—a definition that included both tear gas and defoliants—they listed the gases in their descriptions of chemical and biological weapons alongside mustard gas, chlorine, phosgene, LSD, sarin, and anthrax. Like Meselson's group, they worried that even the use of the most "humane" chemical alternatives could still open "Pandora's box," setting a path for use of more dangerous weapons.[10]

Doctors were also among the first scientifically minded outsiders to visit Vietnam to document the medical effects of the war. In early 1967 four members of the Physicians' Committee for Social Responsibility (an organization to which Sidel belonged) issued a report titled "Medical Problems of South Viet Nam" that described, in stark language, the dietary deficiencies, infectious diseases, and poor medical facilities in South Vietnam. The doctors also described in harrowing detail the war's new kinds of casualties, including vicious burns inflicted by napalm, the jellied incendiary developed during World War II and increasingly employed as an "antipersonnel" weapon in Vietnam. Survivors of napalm wounds faced "a living death," the doctors wrote, while those who perished suffered "death . . . by roasting or by suffocation."[11]

An influential CBW-themed edition of *Scientist and Citizen,* which appeared in the summer of 1967, also cited the work of these doctors. The magazine had started out as a project of the St. Louis Committee for Nuclear Information, the team of scientists and concerned citizens that in the late 1950s had undertaken Barry Commoner's survey of strontium 90 levels in baby teeth. From 1958 to 1963, they had published the periodical *Nuclear Information,* which in 1963 was renamed *Scientist and Citizen.* (In 1969, it would be renamed yet again, this time *Environment.*) In its 1967 incarnation, a collaboration with the Scientists' Institute for Public Information, the publication was infused with post–Manhattan Project notions of ethics and scientific responsibility, including the obligation to educate the public on science-related current affairs. An article introducing members of the new Science Advisory Board, including John Edsall and Milton Leitenberg, proclaimed that: "The scientists of *Scientist and Citizen* have a *special responsibility* to provide their fellow citizens with the information that is relevant to these decisions in an understandable form, free of technical jargon and political bias." John Edsall offered a more radical and somewhat tortured interpretation. "The social mission of *Scientist and Citizen,*" he wrote, "is precisely to enable the general public to escape the tyranny of the expert." Or, as the back cover routinely warned, "It's your world—Don't leave it to the experts." Yet the contents of each issue were filled with expert analyses of major science-related controversies. Articles covered not only the problems of nuclear proliferation, fallout, and civil de-

fense, but also supersonic transport, pollution, pesticides, and other consumer dangers. Contributors ranged from dove to hawk, from pacifist Columbia engineering professor Victor Paschkis to the ardently anticommunist
physicist Eugene Wigner.[12]

With a stark cover featuring a Ranch Hand plane, rocket delivery systems,
and magnified images of microbes arranged in the shape of a gas mask, the
September-October special issue was devoted exclusively to chemical and
biological warfare. Harvard microbiologist John Edsall wrote the introductory essay, in which he reiterated the arguments of the multiple petitions
and letters sent to the administration: despite some possibly humane applications of chemical weapons, the "dangers of escalation" were omnipresent.
He invoked the language of older nuclear debates: "Once either side begins
using weapons of this category," Edsall wondered, "can any dividing line
be drawn at which both sides, locked in conflict, will uphold a binding agreement that says 'At this point we will stop; we will use no weapons more deadly
than this.'"[13]

Elsewhere in the issue, Harvard nutritionist Jean Mayer warned of the
exacerbating effect of crop destruction on malnutrition and disease in
Vietnam, citing the work of the Physicians for Social Responsibility and
describing in detail the medical consequences for the population: elevated
child mortality; degradation of the heart, stomach, lungs, and intestine;
psychological distress; blindness; anemia; and deadly infection. He presented an account of previous attempts to starve enemy forces, from the
Franco-Prussian War to World War II, arguing that "food denial in war
affects the fighting men least and last," taking its toll instead on the civilian
population. Sidel and Goldwyn, the doctors who had written the weapons
"primer" for the *New England Journal of Medicine,* contributed a similar
article describing the chemistry and effects of many categories of chemical
weapons.[14]

The centerpiece of the issue was Arthur Galston's in-depth discussion of
the ecological dangers of defoliation. Like other critics of defoliation, he mentioned the risk of escalation and the warning labels that appeared on domestic herbicides. Galston acknowledged the difference between defoliants
and far more lethal weapons used in war but emphasized that ecological
damage posed enormous and unknown risks. He wrote evocatively: "To
damage or kill a plant may appear so small a thing in comparison to the
human slaughter every war entails as to be deserving of little concern. But
when we intervene in the ecology of a region on a massive scale, we may set

in motion an irreversible chain of events which could continue to affect both the agriculture and the wildlife of the area—and therefore the people, also—long after the war is over."[15]

Galston offered a startlingly accurate assessment of the experimental nature of the sprayings. Though he provided explanations of the general physiological processes of chemical defoliation in plants, he wrote that scientists did "not understand at all well" the changes in plant life caused by external chemicals, and since scientists had only studied a few species carefully in this context, they could hardly predict the ways herbicides might affect a complex ecosystem. Given these uncertainties, the spraying in Vietnam itself constituted a kind of experiment: "When we spray a synthetic chemical from an airplane over a mixed population of exotic plants growing under uninvestigated climatic conditions—as in Vietnam—we are performing the most empirical of operations. We learn what the effects are only after we perform the experiment, and if these effects are larger, more complex, or otherwise different from what we expected, there is no way of restoring the original conditions." Galston condemned the recklessness of U.S. policy, which took the complex ecosystems of Vietnam as expendable laboratories for chemical research.[16]

* * *

Galston's eloquence in the pages of *Scientist and Citizen* was accompanied by diligent efforts to work within all available institutions to try to halt defoliation and gas use. Beginning in the mid-1960s, to complement their letter-writing efforts and circulation of petitions, Galston, Edsall, Meselson, and other biologists and botanists began working within the AAAS, hoping to bring greater institutional backing to their efforts. In 1966 E. W. Pfeiffer, a zoologist from the University of Montana, presented an AAAS resolution calling for an expert study of chemical warfare in Vietnam. Pfeiffer's resolution followed a tortuous path through various committees and votes, culminating in language which, stripped of its reference to Vietnam, nevertheless warned that "the full impact of the uses of biological and chemical agents to modify the environment . . . is not fully known." The effort seemed to elicit a reaction—the Defense Department's Advanced Research Projects Agency contracted with the Midwest Research Institute (MRI) to conduct a review of existing literature regarding herbicide use. The result was a 369-page summary of over fifteen hundred related studies and dozens of interviews. The AAAS's board of directors also corresponded with John Foster, the director of defense research and engineering, who assured attendees at the 1967 national meeting that although questions of detrimental "long-term ecological impacts" of herbicide use had not been answered "definitively," planners had consulted "qualified scientists, both inside and outside our government," who "judged that seriously adverse consequences will not occur." Like McNamara in his dismissal of Hornig's concerns about o-chlorobenzylidene malononitrile and diphenylaminechloroarsine, Foster invoked the support of unnamed researchers to deflect the concerns of civilian scientists.[17]

But despite Foster's assurances, the MRI report contained little detailed information about actual spraying in Vietnam, focusing mainly on summarizing research performed on other vegetation and in other contexts, which was then extrapolated to conclude that the risks of spraying in Vietnam were either negligible or unknown. The AAAS solicited commentary on the report from the NAS and other experts, and the board ultimately expressed deep dissatisfaction with the report and with the failure of the research cited to confirm explicitly Foster's rosy assessment. A majority of AAAS board members wrote in *Science* in 1968 that "many questions concerning the long-range ecological influences of chemical herbicides remain unanswered." They called for a field study of conditions in Vietnam, to be overseen by the United Nations, and a temporary halt to the use of the arsenical Agent Blue until further research on its degradation could be conducted. Other board members, led by Barry Commoner, went further, demanding in a supplementary statement that the use of 2,4-D and 2,4,5-T be halted as well. In *Scientist and Citizen,* associate editor Sheldon Novick complained that "the report as a whole is an emphatic repetition of the fact that we are engaged in a gigantic experiment in Vietnam, and have little idea what its outcome may be."[18]

Frustrated by the slow progress of the AAAS, a small cadre of members, including Galston, Edsall, Pfeiffer, and Berkeley biochemist J. B. Neilands, formed the Scientists' Committee on Chemical and Biological Warfare. The group aimed to promote a strict interpretation of the Geneva Protocol and to conduct field studies on the ground in Vietnam. They made presentations to other scientific organizations, including the Federation of American Societies for Experimental Biology and the American Society for Microbiology. Two members, Pfeiffer and Gordon Orians, traveled to South Vietnam in 1969 and described the "*very* severe" ecological consequences of defoliation.

They warned that the immediate establishment of an international study of long-term consequences of defoliation was critical "if the U.S. scientists wish to maintain—or regain—the respect of scientists in Southeast Asia." But in 1970, Neilands mourned the overall lack of institutional involvement in the antiherbicide campaign: "Within the professional science organizations, only the AAAS has shown any concern about the defoliation of Vietnam and this has required constant prodding from Pfeiffer. Perhaps most distressing of all has been the inability of the AAAS to follow through with any kind of action program." The fervent petition drives and the introduction of resolutions were not yielding results.[19]

∗ ∗ ∗

Gauging the overall opinion of scientists on the problem of Vietnam is difficult. In the summer of 1967, Penn State biophysicist Ernest Pollard wrote a letter to *Science* soliciting volunteers to help with the war effort through "ingenuity regarding weapons," operations research, bombing analyses, or any other form of assistance. He invoked the great contributions of scientists during World War II, offered despite the participants' variety of political viewpoints, and wondered why the current war had failed to evoke similar forms of voluntarism. He wrote, "If indeed it is true that the university scientists as a whole are opposed to the opinion of the majority of the people of the United States regarding support of the war in Vietnam, then obviously we are in a bad situation."[20]

Writing in response two months later, MIT biologist Salvador Luria and Woods Hole marine biologist Albert Szent-Gyorgyi chastised Pollard for promoting President Johnson's disingenuous justifications for war, arguing instead that the war was "a national catastrophe and a moral blight for our country." They replied that yes, thousands of university professors, "including a large percentage of scientists," did oppose the war. Rather than volunteer to help the military, scientists ought to examine their own activities "to make sure that they do not unnecessarily contribute to the waging and prolongation" of the war. A week later, Luria followed his own advice, asserting in the *MIT Tech* that he had "publicly disassociated himself from any research or work on defense projects in protest against the Vietnam War." He refused to pay the portion of his income tax for war funding.[21]

Even if, as Pollard feared, university scientists largely opposed the war, thousands of other nonacademic scientists—industrial scientists, engineers, and researchers, for example—were either supportive or indifferent. Throughout the late 1960s, Lt. Alvin Young, a research scientist in the Bio-Chemical Division at Eglin Air Force Base, compiled outreach rosters and attended meetings of the Weed Society of America, a professional organization with a substantial corporate demographic and a friendly attitude toward military defoliation. In 1967, the same year that Meselson and Edsall collected five thousand scientists' signatures on a request that President Johnson "repudiate the use of chemical weapons in Vietnam," the American Society for Microbiology voted 600–34 in favor of continuing its advisory role to the army's biological warfare lab at Fort Detrick. At the top levels of government, Donald Hornig expressed his relief at the ASM result to President Johnson, observing that "this is a helpful development in view of strong criticism from some segments of the biological community." The vote underscored the differing viewpoints of the majority of working microbiologists and the small group of largely tenured university scientists who opposed the war, or at least its tactics. At the same time, Hornig's report indicated just how powerful those minority viewpoints were—attracting the attention of the president and his science advisor and forcing an organization-wide referendum on military advising.[22]

Not all corporate researchers supported defoliation, and not every report from government-sponsored scientists offered a clear-cut endorsement of operations. In September 1968 two Monsanto scientists "speaking as individuals and not as representatives of Monsanto" contributed a detailed and harrowing article to *Scientist and Citizen* about the effects of Tordon (picloram), the Dow-produced component of Agent White. They revealed that unlike the shorter-lived phenoxyacetic acids, the effects of picloram on vegetation—particularly forest growth—could last two years, and the chemical could remain in the soil for that duration or longer. It could pass intact through the digestive system of a mule. Moreover, it was unclear exactly what the targeted areas were. The two Monsanto scientists refuted claims that Agent White was only used on conifers, observing that military procurement of picloram far exceeded the amount necessary to target Vietnamese pine forests. They cited a warning by New Zealand researchers that picloram sprayed aerially over large areas could lead to widespread contamination and warned, "Large areas *are* being treated aerially in Vietnam, but no studies of contamination by movement have been forthcoming from that field." Noting that no published research existed on the ecological effect of Agent White, a mixture of picloram and 2,4-D, and echoing the views of so many other scientist-critics, the two men emphasized the lack of existing knowledge and the experimental nature of the spraying.[23]

* * *

That fall, another report from an unlikely source fueled further opposition and offered a bridge between insider and outsider scientific critiques of defoliation. Beginning in the winter of 1967–1968, the State Department, long a source of moderate resistance to the use of chemicals, had requested that a small group of scientists, including Charles Minarik, director of the Department of Defense's Plant Sciences Laboratories, and Fred Tschirley, assistant chief of the Crop Protection Research Branch of the Agricultural Research Service, review various aspects of the Ranch Hand program. The combined reports of the scientists generally upheld the current policy as offering greater benefits than detriments, but Tschirley's views, later published for a wider audience in *Science,* included some troubling observations. Although constrained by time and safety, Tschirley nevertheless toured multiple regions where herbicides had been employed and conducted aerial surveys of target areas. Much of his report was dedicated to dismissing fears: that defoliants had caused changes in local climate, for example, or had permanently poisoned the soil. But he did observe, as earlier Advanced Research Projects Agency scientists had at the beginning of the decade, that mangroves were particularly susceptible to Agent Orange. "The trees were not simply defoliated, but were killed," he wrote. He estimated that mangrove areas sprayed in 1962 could take as long as twenty years to redevelop, with possible repercussions for soil erosion and marine life. He acknowledged that little was known about "the effect of killing mangrove on animal populations."[24]

Tshirley's trip coincided with the onset of the dry season—a period of natural defoliation for Vietnam's forests—and he encountered difficulty distinguishing between normally denuded trees and those defoliated chemically or by fires. He was unable to estimate the effects of defoliants on Vietnam's semideciduous forests, relying instead on his own previous U.S. Department of Agriculture (USDA) research on defoliation test sites in Puerto Rico and similar reports from Thailand. Even then, the multiple applications of herbicides in Vietnam surpassed the test sprayings elsewhere. Whereas Tschirley estimated that Vietnam's dense forest canopy could regenerate within a few years of a single dose of Agent Orange, "a second application during the period of recovery would have a wholly different effect." Again, noting the "scanty" research on tropical forest regeneration, Tschirley turned instead to studies of the island of Krakatau, destroyed so thoroughly by volcanic eruption in 1883 that "the only living thing a visitor saw in May 1884 was one spider." The flora of Krakatau did eventually begin to reappear—a development reassuring to Tschirley, who noted optimistically that Vietnam was neither totally defoliated, covered with ash, nor an island.[25]

Tschirley also observed, as some scientists had predicted, the increased presence of opportunistic grasses and bamboo in defoliated areas. Once again, he extrapolated from his research in Puerto Rico to suggest that the sprayed Vietnamese forest would surely recover within a few years were it "not for the probable invasion by bamboo." He could conclude only that "the time scale for succession in a deciduous forest in [the Republic of Vietnam] is unknown," and "the effect of defoliation on animal populations is truly unknown." On the latter point, however, he somberly noted, "I suspect that bombing, artillery, fire, human presence, and hunting have had a far greater effect than has defoliation." His message was clear: war itself was far more devastating for the ecology of Vietnam than the tactic of defoliation alone, but in many areas, the long-term effects of defoliants were unknown.[26]

Tschirley offered one final, interesting note on the "toxicity of herbicides." He wondered if the most lasting risk would come not from the phenoxyacetic acids such as Agent Orange, but from the arsenic-based compounds in Agent Blue. These constituted only a low risk to mammals, he surmised, concluding broadly that "there is no evidence to suggest that the herbicides will cause toxicity problems for man or animals." He offered no discussion of Agents Orange, Purple, or White. As he had mentioned repeatedly early in the report, however, little data of any kind concerning the defoliants' effects existed. The lack of evidence demonstrating toxicity was the same lack of evidence demonstrating safety.[27]

∗ ∗ ∗

Tschirley's report, republished for a wider scientific audience in the winter of 1969, elicited concern from both inside and outside the government, particularly regarding the heavy use of picloram in Agent White. In the aftermath of the report's publication, Galston wrote to *Science*, "While I continue to oppose most aspects of our chemical warfare operation in Vietnam, if it is to continue it would be better to use the readily biodegradable 2,3-D than picloram, which is so resistant in some clay soils that under 5 percent disappears each year." The State Department's Herbicide Policy Review Committee shared this concern over picloram but still endorsed the overall defoliation program. Their qualified views were countered by an air force study of Ranch Hand and a review of crop destruction by the Joint Chiefs of Staff, both of which contained vigorous reiterations of support for the programs. But within six months, all of these areas of concern—the effects of picloram in Agent White and arsenic in Agent Blue; the military effectiveness of defoliation and crop destruction—would be superseded by a new source of worry: the dioxin contamination of Agent Orange.[28]

In 1949 an explosion at a Monsanto plant in Nitro, West Virginia, had exposed over two hundred workers to a chemical by-product of herbicide production, a polychlorinated dibenzodioxin abbreviated TCDD, commonly referred to as dioxin. The exposed workers were stricken with chloracne, a vicious and sometimes permanent inflammatory skin disorder. In 1964, before the major contracts to produce Agent Orange for the U.S. military had been signed, researchers at the Dow Chemical Corporation noticed the dioxin contamination of their own defoliation products. Dow officials responded by setting a safety threshold for dioxin of one part per million and monitoring manufacturing processes to ensure that no herbicides contained dioxin above that level. The following year, representatives from several major chemical companies met to discuss the problem of dioxin, and Dow representatives observed that some of their competitors—particularly Monsanto—were producing highly contaminated products. The existence and nature of this meeting was not publicized, and details of what transpired were not revealed until more than a decade later during the proceedings of a class action lawsuit concerning the effects of Agent Orange on Vietnam veterans.[29]

Nevertheless, in that same year, 1965, the President's Science Advisory Committee encouraged further study of the effects of Agent Orange, and the National Cancer Institute sponsored new research on the effects of defoliants on animals, to be conducted at the Bionetics Research Lab in Bethesda, Maryland. The action may have been due more to the nascent environmental awareness spawned by *Silent Spring* and other works than by the early efforts of Edsall and Meselson in that year. As David Butler, a senior program officer at the NAS, later explained, "The mid-1960s was . . . a time of burgeoning interest in studies of the mutagenic, carcinogenic, and reproductive effects of chemicals. The National Cancer Institute launched an investigation of the tumorogenic, mutagenic, and teratogenic potential of a number of insecticides and herbicides in 1965 and gave the contract to Bionetics Research Laboratories."[30]

The Bionetics study, involving mice fed large quantities of the herbicide 2,4,5-T, was completed in 1966, but the results were not reported to the Food and Drug Administration until the fall of 1968. Additional testing then confirmed the conclusions of the Bionetics researchers: Agent Orange exposure caused high rates of birth defects in mice. In October of 1969, President Nixon's newly appointed science advisor, former Caltech president Lee DuBridge, announced an abrupt shift in Pentagon herbicide policy. Citing the Bionetics study, DuBridge warned of a possible link to elevated birth defects and stated that Agent Orange would no longer be applied in or near populated areas in Vietnam. In a remarkable turnaround from previous official descriptions of Agent Orange as a "common" herbicide in widespread domestic use, DuBridge reassured Americans that their risk of contact was low since "almost none" of the chemical was used in homes or gardens in the United States; rather, it was only applied in nonresidential areas. The USDA would conduct new studies of the chemical and would restrict government sprayings in the interim.[31]

Other reporters and researchers followed up on the Bionetics experiments, in which pregnant mice heavily exposed to the substance spawned a high rate of deformed fetuses. The AAAS quickly passed a resolution calling for an immediate halt to the use of 2,4-D and 2,4,5-T in Vietnam. Four months later, the USDA acknowledged that the birth defects had likely been caused by a contaminant to the herbicide, possibly skewing the results. The contaminant was quickly identified as dioxin, the dangerous substance responsible for the Nitro chloracne cases and detected by Dow researchers in 1964.[32]

Arthur Galston, who understood in detail just how the process of synthetizing 2,4,5-T could create dioxin as an "unwanted" by-product, pushed his activism to a new level. With the selection of DuBridge as science advisor, Galston had an inside line to the White House. He and DuBridge had been colleagues at Caltech and had forged good relations during the contentious debates over McCarthyism and loyalty oaths in the 1950s. Now Galston reached out to DuBridge and to Meselson, who, as a former student of Linus Pauling, had his own ties to Pasadena. The three men scheduled a private meeting with key "military scientific advisors," during which Galston and Meselson explained the inevitability of dioxin contamination, discussed the inherent risks, and pushed for a ban of herbicide use. Galston later credited the meeting with a monumental result: "DuBridge's recommendation to Nixon that the spray operation be terminated."[33] In the spring of 1970, in the face of mounting criticism, the Pentagon announced the suspension of all Agent Orange use in Vietnam, and the Department of Agriculture strictly curtailed the domestic use of the component 2,4,5-T.[34]

* * *

Galston and DuBridge were not the only influential voices who could claim some credit for shaping policy. In the spring of 1969, spurred by recent United Nations meetings and hopes for the upcoming round of détente discussions between President Nixon and Soviet Premier Alexei Kosygin, the Senate Committee on Foreign Relations held secret hearings on CBW. Meselson, called to testify, carefully addressed the politicians' questions concerning the controversial weapons used in Vietnam: tear gas, herbicides, and napalm. Acknowledging that "there are pros and cons to the use of tear gas in war," Meselson reported that many nations considered tear gas banned by the terms of the Geneva Protocol, and that even as a nonlethal chemical weapon, it ran the risk of creating "a highly undesirable escalation." Herbicides, on the other hand, were not explicitly mentioned in the protocol at all. Unlike J. B. Neilands and other members of the Scientists' Committee on Chemical and Biological Warfare, Meselson did not consider incendiaries such as napalm and white phosphorus to constitute chemical weapons because they caused injury through "intense burning" and not "because of a poisonous action." But whatever their interpretive leanings, Meselson urged the committee to work toward a coherent policy. "The important thing," he told them, "is that there be a uniform rule."[35]

Before the year was out, Meselson got his wish. On November 11, 1969, the anniversary of the armistice of World War I, President Nixon formally requested that the Senate ratify the Geneva Protocol of 1925. By then, the agreement had been ratified by more than sixty countries, including the Soviet Union, China, and all other North Atlantic Treaty Organization and Warsaw Pact countries. But where the 1925 agreement had banned all "asphyxiating poisonous gases" and "bacteriological methods of warfare," Nixon added his own qualification, stipulating that these descriptions did *not* include riot-control agents and chemical herbicides. (A month later, the United Nations General Assembly would vote to confirm the exact opposite interpretation.) The process of ratification stretched into the mid-1970s, and after the final 90–0 Senate vote was taken, President Ford maintained the riot-control and herbicide exceptions through executive order. He tried to temper his controversial interpretation with a qualified no-first-use policy forbidding the "first use of herbicides in war except use, under regulations applicable to their domestic use, for control of vegetation within U.S. bases and installations or around their immediate defensive perimeters" and the "first use of riot control agents in war except to save lives, such as, use of riot control agents in riot situations, to reduce civilian casualties, for rescue missions, and to protect rear area convoys."[36]

Despite the achievement of Meselson's stated hope for a coherent policy, these exceptions to exceptions to exceptions pleased few of the critics who had initially pushed for the protocol's ratification. The concerns of the academic biologists and botanists who had criticized gases and herbicides had been largely ignored; their warnings of the ecological dangers of defoliation were explicitly rejected by the language of Nixon's and Ford's clarifications and exemptions. The problem of escalation was addressed only partially: legal boundaries had been set, but they did not allay scientists' fears that the use of nonlethal chemicals would subtly prepare leaders and populations for the introduction of more dangerous chemical and biological weaponry. More concretely, the new policy codified the rejection of all the ecological arguments made by Galston and others and the humanitarian concerns over super tear gas voiced by Donald Hornig. For Meselson, who had hoped for a "uniform rule" that banned tear gas and anticrop chemicals, it was a Pyrrhic victory.

* * *

Despite their allowance under the Geneva Protocol's exemptions, herbicides were still subject to a severe debate rooted largely in the new dioxin revelations. During the temporary ban, a series of influential studies were conducted by the AAAS, the NAS, and a contingent of Pentagon engineers. Meselson was deeply involved in the first of these, arranging for the procurement of $80,000 from AAAS to conduct, finally, a month-long field study of the effects of herbicides on the ground in Vietnam. The AAAS in turn created an Herbicide Assessment Commission headed by Arthur Westing, which dispatched a team to South Vietnam that included Meselson, Westing, John Constable of the Harvard Medical School, and Robert Cook, a Yale ecology graduate student. The commission's preliminary report, issued in December

1970, offered a depressing picture: mangrove populations had not regenerated, even four years after spraying; dead trees and colonizing bamboo was observed in South Vietnam's hardwood forests; and observations of crop destruction targets suggested that civilians had suffered more from the food loss than the Viet Cong.[37]

Meanwhile, inside the administration, science advisors DuBridge and Ivan Bennett Jr. had urged a permanent end to Agent Orange use and crop destruction practices while the Joint Chiefs of Staff argued vociferously for the resumption and continuation of these programs. When the aging DuBridge resigned in August 1970, Nixon appointed Edward David, a 45-year-old Bell Labs computer specialist, as his successor. David, too, pushed for an end to defoliant use. In November 1970, anticipating harsh reports from the AAAS Commission and fearful of jeopardizing other delicate foreign policy concerns, David proposed a compromise policy: that domestic standards for chemical use in the United States be applied to Vietnam. The military leadership balked at the idea and downplayed the risks of dioxin, but Secretary of Defense Melvin Laird sided with David, cementing the ban on Agent Orange, ordering a phasing out of all herbicides, and ending the practice of crop destruction in early 1971. All stockpiles of Agent Orange in Vietnam were to be returned to the United States as of September 13, 1971.[38]

In October 1970 Congress instructed Laird to conduct a review of herbicide use in Vietnam. The Pentagon in turn contracted separately with both the NAS and with the Army Corps of Engineers' Strategic Study Group (ESSG). The NAS study was headed by Anton Lang, a Russian-born plant physiologist and head of Michigan State's Plant Research Laboratory, and conducted by dozens of international scientists who served as committee members and advisors. Their final report, issued in 1974, echoed the earlier work of Tschirley: they saw evidence of mangrove devastation and forests heavily damaged by all forces of war but lacked sufficient data to confirm any of the alleged human consequences. The three-volume ESSG study similarly offered only qualified support for the use of herbicides. After the first two volumes were released to *Science,* Deborah Shapley wrote that the ESSG report's "faint praise" was actually quite damning. (She called on Les Aspin, then a Democratic congressman from Wisconsin fresh from a stint as a Defense Department systems analyst, to decode the "Pentagonese" of the report.) Shapley concluded that despite "the general unanimous pro-herbicide position taken by DOD in public," much internal disagreement existed behind the scenes. Surveys of officials suggested limited enthusiasm and widespread belief that herbicides would not be useful "in future conflicts." Moreover, Shapley concluded that the report's assertion that "at most, the crop destruction program harassed the enemy" was a "Pentagonese" acknowledgment that the program had largely failed. But the same wording could also be interpreted as a Pentagon attempt to downplay the destructiveness of a controversial program.[39]

Shapley also offered an insider's view of the internal tensions among the services and the Defense Department's civilian leadership. These factional disputes included air force resentment at "offering combat support to the Army" (e.g., dropping herbicides for the army's benefit, at great physical cost); navy reluctance to transport chemical weapons that might accidentally be released into a vessel's closed system; and perhaps most importantly, significantly different views on herbicides among the Joint Chiefs and the Office of Defense Research and Engineering, who favored their use, and the more skeptical Office of the Secretary of Defense. As with similar debates over the intensity of bombing campaigns, hopes for resolution lay in technological development. Shapley concluded by quoting the responses of "two former DOD analysts" who noted happily that herbicides would soon be outmoded anyway by sensors and new detection technology. They observed: "sensors can provide surveillance of an area without stripping vegetative cover for friendly use . . . sensors can be delivered or used fairly independent of weather, and . . . an enemy is not likely to know that a sensor is present, whereas he would be aware of defoliation." The implicit suggestion was that detection technologies provided cover for the shift away from defoliants.[40]

✳ ✳ ✳

Since the dissolution of the air force's Ranch Hand unit and the imposition of the ban in Vietnam, Agent Orange has attracted a controversy far greater and louder than the ecological and political concerns put forth by many scientists during the war years. The Bionetics study spawned extensive new research on dioxin and the potential risks of herbicide exposure to humans. In 1973 Matthew Meselson helped develop highly sensitive tests to detect the presence of small amounts of the substance and reported that dioxin had been detected in fish and shellfish samples at multiple sites in South Vietnam. Other researchers confirmed the direct links between dioxin exposure and birth defects and chloracne. And although often overlooked by their U.S. counterparts, after the war, Vietnamese scientists also took up the mantle of researching the human and ecological consequences of defoliation.[41]

In 1977 the first Vietnam veterans began reporting a variety of illnesses and disorders they believed were direct consequences of their exposure to herbicides. The following year a Chicago CBS affiliate station aired *Agent Orange: Vietnam's Deadly Fog*, a television documentary based on multiple cases of sick veterans and veterans whose children had birth defects, as well as interviews with Meselson and Barry Commoner. As Wilbur Scott wrote in a later history of Vietnam veterans: "In one dramatic swoop, Agent Orange went from the private rumblings of a handful of veterans to the center of national attention."[42]

According to Scott, the Veteran Administration's initial reaction was to try to quell fears, and their hastily assembled Agent Orange Policy Group, populated by former researchers from Monsanto and DuPont, offered a skeptical assessment of the veterans' claims. After conducting their own epidemiology studies, the Environmental Protection Agency officially banned 2,4,5-T in 1978. A personal lawsuit filed that year expanded into a class-action affair on behalf of veterans and their families, with seven corporate manufacturers of Agent Orange named as defendants. Arthur Galston provided key explanations of the dangers of dioxins. Much later, Galston would reflect on the difficulties of proving causality in such a case—both in terms of whether exposure to Agent Orange and its dioxins caused the variety of illnesses among veterans and whether illnesses connected to dioxins stemmed in fact from the specific dioxins contained in Agent Orange. Like many scientists at the time, he saw clear links to chloracne and birth defects, which would have affected pregnant Vietnamese but not necessarily male veterans. Nevertheless, after six years of litigation, the lawsuit culminated in a $180 million settlement, which effectively ended claims against the Agent Orange manufacturers. Appeals to the U.S. government, however, as well as Pentagon-sponsored studies of Ranch Hand veterans and other exposed groups, would continue into the twenty-first century, with scientist experts called upon by both sides in the debate.[43]

✳ ✳ ✳

The controversy over defoliant and gas use, the special issues of *Scientist and Citizen*, the field reports published in *Science*, the AAAS prodding of

Pentagon officials, and the heavily critical reaction to the MRI report all reveal two key trends in antiwar scientists' political activity during this period. First was the tendency to emphasize the novelty of the chemicals used and to describe their application as experimental. Such language was consistent with the terms used by Pentagon planners themselves, who had, at the beginning of the decade, referred to Vietnam as a "proving ground" for new weapons technologies. For critics of the war, however, the language of experimentation also evoked images of Nazi doctors and the horrors of World War II. A second related trend was the inclusion of defoliants and tear gas in the category of chemical warfare. Phrases such as "chemical warfare" and "asphyxiating gases" instantly conjured images of World War I; of the "guttering, choking, drowning" men of Wilfred Owen's poetry. Such language targeted the specific and the general—the specific technologies used and the general horrors of war—but not the righteousness of American involvement in Vietnam.

The writings of Barry Commoner and his colleagues typify this second trend. As they wrote in *Science*, "Apart from the morality of the war itself, which is not at issue here, continued use of a weapon with effects that are so poorly understood raises serious moral and political questions for the U.S. government and for the American people." They chose to attack as immoral the technology of the war rather than attack the war directly. Their commitment to curbing CBW was surely genuine, but it seems doubtful that they—and the scientists so busy circulating petitions and penning letters to the president—were quite so indifferent to the cause and nature of the war itself. Criticizing technology was simply a more effective entryway into Vietnam debates for scientists. From the standpoint of credibility and expertise, a chemist expressing sympathy for Ho Chi Minh could be dismissed by war planners alongside hippies and student protesters as ill-informed, naïve, or an example of the "wild-eyed radical." What did a chemist know of Vietnamese politics? A chemist voicing concerns about chemical weaponry and persistent toxins, however, was far more likely to be taken seriously by policymakers. Scientists opposed to the war were trying, consciously or unconsciously, to make the most of the kinds of influence they had, and media outlets reinforced these tendencies by elevating scientists' commentary about technology over broader critiques of the war.[44]

Yet for these very reasons, antiwar scientists found their arguments self-limiting. Criticizing the morality of the technology of Vietnam meant wading into a morass of ethical ambiguity and murky historical precedent. In their

essay in *Scientist and Citizen,* the doctors Sidel and Goldwyn had specifically defined chemical weapons as "agents which produce their effects directly as a result of their chemical properties rather than as a result of blast,
heat, or other physical effects of a chemical reaction." Thus, the destruction produced by oxidizing gunpowder or the burning of napalm were not
included and not explicitly criticized. Why? If part of the risk of nonlethal
chemical weapons was the potential escalation to lethal forms, why not criticize existing lethal weaponry already used in war? For many scientists, the
answer lay in a modification of the classic nuclear nonproliferation arguments:
if unchecked, even nonlethal CBW could set a dangerous precedent for use,
eventually unleashing widespread and uncontrollable dangers far beyond
the scope of conventional weapons. But the antiwar scientists who made
these technical arguments hardly wished to replace the escalated use of
chemicals with escalated bombing campaigns.[45]

Also problematic for scientist critics was the fact that opposition to chemical weaponry had, for historical reasons, followed an unusual path of development. As Matthew Meselson explained to the Senate Foreign Relations
Committee, "The record shows that the governments and peoples of the
world have come to practice and expect a degree of restraint against the
use of chemical and biological weapons not found for any other class of
weapons, except nuclear ones." Meselson attributed this restraint to fears
that both technologies could "open up an unfamiliar and highly unpredict-
able dimension of warfare." But in the aftermath of the Vietnam gas revelations of 1965, Hanson Baldwin, a military journalist and historian, considered the worldwide reaction against gas to be part of "the growth of curious
and inconsistent distinctions between weapons systems." Baldwin explained
this outcome as due to the persistent and frightening images of "blue-faced
men at Ypres, choking to death." Americans dropped fire bombs and atomic
weapons during World War II but refrained from the use of "inhumane"
gas. Besides the reported Italian use of mustard gas in Ethiopia, all of the
fighting parties during World War II had avoided using gas on the battlefield. To Baldwin, the perceived morality of weapons usage arose from the
vicissitudes of historical precedent. Weapons that caused injury through
flame, explosion, or force had a long history of use in warfare and were therefore expected and understood. Weapons that caused harm through invis-
ible mechanisms—toxic gases, bacteria, radiation—were unfamiliar and terrifying. Louis Fieser, the Harvard chemist and inventor of napalm, indirectly
promoted this interpretation, proclaiming in 1967 that despite current ap-

plications, he felt "no guilt" about his napalm work during World War II because it was "better than working on poison gas," which he "didn't feel good about . . . at all."[46]

But examples in other contexts refute Baldwin's claims. In a 1969 article about police departments' use of mace, Seymour Hersh described reactions to the practice as a kind of "index of popular opinion." For example, "white liberals" expressed outrage at the use of gas against civil rights protesters at Selma but went silent when the same chemical was used to restrain white protesters blocking James Meredith's enrollment at the University of Mississippi. The context mattered more than the usage itself. The activism of scientists and the public at large during the Vietnam War reflects this point. Of the controversial weapons technologies and tactics used, almost all— napalm, intensive aerial bombardment, and tear gas—had been employed during World War II and the Korean War without provoking the intense outcry characteristic of the Vietnam era. The use of herbicides and crop destruction in war was new for the United States, but it is plausible that had chemical herbicides been used against Nazi-supporting farms, domestic criticism in 1945 would likely have been far more muted than what ensued in the 1960s and 1970s.[47]

At the same time, the practice of scientists drawing on their credibility as experts to criticize weapons technology had its own strange path of development. It was a form of political intervention honed in the aftermath of the Manhattan Project, and its reappearance during the Vietnam era reflected the influence of that extraordinary precedent. The idea that scientists held a unique authority—an *obligation*—to comment on weapons technologies was evident in Arthur Galston's reflection that with the use of defoliants in Vietnam, "the botanist, probably the last of the scientific innocents, was unexpectedly catapulted into the same ethical hot pot as other scientific colleagues." It was evident in Milton Leitenberg's warning in *Scientist and Citizen* that "as physicists learned to create destruction with nuclear weapons, microbiologists are learning to create disease with biological weapons." And it was evident in the deep ethical concerns described by Sidel and Goldwyn in the *New England Journal of Medicine*. A doctor asked to help develop biological weapons, they wrote, "must carefully evaluate his own attitudes toward the rights and duties as a citizen and as a doctor." They warned of the dangers of inaction, invoking the awful specter of Nazi science and citing an American Medical Association editorial mourning "the failure of German medical organizations and societies to express in any manner their disapproval" of Nazi experimentation. If Arthur Galston saw his early work on plant hormones as the ethical impetus for action, the two doctors felt the weight of the Hippocratic oath in a similar way. They closed their article with an invocation of lines from the famous oath: "Neither will I administer a poison to anybody when asked to do so, nor will I suggest such a course . . ." Their authorship of the article suggested a third obligation: to speak out against others' administration of perceived poisons.[48]

Manhattan Project scientists had not opposed fighting World War II, nor was every scientist who worried about chemical weapons intrinsically criticizing American involvement in Vietnam. But for those who parlayed antiwar sentiment into allegations of immoral technology, the policy implications of their strategy turned out to be extremely limited: beginning at the end of the decade, in the face of expanding scientific opposition and growing evidence of the risks of human exposure to the dioxin-laced Agent Orange, wartime policies regarding herbicides eventually began to change. But the war itself continued unabated, spilling outward into Cambodia and Laos and fueling deeper and more radical modes of protest.

* * *

There were, of course, some exceptions to this trend, as many outspoken scientists did denounce the war itself without manufacturing technological or scientific justifications for their political opposition. And another, broader variant of the argument against CBWs came from scientists connected to the growing environmentalism movement. These researchers were more likely to criticize the overall effects of war, including not just the use of chemicals but the impact of all aspects of battle. For example, Michael Newton, a forestry researcher affiliated with Oregon State University, wrote in a 1968 letter to *Science* that "the philosophical argument against the use of unsolicited biological agents is understandable. But such tactics should not be criticized on the basis of genocidal, biocidal, ecological, or economic considerations because the land and the organisms it supports will recover from such treatment more quickly than from various other instruments of war, and with far less pain. Wouldn't it be more constructive to recommend ways of making the use of all such instruments unnecessary?" Pfeiffer and Orians offered a similar acknowledgment in a 1970 article based on their research trip to Vietnam, noting that "although it has not attracted the concern of

American scientists, the damage caused by raids with B-52 bombers is of considerable ecological significance." Arthur Westing, a botanist at Windham College in Vermont, traveled to Vietnam four times from 1969 to 1973 to research the ecological effects of the war and published his results though the Stockholm International Peace Research Institute (SIPRI) in 1976. He investigated not just chemical use, but also "mechanized landclearing," fire, flood, the use of explosives, and other destructive forces, all of which resulted in the "severe abuse" of the Vietnamese ecosystem, a process he and others termed "ecocide." Westing closed his report with some philosophical reflections, including a criticism of the human tendency to consider the environment only in terms of "anthropocentric concerns"—in other words, the tendency to care about protecting the environment in order to ensure human needs and comforts. He then asked, "But should not living things, and nature as a whole, have some level of immunity in their own right?" Were they not "'noncombatant bystanders' to man's martial foibles? What, indeed, is man's fundamental relationship to the land?" These views were echoed by Barry Weisberg, who in 1970 put together a book titled *Ecocide in Indochina: The Ecology of War,* drawing on Arthur Galston's definition of ecocide as "the willful destruction of the environment."[49]

Although their concerns about ecocide had not seemed to resonate with the government and military planners running the war, scientists such as Westing nevertheless continued their research on Vietnam, following up on the environmental toll in the decades afterward. In 1983 Westing worked with SIPRI to organize an "International Symposium on Herbicides and Defoliants in War," held in Ho Chi Minh City and including seventy-two scientists—ecologists and physiologists—from twenty countries, plus a contingent of fifty-six Vietnamese scientists. The book that resulted from the proceedings, *Herbicides in War,* published in 1984 and edited by Westing, documented the ecological and human consequences of defoliant use. The short-term consequences of spraying were organized by ecosystem. In "dense inland forest" areas, the effects included soil erosion, loss of animal habitats, and replacement of destroyed vegetation by opportunistic grasses. In the mangrove swamps along the southern coast, SIPRI scientists reported that "virtually nothing remained alive after even a single herbicide attack and the resulting scene was weird and desolate." Unlike the inland forest area, these defoliated regions were not even colonized by aggressive grasses or bamboos after the initial devastation. SIPRI described the mangroves as "clearly the ecosystem most seriously affected" by the war; their loss

hastened erosion and resulted in declines in the fish and shellfish populations. The SIPRI-gathered scientists also noted the presence and persistence of dioxin, describing a host of ailments plaguing humans exposed to
defoliants, including allergies, temporary nausea, headaches, and respiratory problems, as well as exacerbation of existing illness and malnutrition.
Defoliation had also resulted in mass human displacement and the increased spread of disease.[50]

Not all scientists who had criticized defoliant use in Vietnam shared this
commitment to ecological preservation. On the concept of ecocide, Arthur
Galston himself later reflected that "this term was coined to evoke the specter
of the parallel crime of genocide, justly condemned after the Nuremburg
trials." But in Vietnam, he noted decades later, defoliation "did not permanently destroy the productivity of the ecosystem" although it did considerable damage resulting in soil erosion, the destruction of mangroves that had
been fish habitats, and the proliferation of "junk" vegetation like bamboo,
which overtook former teak habitats. Galston also noted the unintended consequence of food destruction operations: the accidental destruction of
H'mong crops, which led to exodus from "ancestral homelands" and in many
cases, emigration to the United States. But Galston was blunt and honest:
despite his own deep discomfort with herbicide use, he conceded that it had
not been ecocidal and that, despite all its ecological damage, "the herbicidal
campaign in Vietnam produced certain military advantageous results which
may, in the end, have justified their use in a bitterly contested war." The
costs had been steep, though, wrote Galston—steep enough to make repeat
use highly unlikely. This prospect seemed a source of great relief.[51]

* * *

In the mid-1980s Ranch Hand veteran Paul Cecil wrote in his history of
the operation that its demise in 1971 was a result of "internal and external
political pressures on the US government." Admiral John McCain Jr., the
father of the Arizona senator, blamed two overlapping groups: scientists and
antiwar radicals. J. B. Neilands, the biochemist who had cofounded the Scientists' Committee on Chemical and Biological Warfare, credited "the science community and the Congress" with reform, particular the New York
delegation of Bertram Podell, Richard McCarthy, and Ed Koch. Michael
Gough, a biologist who served at both the Department of Veterans Affairs
and the Congressional Office of Technology Assessment during the peak of

the postwar Agent Orange controversy, surmised that the Pentagon had abandoned the defoliant due to dioxin's link to cancer and allegations of chemical warfare. In a more comprehensive evaluation, William Buckingham attributed the shift in policy to several factors: a reduction in spraying in line with Nixon's stated goal of "reducing the American presence in Vietnam," the evidence linking Agent Orange to birth defects, and the mounting criticism from scientists, particularly the members of the AAAS's visiting teams, whom Buckingham credited with helping "to hasten Ranch Hand's demise."[52]

Whether it was the allegations made by an international chorus of critics that included both scientists and nonscientists, the revelations about dioxin, the meeting between Galston and DuBridge, general public disapproval, or the Congress-ordered Pentagon review that was most responsible for the shift in policy is hard to determine. But one conclusion is clear: scientists' carefully articulated concern during the earlier years of the war—focusing almost exclusively on ecology and the risks of escalation—did not halt the practice of defoliation in Vietnam. It did spur significant political pressure from Congress, but it was not as influential as scientists' later research and concern regarding the links between Agent Orange and dioxin and dioxin to birth defects. Combined with Pentagon acknowledgment of the mediocre efficacy of the defoliation program itself, both contributed far more significantly to the change in policy.

This outcome may have been due in part to the particular character of the debate over the weapons systems employed in Vietnam. Unlike earlier arguments about the hydrogen bomb or the enforceability of a test ban, the controversy over defoliants and gases never included questions of feasibility, for which scientific input could more easily shape political concerns. Instead, the issues raised by scientists about chemical weapons were either moral in nature or rooted in the murky field of ecology, with its difficult and sometimes speculative predictions about the complex ecosystems of South Vietnam, for which little background knowledge existed in the United States. Advocates of defoliation and tear gas use were able to circumvent the broader moral arguments through the ratification and the exemptions of the Geneva Protocol, and they were able to find numerous researchers with more charitable interpretations of the complicated ecological data being collected. Agent Orange, once it was linked to birth defects, was quickly eliminated, but the use of alternative herbicides continued. The conclusions of the in-house studies conducted by Minarik, Tschirley, and others were decidedly less ominous than the outsider studies of Pfeiffer and the AAAS. For every cry of ecocide, there was an internal scientist with a confidently positive assessment for State Department and Pentagon bosses, a welcome confirmation that "the effects of defoliation have not been as disastrous as anticipated."[53] By the time the longitudinal studies detailing the exact consequences could be completed, the war and any opportunities to change its tactics would be long over.

FIVE

Disaster and Disillusionment in Vietnam

A YEAR AFTER he had campaigned for the president with Scientists and Engineers for Johnson, Harvard chemist George Kistiakowsky, Eisenhower's former science advisor, watched with alarm as the war escalated dramatically. It was1965, and the United States was launching an intensive bombing campaign against North Vietnam, deploying hundreds of thousands of U.S. combat troops, and suffering the first heavy American casualties. The day after reports of gas use appeared in nearly every major American newspaper, Spurgeon Keeny, formerly a technical assistant to James Killian and an arms-control advocate within the Johnson administration, appeared at a Harvard seminar on Science and Public Affairs organized by Carl Kaysen. Kistiakowsky met with Keeny after the seminar and informed him that a number of distraught faculty members were organizing among the old Scientists and Engineers for Johnson group. They planned to send a letter to the president expressing opposition to current Vietnam policy. Kistiakowsky confided in Keeny that he sympathized with the activists even though he had declined to take a leadership role in their effort. He offered to meet with McGeorge Bundy to describe and explain Cambridge attitudes toward the war. Reporting the conversation to Bundy the following day, Keeny repeated Kistiakowsky's characterization of the activists as "responsible, sober citizens and not the Alex Rich and Bernie Feld variety." Keeny urged Bundy to meet with Kistiakowsky, both to allow Kistiakowsky an outlet "to unburden himself" and to allow Bundy the opportunity to personally present the administration's position. Keeny emphasized the value of Kistiakowsky's opinion: "I think he could be very influential in keeping some members of the Cambridge community from straying too far off the reservation." A few disgruntled scientists could be marginalized easily as "wild-eyed" cranks, but a broad, organized coalition of Harvard and Massachusetts Institute of Technology (MIT) scholars could pose a serious political problem.[1]

Born in Ukraine in 1900, Kistiakowsky had fled his homeland during the upheaval of the Russian Revolution and ensuing civil war, earning a doctorate in chemistry in Berlin and eventually securing a position at Harvard in 1930. During World War II, he joined the Manhattan Project and witnessed firsthand the devastating power of the Alamogordo test. His honest criticism of nuclear policy during the Sputnik boom years had impressed Eisenhower, who chose Kistiakowsky to succeed James Killian as his top science advisor. Though he had returned to his post at Harvard at the end of the Eisenhower presidency, in 1965 Kistiakowsky was still receptive to personal outreach from key administration officials, and he was himself an important insider liaison to outsider scientists.

Keeny's worries were apt; Kistiakowsky himself would soon stray "far off the reservation." In January 1966 Kistiakowsky wrote a lengthy letter to President Johnson, invoking his status as one of the president's "Advisors on Foreign Policy" and urging de-escalation, possibly through a new strategy of establishing secure noncommunist "enclaves" around Saigon. His suggestions were curtly rebuffed by McGeorge Bundy. Reaching out to likeminded colleagues at Harvard and MIT, Kistiakowsky formed the "Cambridge Discussion Group" as an alternate means to promote de-escalation. Fellow member John Kenneth Galbraith described the informal organization to President Johnson as "a group of your well-wishers and supporters here, several on the scientific side derived from the original Scientists and Engineers for Johnson," who "have been meeting on the problems of Viet Nam." Galbraith assured Johnson that the group's purpose was "assistance and not criticism." Other members included Jerome Wiesner and Jerrold Zacharias of MIT, Frank Long, Carl Kaysen, and Richard Neustadt.[2]

In the spring of 1966, the Cambridge intellectuals began meeting with key Pentagon representatives to propose a special summer session to address possible technological solutions to the war, including the feasibility of a kind of anti-infiltration barrier to prevent Viet Cong entrance into South Vietnam, an idea that had earlier been broached by Harvard Law professor Roger Fisher to John McNaughton. Although initial military reactions to the barrier idea had been unenthusiastic, McNamara himself agreed to meet with members of the Cambridge Discussion Group. Short of meeting with President Johnson directly, Kistiakowsky now had access to the top decision makers in charge of U.S. foreign policy.[3]

* * *

At the same time that Kistiakowsky was consulting with his Cambridge cohort to evaluate new applications of technology, a parallel group on the West Coast was considering a similar proposal. The Jasons, the coterie of young physicists assembled by the Institute for Defense Analyses (IDA) in the aftermath of Sputnik, had by 1966 taken up a variety of Vietnam-related problems. At the Jason spring meeting, William Nierenberg was now organizing a group to study technical approaches to interrupting transit through the Ho Chi Minh Trail.[4]

Whereas the Jasons' work in the early 1960s had been wide-ranging, with an emphasis on the technologies necessary to enforce a nuclear test ban—VELA detection techniques for multiple environments, for example—by 1964 problems of counterinsurgency were appearing with more frequency on the agendas for the famed summer sessions in La Jolla and Cape Cod. IDA had also shifted focus, producing a new round of reports on topics such as "Human Smog as an Ambush Detector" and "Power Sources for Remote-Area Counterinsurgency" and organizing an IDA-wide "show-and-tell meeting on counterinsurgency." At the Jason meeting in the spring of 1964, the familiar sessions on missile technologies and deterrence took place, but a new panel headed by William Nierenberg addressed issues of counterinsurgency and limited war. The group listened to reports from White House representatives and Advanced Research Projects Agency (ARPA) personnel explaining Project AGILE and "remote area conflict." Seymour Deitchman, the physicist and limited-war advocate from the Pentagon's Defense Research and Engineering Department, presented a personal account of his recent trip to Vietnam. Political scientist and Indochina expert Bernard Fall was the rare nonphysicist member of the panel.[5]

Berkeley physicist Kenneth Watson later recalled that the Jasons' involvement in Vietnam began in the summer of 1964, "before the actual war," and was initially "a marginal thing." Bernard Fall offered lectures on Vietnamese culture and history throughout the summer session, but to Watson "it was all background." For other Jason members, however, 1964 marked the start of more serious involvement. Throughout that spring, Jason administrator

David Katcher was forwarding reports on counterinsurgency in Southeast Asia to panel members Seymour Deitchman and Murray Gell-Mann.[6]

The gradual inclusion of counterinsurgency and limited-war topics coincided with a decrease in enthusiasm among Jason members. In September 1964 Marvin Goldberger of the California Institute of Technology (Caltech) wrote to the Jasons on the occasion of the group's sixth anniversary. He acknowledged that "it is hard to put aside one's regular academic research and teaching duties during the course of the year" but nevertheless worried that it was becoming increasingly difficult "to get people to take initiative to get involved and stay involved." Goldberger and others discussed the creation of new programs to attract younger scientists, or "junior Jasons," such as postdoctoral fellowships, partnerships with Jason mentors, and consulting opportunities. Goldberger seemed unsure whether the problem was a lack of youthful enthusiasm or deeper ambivalence about the relevance or appropriateness of the group's projects. He offered a last-ditch pep talk: "Jason has nothing to be apologetic about. We have made definite contributions and these are widely acknowledged in the defense community."[7]

Whether actual ambivalence existed or not, Jason members acknowledged among themselves the controversial aspects of their work. As the Nierenberg Panel prepared its final report on counterinsurgency, panelists debated how widely to distribute the document. Katcher reported that Harold Hill of ARPA wanted the Jasons' explicit approval of distribution plans because "the nature of the work would make the group vulnerable to criticism." In Katcher's view, though, "As far as I can tell, no one is ashamed of this paper and would want it hidden from view."[8]

Katcher did not attribute the Jasons' malaise to the shift toward counterinsurgency and limited war. In a 1966 letter, he linked falling productivity to boredom, noting behavior that was surprisingly "passive for a group of individuals selected because they aren't." He hoped organizational shakeups might help, whether in the direction of decentralization or greater institutionalization. Other explanations were in the air, however. In July 1965 Jason members and IDA's Jack Ruina discussed ways to encourage the creation of a British equivalent of Jason. He acknowledged that nuclear technology and strategy had drawn the first wave of promising young physicists to Jason, with "the problem of antimissile defense" serving as "one of the powerful forces that motivated busy young scientists to take time out for Jason work." But in 1965, antimissile defense was arguably no longer the top area of research for the Jasons. Kenneth Watson later recalled that from 1965 to 1969, it was work connected to the war in Vietnam that became "a predominant Jason activity" and "a major Jason effort."[9]

Watson's recollection confirms the impact of Vietnam on the organization's sense of work and mission, but whether it truly "predominated" is difficult to gauge. Of the more than forty Jason publications in 1965–1966, the majority concerned problems of nuclear weaponry: missile penetration, effects of nuclear explosions, and aspects of antiballistic missile technologies. Six items issued under the heading "Remote Area Conflict" were clearly connected to the war in Vietnam: "Tactical Nuclear Weapons in Southeast Asia" (S-266); "Interdiction of Trucks from the Air at Night" (P-289); "Manned Barrier Systems: A Preliminary Study" (P-322); "Air Sown Mines" (P-315); "Tactical Nuclear Weapons in SE Asia; Questions Requiring Further Study" (N-406); and a top-secret assessment, "Air Supported Anti-Infiltration Barrier (C)" (S-255), prepared by Deitchman, Gell-Mann, Nierenberg, and a dozen others. The first and last of these reports would later draw enormous attention and controversy, though they constituted less than 5 percent of the Jasons' output that year.[10]

In the spring of 1966, the Cambridge Discussion Group and the West Coast Jasons joined forces. On April 15 MIT's Jerrold Zacharias wrote to Gell-Mann to inform him that he and his Cambridge colleagues had been meeting with Defense Department officials to discuss "a special study of the military and technological options open to the U.S. in Vietnam." The catalyst, according to Zacharias, was frustration with the escalation of the war and the heavy death toll. "Our hope," he wrote, "is that by re-examining the present military tactics, especially in the light of technological opportunities that may not have been adequately considered, military alternatives might emerge that would be less costly and more likely to lead to a political solution." He invited Gell-Mann to join the group for an exploratory meeting, to be held in the penthouse of the MIT Faculty Club in early May.[11]

An attached draft proposal clarified the group's goals more precisely. In order to "enhance the probability of achieving military objectives," the group should evaluate a variety of military options, including further escalation, continuation of current policy, or de-escalation (in various forms), drawing on potential new innovations in communications, transportation, reconnaissance, weapons systems, and "geographical barriers (the possibility of 'sealing off' South Vietnam)." Ideally, a six-member steering committee would lead a team of up to fifty scientists and engineers during a multiweek summer study. Recipients of this proposal included Luis Alvarez, Seymour Deitchman, Richard Garwin, Murray Gell-Mann, James Killian, George Kistiakowsky, I. I. Rabi, and Jerome Wiesner, among others.[12]

Zacharias's invitation appealed to scientists who, like him, hoped to contribute to a de-escalation of the war. For those with deeper objections to the war itself, the prospect of the summer study offered only small hope and stirred other complex concerns. Such views were conveyed with devastating clarity by George Rathjens, then head of IDA's Weapons System Evaluation Division. In a letter to Kistiakowsky in early April 1966, Rathjens expressed interest in the summer study but concluded with a painful assessment of his frustration and his ethical reservations about both the war and his tenure at IDA:

> I am extremely upset about the whole Viet Nam business. This has been one of the things that has made the IDA job so difficult for me. . . . [Last summer] if I had then foreseen our getting involved in Viet Nam as we have (and my reacting as I have) I probably wouldn't have done it. I now have the feeling that I am to a substantial degree an instrument for a policy with which I am very much in disagreement but which I have damned little chance of influencing. The bombing of the North just about brought me to the point of resigning despite my commitment to stay two years, but I have stayed on arguing with myself that I have, or may have, more opportunity to influence things than if I left abruptly. But the hell of it is that I'm not sure whether this is really right or just a rationalization for doing the easy thing. Events since the resumption of bombing, and most recently the Rostow appointment, haven't made things any easier. I am writing all this because I want you to know that I do feel very strongly about this business and would welcome the opportunity to help in any way I can in your efforts. Unfortunately, I am afraid I am least qualified to help with respect to the political and moral aspects of the problem where I think we are most clearly in error, and qualified, if at all, only on the military side . . . I feel awfully frustrated and uncomfortable about not being able to find some mechanism for trying to affect what we're doing. . . .[13]

Kistiakowsky replied with appreciation and honesty. "As you know I feel about Vietnam just like you do," he wrote. But he planned to attend the summer study, explaining, "I personally have considerable misgivings about anything useful coming out of that kind of undertaking but will participate to the best of my ability because I know of no obviously better alternative."[14]

With Kistiakowsky's qualified enthusiasm, Nierenberg's Jason panel and members of the Cambridge group (now dubbed "Jason East") quickly consolidated their efforts, arranging for two two-week IDA-sponsored summer sessions to be held at the Dana Hall private school in Wellesley. McNamara requested that among the other projects, the scientists specifically address the prospect of "a fence across the infiltration trails, warning systems, reconnaissance (especially night) methods, night vision devices, defoliation techniques and area denial weapons." But the scientists' hopes for altering the course of the war went beyond plans for a barrier. As one attendee wrote to his colleagues, "The substantive question is, considering the overwhelming philosophical power assembled in Wellesley, whether or not some effort should be made to enlist outstanding physiologists, biochemists etc. in order to explore the possibility that a 'Manhattan District' effort could not in fact produce an effective system in time to stop this war in Vietnam." Such a sweeping project never materialized; instead, the summer sessions were devoted to studies of the bombing campaign and the design for a potential anti-infiltration barrier.[15]

✳ ✳ ✳

The Dana Hall summer meeting resulted in a series of Jason reports related to Vietnam concerning the bombing campaigns, the construction of the "electronic barrier" to cut off the Ho Chi Minh Trail, and the use of tactical nuclear weapons (TNW)—all controversial subjects that would later attract considerable attention from the growing antiwar movement.

The Jasons' first major study assessed the effect of Operation Rolling Thunder, the campaign of aerial bombing in North Vietnam that adhered strictly to McNamara's concept of gradual escalation. As historian John Lewis Gaddis has written, "The bombing campaign against North Vietnam was intended to be the most carefully calibrated military operation in recent history." Over the objections of military leaders who preferred more intensive bombardment, Rolling Thunder began under heavy restrictions, many of which would be gradually lifted over time. Gaddis has estimated that "the scale and intensity of the bombing progressively mounted, from 25,000 sorties and 63,000 tons of bombs dropped in 1965 to 108,000 sorties and 226,000 tons in 1967." But in his final estimation, "none of it produced discernible progress toward what it was supposed to accomplish: a tapering off of infiltration into South Vietnam, and movement toward negotiations."[16]

It was in the context of this failure of Rolling Thunder, combined with the large-scale introduction of ground troops beginning in 1965, that the Jasons undertook their evaluation of the effectiveness of the bombing campaign in North Vietnam. The effort would eventually result in two reports, the second of which, a weighty four-volume affair, offered the more thorough analysis and assessment. But it was the first volume, completed in the midst of the escalation in August 1966 and clearly intended to affect policy quickly, that created controversy and exacerbated tensions between the scientists and military leaders.

Titled "U.S. Bombing in Vietnam: The Effects of U.S. Bombing on North Vietnam's Ability to Support Military Operations in South Vietnam and Laos: Retrospect and Prospect," the report drew on the "extensive official analyses" provided by the U.S. military and intelligence communities. Within it the scientists analyzed the outcome and effectiveness of the bombing campaign as well as the underlying logic and assumptions behind it. The report reached five major conclusions: that the bombing had had "no measurable effect on Hanoi's ability to mount and support military operations in the South at the current level"; that aid from China and the Soviet Union had offset the damages caused by bombing; that escalating and expanding the scale and scope of the bombing would be unlikely to alter North Vietnam's ability to support military actions in the South; that it would be difficult if not impossible to define a damage level at which North Vietnam would capitulate; and that in the face of a year of bombing, Hanoi's infiltration into the South had actually accelerated.[17]

In providing the context for their work, the Jasons noted astutely that they were intervening in an ongoing disagreement between the Pentagon's civilian leadership and top military brass. The bombing campaigns so far had followed McNamara's strategy of graduated escalation, the "graduated military pressures directed systematically against [North Vietnam's] ability to support the insurgencies in South Vietnam." The Jasons acknowledged that this strategy had been adopted over the objections of the Joint Chiefs of Staff, who had called for a campaign using "the full limits of what military actions can contribute." During the bombing pause of 1965, the entire campaign had been evaluated, with the general military and government consensus that it had failed to deter North Vietnam's activities in the South. The Joint Chiefs blamed the failure on the limited nature of the bombing, arguing again for a vastly increased campaign. But when the evaluation period ended, only a limited resumption of bombing began, on January 31. As the Jasons were writing the report, considerable disagreement still existed. The Joint Chiefs proposed mining the major harbors and targeting petroleum resources, transportation systems, and economic and industrial sites while the intelligence community was, in the words of the Jasons, "pessimistic about the prospect of achieving a level of interdiction that could significantly reduce the flow of essential war materials through North Vietnam or even prevent the flow from reaching higher levels than in recent months." The North Vietnamese, intelligence officers suggested, were too good at compensating, improvising, and finding alternate routes for southern assistance for *any* bombing campaign to be successful.[18]

The Jasons largely sided with the intelligence experts, agreeing that further bombing would be disruptive but likely insufficient to achieve Rolling Thunder's original goals. The Jasons observed accurately that "the Joint Chiefs of Staff have never fully accepted the strategy of 'graduated escalation' that was finally adopted for the air attacks on North Vietnam." But they considered this analysis misguided. North Vietnam was not a "complex industrial society" vulnerable to attacks on its infrastructure; rather, it was "relatively primitive," so that the assumptions underlying the proposed campaign of the Joint Chiefs would have "little applicability."[19]

Despite their agreement with the conclusions of the intelligence reports, the Jasons argued that the analysis and the methodology of the intelligence community and Pentagon planners were also deeply flawed. The scientists were particularly skeptical of the assumption that key aspects of the war could be calculated and predicted with any meaningful reliability. For example, they wrote, intelligence analysts tended to evaluate the bombing campaign in isolation without considering it in concert with other operations, such as the trajectory of the war in the South. As a result, "The fragmented nature of current analysis and the lack of an adequate methodology for assessing the net effects of a given set of military operations leaves a major gap between the quantifiable data on bomb damage effects . . . and policy judgments about the feasibility of achieving a given set of objectives." The Jasons concluded damningly that "there is currently no adequate basis for predicting the levels of US military effort that would be required to achieve the stated objectives—indeed, there is no firm basis for determining if there is any feasible level of effort that would achieve these objectives." In a passage later quoted by the Senate Committee on Foreign Relations, the

Jasons noted pointedly that bombing planners had ignored centuries of historical lessons and clung instead to flawed and overly optimistic assessments of the bombing's success:

> Initial plans and assessments for the Rolling Thunder program clearly tended to overestimate the persuasive and disruptive effects of the US air strikes and, correspondingly, to underestimate the tenacity and recuperative capabilities of the North Vietnamese. This tendency, in turn, appears to reflect a general failure to appreciate the fact, well-documented in the historical and social scientific literature, that a direct, frontal attack on a society tends to strengthen the social fabric of the nation, to increase popular support of the existing government, to improve the determination of both the leadership and the populace to fight back, to induce a variety of protective measures that reduce the society's vulnerability to future attack, and to develop an increased capacity for quick repair and restoration of essential functions. The great variety of physical and social countermeasures that North Vietnam has taken in response to the bombing is now well documented in current intelligence reports, but the potential effectiveness of these countermeasures was not stressed in the early planning or intelligence studies.[20]

In this area, as elsewhere, the Jasons offered criticism rooted in commonsense scrutiny and observation, devoid of any particularly complex scientific or technical analysis.

✳ ✳ ✳

Reactions to the report were predictable. Robert Ginsburgh, the assistant to the Joint Chiefs of Staff and a member of the National Security Council's staff, agreed with some of the report's conclusions, particularly in its descriptions of the limited effects of the bombing. But he considered the Jasons' overall predictions ill-founded and overly pessimistic. How could anyone know that mining Haiphong and other harbors would not have a substantial effect? Ginsburgh forwarded the Jasons' report to Rostow with an accompanying memo noting that despite the Jasons' arguments, it would be "very difficult . . . to prove conclusively" what the effects of an expanded bombing program would be. Similarly, Ginsburgh questioned the Jasons' assertion that it was impossible to determine the limit at which bombing would finally halt North Vietnamese military activity. To Ginsburgh, asserting that the limit was unknowable "is the kind of statement that could be made right up to the time that [North Vietnam] decides to sue for peace." Similarly, the Jasons' claim that North Vietnamese "will" seemed undiminished "is the kind of statement that could be made right up to the moment when Hanoi's leaders change their mind." Ginsburgh considered the report an argument for de-escalating Rolling Thunder, which he informed Rostow would be "a grievous mistake."[21]

Ginsburgh's criticisms epitomized the Joint Chiefs' attitudes toward expanded bombing. All through the fall, the Pentagon was awash in reports and discussions concerning the bombing, including an additional RAND report by Oleg Hoeffding offering conclusions similar to those of the Jasons. But Rostow was simultaneously receiving memos from Westmoreland lauding the "significant impact" of Rolling Thunder, even while the general complained that the policy of "creeping escalation" was far less effective than his proposed "shock action" of intensive attack. Westmoreland was frustrated by gradual escalation, but his protests failed to convince McNamara, who felt bolstered by the Jasons' report, even with its deep criticisms of the bombing. As a later Senate report noted, this first Jason study had a "powerful and perhaps decisive influence in McNamara's mind."[22]

The Jasons would reiterate their arguments in the expanded second study of the bombing, released in December 1967. In a devastating assessment, they asserted that the bombing "has had no measurable effect on Hanoi's ability to mount and support military operations in the South." Indeed, China and the Soviet Union had more than compensated for the costs of the bombing so that from an economic perspective, North Vietnam had actually *gained* from the bombing. The influx of money and supplies, in turn, improved North Vietnam's military capabilities, making it "a stronger military power than before." At best, the bombing improved morale in South Vietnam but only in "transient" ways. In no uncertain terms, the Jasons were arguing against the effectiveness of further bombing.[23]

※ ※ ※

The second major report born out of the Dana Hall meetings flowed naturally from the conclusions of the first and concerned the electronic barrier plans of Kistiakowsky and Zacharias. The plan—a system of sensors that could trigger military strikes against people and supplies crossing from North Vietnam to South Vietnam—offered hopes of de-escalation through deterrence, an alternative to the devastating bombing of Rolling Thunder. It also promised to minimize American casualties. For scientists such as Kistia-kowsky, who had reservations about the war, the barrier was a path to de-escalation, an attempt to save the McNamara approach and preserve the arms-control liberalism that had gained ascendancy during the Kennedy administration.[24]

The Jasons' report proposed the electronic barrier as a means to solve two key problems: the movement of supplies and the movement of troops from North Vietnam into South Vietnam. In its simplest form, the barrier consisted of a variety of mines placed throughout key areas and the "pro-fuse use of simple sensors" that upon detection of any interloper would trigger air strikes. In principle this system was not new, but the novelty of the Jasons' approach lay in its extent and intensity, specifically "the very large scale of area denial, especially mine fields kilometers deep rather than the conventional 100–200 meters." Supplies tended to be transported by trucks, boats, animals, or individuals on bicycles or on foot, through many different secondary road systems. Due to the high degree of "redundancy and flexibility" in available trails and transportation routes, an effective bar-rier would have to be "applied over sizeable areas," and the scientists offered several possible geographic options, mostly near the Laotian border.[25]

More importantly, they provided detailed descriptions of the kinds of sen-sors and weaponry necessary for a functional barrier. In the case of the an-titroop barrier, the mine and sensor field, potentially a strip of 500 square kilometers, would be "constantly renewed" with gravel mines and "button bomblets." As one historian has described it, a gravel mine "looks like a piece of ravioli" and is intended "to blow off the foot that steps on it." Button bom-blets, in contrast, were tiny "aspirin-sized" mines that when triggered emitted noises to alert acoustic sensors but otherwise would not injure "a shod foot." For the most lethal "anti-personnel" swath of the barrier, the Jasons pro-posed that gravel mines be "sown" at a density of 50,000 per square kilo-meter. Every thirty days, the minefields would be "reseeded."[26]

While the gravel mines would deter crossing on their own, the range of acoustic sensors, photo reconnaissance systems, and traversing P-2V planes equipped with infrared detection would also trigger additional airstrikes, composed of gravel mines and SADEYE/BLU clusters. The SADEYE "Bom-blet Dispenser Weapon" was deemed particularly effective: each cluster could be rigged to pass effectively through jungle canopy and contained roughly six hundred "anti-personnel/anti-vehicle" bomblets, themselves filled with steel pellets. Due to the clusters' indiscriminate lethality, the Jasons considered the SADEYE the "canonical" weapon of the barrier, "on the basis that area coverage with high kill probability will be needed to compensate for uncertainties in target location." In other words, the SADEYE's large and deadly range of impact rendered the minor imprecision of sensor data irrelevant.[27]

For the antivehicle and supply barrier, acoustic detectors would be placed along targeted roads at mile intervals and would in turn trigger SADEYE cluster air strikes. Additionally, every evening, air patrols "would distribute self-sterilizing Gravel over parts of the road net," ensuring that "road-watching and mine-planting teams" could still access the area.[28]

Rather than requesting new areas for scientific research or hypothetical new detection tools, the Jasons focused instead on developing a system that "could be largely operational . . . using nearly-available weapons, aircraft, and equipment" though "some component engineering will be necessary." In a section headed "Some Comments on the Orientation of the Study Team and the Task," the authors noted that they could have devoted more time to new technologies useful for the barrier but instead thought it was more valuable to focus on "the more prosaic task of trying to see how one could assemble hardware that will soon be available, with some minor modifications, into a system that could begin to function within about a year from go-ahead." Although the group hoped that over time the barrier would employ new sensor technologies and new computing power (described as "information processing/pattern recognition techniques"), the focus of the report adhered to the spirit of Kistiakowsky's words to McNamara during the summer of 1966: "We do not propose to become involved in a broad effort at inventing new gadgets."[29]

✳ ✳ ✳

In researching the barrier, the Jasons had been briefed by numerous military personnel, from generals to specific field experts, as well as representatives from the Central Intelligence Agency, ARPA, and the Pentagon's Defense Research and Engineering office. Despite the Jasons' self-identification as independent, candid outsiders, their report frequently invoked many of the administration's political assumptions and analytical approaches, such the as discussion of the importance of "deniability," the acknowledgment of the need for secrecy about the geographic scope of the war, and the euphemistic phraseology and quantitative analysis characteristic of McNamara's

Pentagon. In a section titled "Political Constraints," the Jasons wrote that in Laos "everything we do must satisfy the principle of deniability, to give the Soviet Union the opportunity to close its eyes to our operations . . . To this end, the North Vietnamese have never publicly admitted their infiltration operations in Laos, nor have we officially admitted the air or ground reconnaissance operations in all their scope." In describing the potential effectiveness of the barrier, the Jasons calculated confidently that "the system is designed for probability ≪1 of a small group penetrating the denied area; for probability ~1 that all moving targets on roads or well-used trials are detected; and for ~0.3 kill of the moving targets, on the presumption that the enemy will not continue to 'run the gauntlet' at that price. All probabilities are for the basic system design in the absence of countermeasures."[30]

But countermeasures were of course almost guaranteed, as the scientists themselves acknowledged in a separate section. These might include moving troops individual by individual rather than as a unit; using groups of local tribesman as advance porters to "sweep" areas; deploying decoys; physically moving acoustic sensors; constantly firing into the minefield to "spoof" the system; or constructing a network of foxholes and bamboo bridges. The quantitative analysis lent a veneer of scientific objectivity to rosy assessments of the barrier's effectiveness while the probable countermeasures were detached from the calculation and addressed only qualitatively. Such calculations, when marshaled in support of expanded bombing campaigns, had been deeply criticized by the Jasons in their report on Rolling Thunder.[31]

* * *

Only twenty copies of the Jasons' barrier report were produced, but throughout the following year they circulated at the highest levels of government. Robert Ginsburgh, the National Security Council staff member who had criticized the Jasons' earlier report on the bombing, now suspected the Jasons of tailoring their technical assessments to political aims and inflating their assessment of the barrier while offering unduly harsh analysis of the effectiveness of Rolling Thunder. In a memo to Rostow, he noted drily: "It seems to me that if the Jason Group had applied to the barrier concept the same rigor that they applied to the bombing of North Vietnam the report would have been decidedly less enthusiastic about the infiltration barrier. Conversely, if they had applied the same standards to the bombing that they used for the barrier they would have concluded that the bombing was considerably more worthwhile than they indicated."[32]

Despite Ginsburgh's complaints, in September 1966 McNamara officially approved the barrier project. As one participant later recalled, this final moment of decision making occurred at Zacharias's summer house in Cape Cod: "The occasion was highly informal—maps were spread on the floor, drinks were served, a dog kept crossing the demilitarized zone as top secret matters were discussed." McNaughton and John Foster, the director of defense research and engineering, were both present.[33] Kistiakowsky later recalled that although he had recommended further study before implementation, McNamara enthusiastically preferred to "gamble immediately on putting our rather sketchy proposal into effect." Soon afterward, McNamara named Lt. Gen. Alfred Starbird as director of the Joint Task Force charged with setting up "an infiltration interdiction system." McNamara instructed Starbird to work closely with John Foster and to keep John McNaughton and the Joint Chiefs fully informed. He also advised Starbird that since further experimentation and development of new features would be required, additional input from scientists would be necessary. McNamara wrote explicitly, "I expect you to make use of an advisory group of non-government experts, including Dr. George Kistiakowsky." The advisory group, known as the Defense Communications Planning Group (DCPG), was tasked with the necessary design and engineering work required to make the barrier operational.[34]

This last instruction rankled Westmoreland. After meeting with Starbird shortly after his appointment to the barrier project, Westmoreland recorded his recollection of their conversation: "My general comment was that I was very much in support of the development of new weapons and devices by scientific and engineering communities, but I did not think it wise to have the scientists deeply involved in tactical employment."[35] The following month, Westmoreland expressed further skepticism; in his view, McNamara expected "that great dividends might accrue from [the barrier] which I am very doubtful of." Without mentioning the Jason scientists specifically, he wrote of his "suspicion that somewhere along the line certain parties feel that if a barrier could be established to stop infiltration, the bombing in the North could be stopped. This is completely unrealistic thinking."[36] Moreover, as he wrote to General Starbird, diverting precious resources to the untested barrier would likely "degrade" his command's "overall mission capability."[37]

✳ ✳ ✳

Westmoreland was not the only skeptic. Kistiakowsky himself harbored reservations from the opposing side of the political spectrum. In late 1966 he was appointed as General Starbird's senior advisor and commenced his formal involvement in war planning. In a 1968 recollection of his work with the DCPG, Kistiakowsky described his initial reaction to his appointment:

> Upon reading the directive I telephoned McNamara from Hornig's office and told him that I hesitated accepting this assignment because I was opposed to the administration's Vietnam policy. I wanted to work towards its modification and felt that accepting the job within DCPG . . . would cost me the freedom that I would have otherwise about speaking on Vietnam policy matters. I asked McNamara. therefore, whether he agreed with our summer study conclusions that the barrier could and should be used as a means for de-escalating the war and specifically making the bombing of North Vietnam unnecessary, because *it was only with this understanding that I would take the job.* I received assurances from McNamara that if the barrier were successful he, McNamara, would make an effort to use it for the de-escalation of the war through cessation of bombing. With this assurance I approached a number of Jason West summer study members and asked them to become members of the DCPG.[38]

McNamara offered Kistiakowsky his assurances, and, perhaps to mollify Westmoreland, John Foster provided the general with his own scientist, the UCLA chemist and air force advisor William McMillan. McMillan proved sympathetic to Westmoreland's commitment to expanded bombing, and Westmoreland found his bluntness refreshing. McMillan conducted an early study of the current state of the "air-interdiction campaign" in March 1967 and concluded that while analysis was difficult because "reliable figures" were hard to come by, the painful truth was that the present system was "simply not very effective."[39]

Thus, with Kistiakowsky reassured and Westmoreland at least tentatively supportive, the stage was set for implementation of the barrier. That summer, Rostow informed President Johnson that installation would begin in November, followed by "several months" of adjustments "to perfect the system." But in July 1967, aspects of "the System" were already being tested. Kistiakowsky, closely involved in the process, recorded his hopes for the project in terms that justified Westmoreland's suspicions. The barrier could be "a tool for de-escalation," he wrote. He acknowledged the possibility that North Vietnam might simply accept the higher attrition caused by the barrier and continue infiltration, but he maintained that "the System could provide Hanoi with a face-saving device to reduce the infiltration into the South without seeming to have been forced to yield to a bombing campaign." To achieve this preferable outcome, Kistiakowsky urged that the barrier be used in connection with a drastic reduction in aerial bombing while communicating to North Vietnam that any increased infiltration attempts would be considered acts of escalation. As he witnessed the limited successes of the early testing, however, Kistiakowsky understood that the barrier's functionality might initially be very limited. Therefore, he desperately hoped that mere description of the system would prove an effective deterrent and that a bombing halt would not be postponed while waiting for improved barrier results. Immediate de-escalation was critical, he wrote, even if it meant that "somebody will have to assert convincingly that [the barrier] will work without really knowing that this is so."[40]

Kistiakowsky's exuberance was shared by his colleagues. In a later oral history, Richard Garwin, who had traveled to Vietnam to assist personally with the implementation in February 1968, described the centrality of deterrence and de-escalation for the scientists' barrier plans: "Sensors don't keep anybody from coming through, so the idea was that you would have such an effective capability of striking trucks . . . that they wouldn't come at all. It's like a perfect mine field or a fence; there's no sense coming, you won't get through; so you don't hurt anybody." It was the logic of nuclear deterrence applied to the ground war: a sufficiently lethal and destructive barrier would ideally remain untriggered, deterring crossings and thus eliminating casualties and hastening peace talks.[41]

Within the administration, however, few shared the conviction that de-escalation could proceed without the barrier's demonstrated functionality. Richard Neustadt, the political scientist and Johnson advisor, was among those who emphasized the necessity of proving that the barrier was effective. Although only "marginally involved" in the barrier-related summer sessions, Neustadt considered the work integral to his larger goal of securing Johnson's reelection in 1968 and then "getting the war off the President's back so his third term isn't burdened like his second." In a memo written a year after Dana Hall, Neustadt clarified his assessment of the barrier's potential political importance as "both the symbol of our purpose and the center of our effort." It would reduce the American military presence in Vietnam, which could be replaced by an "international police-force," thus dramatically de-escalating the war. Neustadt understood, however, that political goals depended on technological success. The barrier had to *work*. Otherwise, U.S. claims would lack credibility and any political benefits would vanish. Neustadt thus pinned his hopes on Kistiakowsky and the other Jasons, whom he labeled "our scientist-weaponeers," and their ability to create a functional barrier.[42]

* * *

Not every Jason member shared Kistiakowsky's brand of antiwar sentiment and corresponding action. Steven Weinberg, Jason member and future Nobel Prize winner, later recalled that the war in Vietnam posed ethical problems for the scientists, to which they responded in three ways. "Some members looked at [the war] as a purely military problem, to which the expertise of JASON members might make a useful contribution," Weinberg wrote. "Some thought of it as nasty business, which could best be ended by winning the war. Others simply wanted nothing to do with it. I was in the last group."[43]

Freeman Dyson, the Princeton physicist and chairman of the Federation of American Scientists during the years of the test-ban debates, later wrote about his decision-making process in his book *Disturbing the Universe:* "I was invited to join the Barrier project and considered with some care the ethical questions that it raised. According to my general principle of preferring defensive strategies, the Barrier was theoretically a good idea. It is morally better to defend a fixed frontier against infiltrators than to ravage and batter a whole country. But in this case, if one believed that the war was wrong from the beginning, a shift to a defensive strategy would not make it right. I refused to have anything to do with the Barrier, on the grounds that the ends it hoped to achieve were illusory."[44]

* * *

While the DCPG worked on the barrier project, yet a third Vietnam-related Jason report was in the works. This stemmed only indirectly from the Dana Hall summer session. For antiwar scientists like Dyson and Weinberg, who had opted out of other Vietnam studies, the frequent contact with the military leaders and Pentagon planners waging the war instilled a significant fear of the possible use of smaller "tactical" nuclear weapons (TNW). In a later memoir, Dyson wrote that TNW use was indeed broached at high-level meetings during the war and he cited the circulation of a memo titled "Situations in Which the Use of Tactical Nuclear Weapons is Plausible." He recalled a meeting in 1966 during which a senior official (whom Dyson tactfully referred to only as "official Z") suggested, "I think it might be a good idea to throw in a nuke now and then, just to keep the other side guessing." Dyson was horrified at the proposal and sought counsel from three other Jason scientists who were present: Robert Gomer, Steven Weinberg, and S. Courtenay Wright. The four "decided that something must be done" and "concluded that the only way we might exert some real influence was to carry out a detailed professional study of the likely consequences if Z's suggestions were followed."[45]

The extent to which war planners seriously considered using nuclear weapons is the subject of debate, with most scholars agreeing that whatever idle chatter among military personnel might have occurred, Johnson and McNamara were firmly opposed. Nevertheless, Seymour Deitchman, the physicist and limited-war expert, later remembered "recurring talk around the Pentagon" in the spring and summer of 1966 about "using nuclear weapons to block passes between North Vietnam and Laos, especially the Mu Gia Pass, a key part of the supply route heading south." Both RAND and the Research Analysis Corporation had conducted war games addressing targeting and strategic problems relevant to TNW use, and contingency plans existed, including nuclear weapons use, should Chinese forces enter the war. But as a later assessment in the *Bulletin of the Atomic Scientists* concluded, Dyson and Weinberg "were undoubtedly responding to loose talk about using nuclear weapons from lower-level officials" rather than the serious consideration of top decision makers.[46]

Whatever the actual risk, the four scientists determined to present, in military-style language scrupulously scrubbed of any moral or ethical taint, the case against TNW in Vietnam. Their goal was to influence, as effectively as possible, an audience of "official Z" and his peers, for which they determined a kind of "value-free" approach would work best. They stated this aim clearly in the opening passage of the report itself, writing, "The purpose of this study is to evaluate the military consequences of a US decision to use tactical nuclear weapons (TNW) in Southeast Asia," and rather than relying on "intuitive judgment" or "moral" reactions, the scientists promised to provide "detailed analysis" and technical assessment to reach their conclusions. It was, in Dyson's later words, an attempt to present "the narrowest military point of view disregarding all political and ethical considerations."[47]

Despite their careful language, the scientists' own personal motivations were entirely rooted in their values and their moral commitment to preventing nuclear war. Weinberg later wrote, "It was clear from the beginning that the report should not go into ethical issues. For us to raise such issues would cast doubt on the impartiality of our analysis." But his motivation was almost entirely ethical: "The analysis was honestly done, but I have to admit that its conclusions were pretty much what we expected from the beginning, and if I had not expected to reach these conclusions then, for the ethical reasons that we left out of the report I would not have helped to write it." Indeed, the report's conclusions were stark: tactical nuclear weapons would not be "cost-effective," would offer few if any improvements over conventional weapons, and could potentially escalate the war to include "bilateral use" in Vietnam and possibly "general war" with the Soviet Union or China. Politically, the consequences could be "catastrophic."[48]

As promised, the scientists reached these conclusions through a patient review of the many scenarios in which TNW might be employed and the possible consequences and outcomes. For example, "bridges, airfields, and missile sites" might be "good TNW targets," but in many cases alternative high-power but nonnuclear weaponry could offer similar results. Using TNW to block roads and trails would also be effective, but only temporarily, since downed trees or other roadblocks could be cut through or cleared eventually. As an antipersonnel weapon, TNW would only be effective against "large masses of men in concentrated formations," not small, clandestine groups.[49]

The scientists also made heavy use of the war games and simulations of RAND and other institutions, which they criticized for being too short in duration and involving improbably large battle forces, so that the results tended to "exaggerate the effectiveness of TNW." Even so, the games revealed that "the outstanding difficulty in the use of TNW lies in locating troop targets accurately and striking before the location becomes obsolete." In other words, "target acquisition, rather than firepower" was the challenge facing U.S. military forces. In an analysis that could extend far beyond the planners of war games to the actual planners of war, the scientists noted that the simulations did not "credit the enemy with the ability to hide and maneuver in the jungle, an ability that he has already demonstrated in

Vietnam" and "are played on much too short a time scale; the proper time scale for war in Southeast Asia is almost certainly years, rather than days or months, with or without TNW."[50]

In the report's most alarming section, the scientists addressed the risk of TNW use eliciting a nuclear response from the Soviet Union or China. They described how the Soviet Union might provide nuclear weapons to Vietnamese forces, the ways in which these weapons could be transported and used against U.S. forces, and the possible consequences: "If about 100 weapons of 10-KT yield each could be delivered from the base perimeters onto all 70 target areas in a coordinated strike, the U.S. fighting capability in Vietnam would be essentially annihilated." Even if "only a few weapons could be delivered intermittently, US casualties would still be extremely high and the degradation of US capabilities would be considerable." Furthermore, should U.S. use of TNW in Vietnam prompt the Soviet Union to provide similar weapons to the North Vietnamese, a dangerous precedent might be set, resulting in Soviet provision of nuclear weapons "to her friends in South America or Africa." Thus, using TNW in Vietnam might lead to "nuclear guerrilla operations in other parts of the world." (The scientists went so far as to list places "where dissident groups armed with TNW could do particularly grave damage": Panama, Venezuela, the Middle East, and South Africa.) More critically, as the scientists worked their way through different nuclear scenarios, all paths seemed potentially to lead to nuclear disaster. "The ultimate outcome is impossible to predict," they wrote. "We merely point out that general war could result, even from the least provocative use of [nuclear weapons] that either side can devise." Even in the best outcome, with no corresponding Soviet or Chinese nuclear response, the United States would suffer from world condemnation for "crossing the nuclear threshold." The scientists closed with a strong, clear warning: "In sum, the political effects of US first use of TNW in Vietnam would be uniformly bad and could be catastrophic."[51]

＊ ＊ ＊

Official response to the report is difficult to gauge. Dyson wrote in his memoir: "We handed the report to our sponsors in the Defense Department. That was the last we saw of it." Seymour Deitchman later recalled that both John McNaughton and McNamara were briefed on the scientists' conclusions, but his time frame of summer 1966 does not match the publication date of the report: March 1967. In 2002, McNamara told the *Bulletin of the Atomic Scientists* that he had no recollection of the report or a related briefing, but he acknowledged that such a meeting might have occurred.[52]

Nuclear weapons—tactical or otherwise—were never deployed in Vietnam, but the prospect surfaced again briefly in early 1968. That winter, during the siege at Khe Sanh, Westmoreland reportedly set up a secret but short-lived "study group" to look into nuclear options. In February McNamara personally reassured concerned scientists Killian, Kistiakowsky, and Rabi that the Joint Chiefs had never discussed using nuclear weapons in South Vietnam, and "because of terrain and other conditions peculiar to our operations in South Vietnam," nuclear warfare against North Vietnamese forces was "inconceivable." In June William McMillan, the UCLA chemist who served as Westmoreland's science advisor, an Air Force Scientific Advisory Board member, and a President's Science Advisory Committee consultant, discussed with a Reuters reporter the possibility of using nuclear weapons in Vietnam. He explained that he wouldn't recommend their use "on a scientific basis" because they were "not suited to the kind of warfare being waged." But, he added, "I certainly would not suggest the use of nuclear weapons by the United States in South Vietnam—and I think it is a very low key possibility that we will ever use them against the North. But I can see circumstances under which nuclear weapons could be used." Whether the Jasons' report on tactical nuclear weapons affected the views of Westmoreland, McNamara, or McMillan remains unknown, but their statements suggest at least the plausibility of Dyson's hopes for his effort. "I have no way of knowing whether anybody ever read our report," he wrote later. "I have no way of knowing whether there was ever any real danger that Johnson would use nuclear weapons in Vietnam. All I know is that if Johnson had ever considered this possibility seriously and had asked his military staff for advice about it, our report might have been helpful in strengthening the voice of those who argued against it."[53]

✳ ✳ ✳

While the nuclear recommendations of Dyson and his colleagues seemed to vanish into the depths of the Pentagon bureaucracy, press speculation about the Jasons' barrier project began to appear in force beginning in the fall of 1966. As information leaked out of the Senate Preparedness Subcommittee, public accounts edged closer to the truth of the sensor-and-airstrike system. Andrew Hamilton of the *New Republic* reported on McNamara's "Practice Nine," a study of possible barrier technologies that, in the words of a quoted official, drew on "everybody and his brother." Hamilton estimated that the proposed system of barriers and fencing would cost $1 million per mile to build and operate and offered a pessimistic assessment of its chances for success. On September 7 the *Washington Post* carried the first detailed descriptions and a critical analysis by Joseph Kraft, who considered the barrier an indication that the U.S. military was now taking a purely reactive approach to the war (rather than "taking the offensive on raids"), which could lead to "terrible casualties." While supporting the barrier in theory, he expressed skepticism as to its effectiveness.[54]

Kraft's article appeared on the same day as a press conference on the barrier held by McNamara. Shortly afterward, McNamara further clarified the nature of the barrier in a memo to President Johnson, calling the word itself a "misnomer," as it was an "anti-infiltration system" rather than a kind of Maginot Line. The overall project did include "an obstacle line" of barbed wire, mines, and fortifications south of the demilitarized zone, but its larger component was the "unique" sensor-and-air-support system. McNamara cited both the optimistic estimates of Kistiakowsky and the applicability of the system to other limited war operations. As he put it, "If effective at all, the components should be useful in other parts of Southeast Asia (or the world) where selective detection and strikes are desired." In this regard, McNamara was reconfirming the mutual support for "flexible response" he shared with Kistiakowsky and many of the other Jasons.[55]

But McNamara's note to the president contained other messages. Despite his citations, he expressed some skepticism about "Dr. Kistiakowsky's forecasts" and cautioned, "There is the risk, too, that expectations for impressive early results will create clamor to substitute the new anti-infiltration system for other military measures." He did not specify the bombing of North Vietnam as among these measures, but such was the hope of many of the Jasons, particularly Kistiakowsky himself. McNamara also noted pointedly that while Gen. Earle Wheeler endorsed the project, military support among Westmoreland and the Joint Chiefs was lukewarm at best. McNamara clearly understood the tensions inherent in the contradictory concerns and prognostications of the Jasons and the Joint Chiefs. The views of Kistiakowsky and Westmoreland were irreconcilable.[56]

✳ ✳ ✳

In the meantime, the Jasons and the DCPG continued to work; in the spring of 1967, Val Fitch and Leon Lederman submitted a study of "Air-Sown Mines for the Massive Barrier," recommending "a pencil-shaped, fin-stabilized device which would be capable of soil penetration to a predetermined depth, so that a plunger-activator projects just slightly above the trail surface." Regular meetings with McNamara took place in the form of the "DCPG Advisory Committee," whose bicoastal members included physicists David Caldwell of the University of California, Gell-Mann and Fredrik Zachariasen from Caltech, Harold Lewis of the University of California, and, from the east coast, Marvin Goldberger of Princeton, IBM's Richard Garwin, Henry Kendall and Jack Ruina from MIT, and Harvard's Kistiakowsky, among others.[57]

But as McNamara's memo to the president had intimated, the Jasons were no longer in control of the project's trajectory, if they ever had been. Plans for the barrier were expanding beyond the Jasons' commitment to using existing technology, drawing on experts outside of the original Jason participants. Flush with money, including massive funding for further research and development, the DCPG was endowed with its own "Engineering Directorate" that drew in engineers and technicians from the military labs and the private sector. Contracting to the country's major defense engineering firms began in earnest. According to *Electronic News*, private sector companies involved in sensor production included General Dynamics, Defense Electronics, Texas Instruments, and Teledyne, among others. Many of the key military agencies devoted to limited-war research were involved: the Rome Air Development Center, the Elgin Air Force Base Limited War Group, and the Army Limited War Laboratory at Aberdeen. These sites worked on acoustic, chemical, and seismic detectors.[58]

Another major feeder was the MITRE Corporation, the Sputnik-era MIT spin-off with a decade of experience conducting defense research for the air force. Paul Dickson, who documented the creation of the DCPG in his history of the electronic battlefield, recalled the excitement and prestige felt by the new, non-Jason personnel. "DCPG alumni . . . talk about the organization in reverential and enthusiastic terms," he wrote, quoting one "typical" MITRE veteran's recollection that "as an engineer it is what you dream about." In Dickson's estimation, participants swooned at the prospect of "being present at a turning point in the technology of warfare" and "being a charter member of a new technological fraternity." As experts in sensors, acoustics, and signal processing convened, MITRE's Casper Woodbridge remembered the thrill of feeling part of "a new community, the sensing community." To an increasing extent, the old guard of Manhattan Project veterans and Jason physicists existed only on the peripheries of this "new community."[59]

* * *

As the barrier implementation efforts foundered during the early months of testing in 1967, Kistiakowsky continued to push for de-escalation during his private meetings with McNamara. McNamara, in turn, warned Kistiakowsky that no change in policy could occur until "the barrier proved itself militarily." The Jasons' second barrier study in the summer of 1967 argued, in Kistiakowsky's words, "that to make the barrier effective politically it was necessary to stop the bombing of North Vietnam very soon after putting the barrier in operation." But McNamara still resisted immediate de-escalation. He stalled Kistiakowsky by requesting an additional assessment of the effectiveness of the bombing in North Vietnam, resulting in the second Jason report on the topic—the massive, four-volume study reiterating earlier conclusions.[60]

Meanwhile, the bombing continued. For Kistiakowsky, the failure to implement the Jasons' barrier plan alongside bombing reduction constituted a betrayal. Since his work with the Cambridge Discussion Group beginning in early 1966, he had urged de-escalation and a change of course in Vietnam, to no avail. To him, the barrier had come to represent the last, best hope for scientists to work toward these goals within existing political and military channels, and by the fall of 1967, Kistiakowsky had transferred all of his political energy to its design and implementation. Privately, however, he sensed its futility. When fellow chemist George Pimentel recruited him for an antiwar effort in November, Kistiakowsky declined on the grounds that he was occupied with a major "classified project" whose success he didn't want to jeopardize. But he wrote, ominously, "It is, of course, not only possible but probable that the project I am involved in will be essentially a failure, in which case it is becoming clear to me that I will have no alternative than to go into opposition."[61]

Kistiakowsky was not alone in his growing skepticism. In a startling letter written to Kistiakowsky in the spring of 1968, Fredrik Zachariasen of Caltech reviewed the wholesale disillusionment that had plagued the entire DCPG Scientific Advisory Committee in the fall of 1967. He confirmed that the prospect of a mass resignation had not only been discussed but largely endorsed by members:

> Beginning in the fall of 1967, I, and most of the other members of the committee, were becoming more and more concerned about the US policy of slow, steady escalation of the war in Vietnam and in particular about what this implied about the use to which the DCPG project would be put. I recall telling you, one morning before one of our meetings, that I was not interested in helping you with a project that would be used merely as an additional escalatory step, and that I felt that the administration had no interest in using the DCPG effort in the way all of us had originally hoped it would be used, as part of a general de-escalation of the war. You said you felt the same way, and I know that all but at most one or two of the other members of the committee did too.
>
> In later discussions, both at DCPG and at the Cosmos Club, dissatisfaction with US Vietnam policy was expressed by everyone present. You said, at various times, that you thought you should eventually resign from the committee, and several of us, myself included, thought it would be better if the whole committee resigned together. You, however, felt that this was a decision for each of us separately and did not think that anything should be done by the committee as a group. Nevertheless, a number of us, again including me, told you that if you resigned we intended to do so too; and that we still felt all of us should quit together.[62]

Both Kistiakowsky and MIT's Henry Kendall argued against a mass resignation, although Kistiakowsky agreed to warn McNamara of such a possibility in an attempt to influence policy. In December, however, McNamara abruptly left his post as secretary of defense. Kistiakowsky, who had considered McNamara his "only channel for offering advice on policy matters," remained loyal to the end, defending McNamara to critical colleagues and assuring his former boss of his respect and sympathy. But as promised, Kistiakowsky informed McNamara that with his departure, and with clear evidence that no change in policy was likely, "most of the members" of the DCPG advisory committee planned to resign quietly. Kistiakowsky cited his own frustration at the committee's complete lack of policy influence and wrote that he found it "intellectually unacceptable to continue participating in a project that in the framework of present policies will tend to further expand the war."[63]

At the DCPG committee's next meeting on January 13, 1968, Kistiakowsky announced his resignation to his colleagues. He would retain his posts as a President's Science Advisory Committee "member-at-large" and Arms Control and Disarmament Agency advisor, but his Vietnam work was over. Despite his words to McNamara, he rebuffed calls by Gell-Mann and others to lead a group effort and instead urged that any resignations be individual matters of "conscience." He also suggested that any accompanying statements emphasize opposition to general policy rather than specific failures of the DCPG. Perhaps Kistiakowsky still held out some tiny hope that sensors and minefields could yet lead to de-escalation.[64]

* * *

Caltech's Zachariasen, for one, was caught "by surprise" at the quickness of Kistiakowsky's decision, and his immediate reaction was to resign as well. The remaining members of the advisory group later met to discuss their response, and, in Zachariasen's words, "It became clear that in fact very few wanted to quit immediately, and most preferred to wait for a little while, though nearly all were still completely disenchanted with US policy and, they said, had no intention of remaining on the committee for very long." Zachariasen worried that his individual decision would have little impact unless the entire committee resigned. Soon after, Johnson's announcement that he would not seek reelection cemented the group's decision to stay, as it held out "the possibility of a real change in policy."[65]

In the interim, Kistiakowsky selected an audience of roughly one hundred acquaintances, many of them members of the National Academy of Sciences, and distributed to them a carefully worded letter describing his views on de-escalation and "hinting" at his resignation. In a short time the letter had been widely circulated among physics and chemistry departments at nearly every major research university. Kistiakowsky's words and deeds prompted an outpouring of praise, hand-wringing, and political reflection among scientists, largely expressed in private correspondence to Kistiakowsky himself.

Kistiakowsky estimated that he received roughly seventy letters of support and one angry denunciation. A survey of the correspondence reveals the deep respect with which Kistiakowsky was regarded and the power of his words and actions to spur serious self-reflection among his colleagues, nearly all of whom wrote of their inner turmoil and felt compelled to articulate exactly why they had chosen to remain in their posts. This is particularly apparent in the letters from fellow Jasons and DCPG members. Physicist and DCPG committee member David Caldwell wrote that he agreed with Kistiakowsky and had considered resigning but felt his effectiveness would be equally minimal as an outsider. Leonard Sheingold of the DCPG concurred, as did Jack Ruina, who lauded Kistiakowsky's analysis of the war as "excellent" while defending his own decision to remain on the committee. William Nierenberg, a Jason leader on Vietnam work, likewise wrote in support of Kistiakowsky's words and actions.[66]

In the most prestigious halls of academia, Kistiakowsky's letter elicited similar reactions. Nearly every respondent expressed hopes for de-escalation and offered political observations about the state of the war and prospective candidates in the upcoming presidential election. Caltech's Max Delbrück lauded Kistiakowsky's actions but chided him for his faith in McNamara, whom Delbrück considered guilty of "dreadful errors of policy." His colleagues Norman Davidson, Jesse Greenstein, George Hammond, and John Roberts all wrote in support, ranging from the effusive to the qualified. Other high-profile supporters included Mark Kac of Rockefeller University, George Fraenkel of Columbia, Kenneth Pitzer, Emanuel Piore, and Wolfgang Panofsky.[67]

Piore and Panofsky, in particular, homed in on the relative merits of acting as either insiders or outsiders. As Piore put it, "I'm ambivalent about how best to apply pressure—whether completely to dissociate oneself and possibly get some satisfaction that one has made the break, or stay with it and try to apply pressure internally." Wolfgang Panofsky echoed these views, admitting to doing some "soul-searching" about his advisory work and wondering "whether I could be more helpful outside or inside the present system, and I am afraid the conclusion is that neither appears too promising."[68]

In nearly all the dozens of letters received by Kistiakowsky, his elite colleagues expressed frustration, bitterness, anger, and deep ambivalence about their own work and their political options. In perhaps the most telling example of the stark shift in political outlook, Berkeley chemist George Pimentel, whose own efforts to persuade President Johnson to de-escalate had been personally rebuffed by Walt Rostow, wrote Kistiakowsky to propose the creation of a new organization: "Scientists and Engineers against Johnson."[69]

* * *

Despite this response, Kistiakowsky considered his resignation an act of personal conscience, not a public statement of political opposition. This stance was evident in his objection to a mass resignation of the DCPG advisory committee (although he had invoked the possibility of such an act in his letter to McNamara) and in his own explanations to colleagues and friends. When he wrote to a colleague that his resignation "doesn't mean that I have grown long hair and have become a 'peacenik,'" he firmly separated himself from the youthful antiwar movement. Instead, he explained that he had circulated his letter without publication because he didn't want to "reduce my usefulness to a candidate for the presidential office that I hope to work for (most likely Rockefeller)." Sensitive to the classified nature of his work and reluctant to reveal too much, he felt he had "to walk a tight rope" in his public actions.[70]

Kistiakowsky's efforts to contain the publicity predictably failed, however. Over his "remonstrations," in March Daniel Greenberg wrote a one-page account of the resignation in *Science,* describing Kistiakowsky as "one of the federal government's most influential science advisers" at "the heart" of the so-called "scientific establishment" and someone "noted for his prudence and conservatism." Greenberg therefore interpreted the resignation as potential evidence of "massive disaffection" among elite scientists and an indication of the Pentagon's "increasing difficulty . . . in attracting top-level scientific talent to work on military matters." More ominously, the resignation came "against a background of worsening relations between the academic world and the military." The story was picked up by the *New York Times,* which emphasized Kistiakowsky's prestigious scientific past and recent status as a National Medal of Science recipient. It also noted widespread "speculation over the political ramifications of the action."[71]

* * *

Jason work connected to Vietnam did not end with the Dana Hall summer study and the creation of the DCPG. In August 1966 Kenneth Case, Leon Lederman, and Malvin Ruderman wrote to Marvin Goldberger at Princeton's

Palmer Physical Laboratory proposing the development of new nonlethal weapons for use among police forces and U.S. and UN operations in a variety of countries, including Vietnam, where military encounters included "close engagement of enemy and friendly troops and also engagement in the presence of villagers whose minds and hearts we would like to leave intact." The scientists acknowledged that "many moral and political, as well as military, arguments have been raised against such a program" on the grounds that "fluctuations in applied dosage from such weapons and also in the human sensitivity to them preclude their really being humane, effective, and limited to specific areas." But they still believed in the potential of "effective, controllable, non-lethal weapons" that could reduce civilian fatalities. They described one particular plan for a kind of anesthetic dart gun, a "humane" alternative to "conventional projectiles [and] explosives." Such projects might not only preserve the hearts and minds of Vietnamese villagers but could also draw in the alienated scientists who "previously had been repelled" by weapons research. "We feel that the competent scientific community has not been engaged," they wrote. But if "humane approaches" that were "separated as far as possible from conventional CW and BW work" were proposed, perhaps top scientific minds could be attracted to the program.[72]

Despite the optimism of key members, morale problems continued to plague the Jasons as the war progressed. Early plans for the 1967 Falmouth summer study included sessions on pacification, night operations, nonlethal weapons for crowd control, war sensors and communications, and American actions in Thailand as well as the traditional meetings on antiballistic missile technology and other nuclear weapons-related problems. After a period of "soul-searching" among ARPA and Jason leaders, however, the Jason Steering Committee met in March and April of 1967 and agreed to "start over" with a new list of "important but not urgent" topics for Jason subgroups to choose from and wider allowances for the Jasons to produce either original studies or critiques of existing programs and proposals. Once again, committee members emphasized lagging productivity and the importance of recruiting and retaining new members; specifically, that "an effort should be made to keep ingesting young Jasons and keeping them from becoming senior scientific statesmen too fast." Throughout 1967, in addition to studies on weather modification, antiballistic and missile guidance systems, naval warfare, lasers, turbulence, transportation problems, and other topics, Jason continued its Vietnam work, drawing on the expertise of Marvin Goldberger,

Murray Gell-Mann, George Kistiakowsky, Herbert York, Richard Neustadt (first as a non-Jason consultant, later a Jason consultant), and Adam Yarmolinsky (a Jason consultant).[73]

As information about the barrier began to appear in the press, reports of the existence of the secret Jason group began to percolate out into the academic communities of the members, attracting unwanted attention and difficult questions. In November 1967 the Jason scientists were instructed to direct any questions on the connections between Jason and IDA to James Cross, IDA's secretary and soother of university-IDA trustee relations. IDA also assured Jason members that they had the right to take on whatever consulting projects they chose without university interference, and as for storing classified documents in special cabinets on campus, this was no different from "locking one's desk or other container with valuables." Two months later, John Martin wrote worriedly to Gell-Mann that "two gentlemen by the names of Kraslow and Loorie have been making inquiries about the Jason-sponsored work at Dana Hall in '66" and were aware of the Jason members' identities and at least "some ideas of the matters which transpired at that time." For the first time, the Jasons were beginning to face public scrutiny of their war-related work.[74]

✳ ✳ ✳

On the ground in Vietnam, the early results of the Jasons' Dana Hall reports had taken physical shape in the two-part "Muscle Shoals" system, later renamed "Igloo White," consisting of MUD RIVER, the antivehicular barrier located largely in Laos, and DUMP TRUCK, the antipersonnel barrier to be located along the western half of the demilitarized zone into Laos. Results were far from the near-perfect success rates anticipated by the Jasons. A Pentagon report from April 1968 complained that "the value of any intelligence estimate based on sensor information is questionable." Particularly near Khe Sanh, enemy forces seemed to be eluding detection. Frequent artillery use and passing planes overstimulated the local sensors. Reviews of MUD RIVER conducted in the spring of 1968 noted that almost half of the targets relayed by sensors were never investigated, often due to poor weather conditions, and of the targets confirmed, only half were pursued, with a further 67 percent success rate. As one official complained, "It costs us more than $55,000 for each enemy vehicle destroyed or damaged."[75]

Pilloried in the press and deemed expensive and ineffective by military analysts, the barrier was never fully installed. By the spring of 1968, its function had already shifted to a kind of data collection; in the words of Earle Wheeler, the chairman of the Joint Chiefs of Staff, the system had "evolved from one which initially produced 'real time' intelligence on truck targets to one which now provides an in-depth analysis of traffic movement patterns, choke points, truck parks, and bypasses within the Laotian Panhandle transportation complex," enabling more effective air strikes "on the most lucrative truck targets in Laos." Moreover, after "antipersonnel" sensor technology had been employed during the battle of Khe Sanh, the Pentagon approved plans for use of the barrier's technical "assets" to support "ground operations in South Vietnam." By the summer of 1968, the stage was set for testing "eight different combat applications" of former barrier technologies, with an eye toward integrating DCPG "surveillance techniques into ongoing ground tactical operations." In 1967 Kistiakowsky had worried that the barrier would not be accompanied by a reduction in bombing. Now, rather than offering a stepping-stone to de-escalation, the fruits of the Jasons' early work on the barrier were being transferred to the ground war in South Vietnam.[76]

As Dickson observes, "The McNamara Line did not so much die as it became the 'Westmoreland Umbrella,'" defined as the "flexible and mobile" application of new and updated technologies throughout "South Vietnam and into Laos, Cambodia, and Thailand." Despite the travails of its advisory committee, the DCPG itself continued for another five years after the Dana Hall reports, overseeing the development of new kinds of sensors, mines, and laser and TV-guided bombs as well as the refurbishing of older technologies for application in Southeast Asia. These included "spider bombs," the air-dropped weapons whose swath of wire triggers could produce an "antipersonnel" detonation covering two hundred feet; "daisy cutters," the updated World War II-era blockbuster bombs that could destroy enough jungle vegetation to create "a bald spot about the size of a football field"; and "cluster bombs" refitted to carry tear gas or flechettes, which Dickson describes as "steel darts whose prime function is to shred human flesh or pin people to trees." The most influential technical developments promoted by the DCPG, however, were advances in sensor technology, signal processing, and computing—areas of research that had roots in earlier VELA work, were expanded during Vietnam, and would form the controversial underpinnings of the Reagan-era Strategic Defense Initiative.[77]

* * *

In 1984, at a celebration in honor of the twenty-fifth anniversary of Jason, Gordon MacDonald, who had been Maxwell Taylor's deputy at IDA during Vietnam, asserted that the war had not been "a major consideration among those who founded Jason" and that work related to Vietnam was largely "sporadic and, for the most part, the result of individual initiatives." The exception, however, was the DCPG. MacDonald presented a series of lessons learned from the experience. First, in order for a "study" to have effective results, there must be "a customer with the authority and access to resources to implement the findings of the analytic effort." In this case, the customer was McNamara, who had been involved from the very beginning and "gave it the personal attention that was required." Additionally, the group performing the study should be familiar with the topic and work according to its strengths—in this case, although the Cambridge group was less familiar with limited warfare, the California Jasons had greater expertise, and the group was able to produce a largely technical study that avoided administrative or managerial recommendations. Their work wasn't rigidly technical, however—they allowed for plenty of flexibility in the development and implementation of the actual system. Thus, in MacDonald's view, the very aspects of the DCPG that had so disillusioned Kistiakowsky were actually strengths: the Jasons' lack of political influence, the evolution of the project to include new technologies with expanded wartime applications, and the power of a single administration official, McNamara, to set the course of development and application.[78]

Such a view of the DCPG, offered a decade and a half later at the height of the Reagan defense boom, probably did not reflect the actual ambivalence among Jason members in the late 1960s. It is hard to gauge exactly how problematic Vietnam was for Jason members. "Ninety percent of us are doves," one member reported anonymously to *Science* in 1973, but the practical meaning of the term was left unexplained. In a series of oral history interviews conducted years and sometimes decades later by Finn Aaserud for the American Physical Society, recollections were mixed. Kenneth Watson, for example, acknowledged that antiwar sentiment "may have contributed to people's leaving," but there was no "great exodus." Internal documents from the period, however, suggest that the war and the fate of the DCPG project caused substantial angst among many Jasons. Without mentioning Vietnam explicitly, minutes from the January 1968 Steering

Committee meeting reveal a new level of turbulence, exemplified by a wave of resignations and withdrawals: three members had resigned, with Freeman Dyson tentatively offering to do so as well, while Luis Alvarez and Donald Glaser had retreated into "inactive" status. When the same group met three months later, they agreed that more recruitment was necessary and hoped that committee members could "suggest possible candidates with the emphasis on 'bright applied types.'" Something had changed—the original conception of the Jasons as young, iconoclastic geniuses was fading, replaced by a new vision of loyal talent eager to work within the defense establishment; experts who had already eschewed basic research in favor of projects hitched to that telltale modifier: "applied."[79]

Although early schedules for the spring 1968 meeting included planned sessions on the barrier and the North Vietnamese air-defense system among the standard antiballistic missile system, satellite systems, over-the-horizon detection, and command-and-control topics, by February the Jasons were reaching out to the Department of Transportation, expressing interest in projects related to air-traffic control, noise abatement, and other less explicitly defense-related topics. In March 1968 the Jasons participated in a day of special sessions to discuss the larger political implications of their work, and "major events followed quickly." The revised spring schedule now included an informal Saturday morning discussion about Jason and the war, to be followed by an executive session on the same topic with Maxwell Taylor. Attendees reportedly expressed "a wide spectrum of opinions" centered on "a sense of moral outrage at the war." The summer session followed suit, as Steering Committee members planned for a "postmortem" after inevitable "JASON-and-the-war conversations" among members. That summer and fall, the Steering Committee met to discuss possible reasons for poor attendance at the 1968 summer meeting. Ten members, all of whom had attended in 1967, had opted to skip the following year. The committee proposed that perhaps coincidence, competing meetings, or subpar remuneration were to blame. They never mentioned a possible lack of desire or ideological dissatisfaction. The following month, they voted to increase compensation for summer meeting attendance.[80]

By the spring of 1970, it had become clear that whatever administrative or organizational problems had been troubling the Jasons, deep-rooted political and ideological concerns had not been addressed. In May Edwin Salpeter submitted a letter of resignation to Hal Lewis, cc'd to the entire Jason membership. He wrote: "As you know, for several years now I have had misgivings about DoD consulting while the Vietnam war continues, but on the whole I felt (as Herb York expressed in a letter last year) that one can do more good by 'working from within.' In principle I still feel this way, but in practice I consider the present Nixon-Agnew administration's behavior so reprehensible that I am simply unable to 'do business as usual.' For this reason I herewith wish to resign from JASON. I hope reason will eventually prevail in the executive branch so we can all reevaluate the situation." Salpeter was joined in his action by other members for whom the Cambodian invasion of 1970 proved a tipping point. The science journalist Ann Finkbeiner has estimated that as many as nine Jason members resigned in the years immediately following the barrier debacle. Others never formally resigned but simply stopped participating in meetings and projects.[81]

In his response to Jason members, Lewis acknowledged that "many members of JASON have long felt anguished about the fact and the prosecution of the war in Viet-Nam, an anguish which has been exacerbated by the Cambodian adventure, and which I share." As in 1968, he suggested setting aside a day of the upcoming summer study "to talk about Viet-Nam, Cambodia, JASON, the Universities, etc., and to decide whether JASON, as an entity, should do anything."[82]

Besides the shift in membership, the Jasons did not really "do anything" on this front during the early Nixon years. Rather, the organization continued its defense-related work, with some additional attention to nonmilitary topics. Meetings in 1970 and 1971 included top-secret sessions on Cambodia, communications, Soviet expansion in the Mediterranean, nuclear explosions as possible earthquake triggers, reactor technology, antihijacking technology, housing and transportation technologies, lasers, and the supersonic transport. Internal correspondence focused on upcoming administrative changes in 1973 when the Jason contract with IDA was set to end, to be replaced by new institutional ties to the Stanford Research Institute.[83]

✳ ✳ ✳

Although the Jasons' parent organization, IDA, had long been a target of antiwar protests (resulting, for example, in the termination of its special housing arrangements at Princeton in the early 1970s), only a few partial reports of the Jasons' wartime activities had dribbled out in the years since the barrier. In 1971, however, the publication of the Pentagon Papers thrust the organization into a withering public spotlight. For the first time, detailed descriptions of the Jasons' work were revealed, including summaries of the Jasons' bombing reports, particularly the sweeping and detailed criticisms of the 1967 multivolume study, as well as the initial planning and development of the electronic barrier, including its vast and lethal minefields. Some titles of Jason reports were also released, including the infamous "Tactical Nuclear Weapons in Southeast Asia," without accompanying synopses or explanations.

Overall, the depiction of the Jasons in the Pentagon Papers was neither overstated nor inaccurate, and in the larger sweep of revelations about the prosecution of the war and Pentagon decision making, the participation of scientists might have seemed a minor footnote. But among the growing New Left, rooted as it was on college campuses, the account of the Jasons' work both confirmed suspicions of the reach of the military-industrial-academic complex and, at two of the most politically active campuses in the country, Berkeley and Columbia, it offered up convenient local targets of protest. In a bitter and sometimes violent attack, New Left critics rejected the Jasons' defensive claims that they had been working for de-escalation. Viewed in another, much more critical light, the Jasons appeared to be prestige-hungry collaborators who had created a lethal, civilian-killing component in an immoral war, all while cloaking themselves in the guise of dovish intellectuals trying, ineffectively, to work from within. Covering the controversy for *Science,* Deborah Shapley quoted Fred Branfman of the research organization Project Air War, who charged that the electronic battlefield, built upon personnel bombs that "cannot distinguish between soldiers or civilians," constituted a war crime. Branfman acknowledged that the Jasons may "have had a more beneficial effect than some others." But in his view that simply meant "they are lesser, rather than greater, war criminals. They are dramatic examples of how it is possible to be a moderate, well meaning, decent war criminal." He called on the Jasons either to resign from their posts or, following the precedent of the Pentagon Papers, to sabotage Defense Department operations through further document leaks.[84]

Jason scientists faced sudden hostility from academic audiences the world over as students and professors at international conferences and symposia protested their appearances and picketed their presence. French scientists rejected Murray Gell-Mann's attempt to speak at the College de France, and Jason physicists at a symposium in Trieste faced a crowd of three hundred protesters in a confrontation that ended with the deployment of French riot police. Protesters accosted Sidney Drell in Rome and in Corsica. The

Jasons' most vigorous opponents, however, were the members of the Scientists and Engineers for Social and Political Action (SESPA). In New York, the local SESPA chapter picketed and leafleted every week outside Columbia University's Pupin Hall, home of the physics department and five Jason members. The Jason members, in turn, complained of harassment: hate mail, irate phone calls, and vandalism. On April 24, 1972, the "New York Anti-War Faculty," a coalition of professors and students from twenty New York universities, occupied Pupin Hall for four days to protest the Jasons' war work.[85]

Unprepared for this level of hostility and anger, most Jasons responded with deep defensiveness; they closed ranks and publicly portrayed their opponents as thuggish and misguided. Columbia physicist Malvin Ruderman told *Science*'s Deborah Shapley, "It's impossible to resign under this kind of tactic. Nothing could be better designed to draw us together." Under the cover of anonymity, however, a handful of members confided to Shapley in early 1973 that their organization had made grievous mistakes. One Jason stated bluntly, "Obviously we blew it. When McNamara came to us in 1966 we should have told him to shove it and made a public statement." Others explained only that they had failed to achieve their goal of de-escalation or had antagonized the military brass and found themselves with little meaningful influence. Some members tried to emphasize the group's laudable work on arms control. One Jason even violated security regulations to assure Shapley that "Tactical Nuclear Weapons in Southeast Asia," despite its title, gave "all the reasons why you wouldn't use nuclear weapons in Vietnam." Though Shapley's 1973 article afforded the Jasons an important forum through which to defend their actions, Shapley herself characterized the group as stuck in a predicament of its own making. "Like the Jason of mythology," she wrote, the Jasons had "sewn a field with dragon's teeth which have sprung up into a host of hostile soldiers."[86]

Among those hostile soldiers were the members of the Berkeley SESPA chapter, led by the young physicist Charles Schwartz. In December 1972 Schwartz spearheaded the group's publication of a forty-page exposé of the Jasons' Vietnam work titled "Science against the People." It was, as the cover proclaimed, "The story of Jason—the elite group of academic scientists who, as technical consultants to the Pentagon, have developed the latest weapon against people's liberation struggles: 'automated warfare.'" The authors quoted at length the sections of the Pentagon Papers that mentioned the Jasons, placing the group squarely in "the technological wing of the military-industrial complex." Whatever the Jasons' intentions had been, they had offered their services to the Pentagon, and the result was the construction of the barrier, an "intervention contributing decisively to the prolonging of the Indochina war."[87]

Schwartz and his colleagues interviewed five Jasons in the summer of 1972 and published summaries in the SESPA report. The activists offered clear admiration for the brilliance of their subjects, noting that three of the scientists—Charles Townes, Donald Glaser, and Luis Alvarez—had received Nobel Prizes for their discoveries. But they found the Jasons' justifications for their actions thoroughly unconvincing. Marvin Goldberger told SESPA that he had been disillusioned about the war since 1966 and had hoped the barrier would be a means to end the war. Glaser noted that at its best, Jason could help the Pentagon avoid irresponsible decisions and investments in wasteful systems. With the barrier, however, he felt that military planners "used us technically but didn't listen to us."[88]

SESPA responded that traditional categories such as "dove" and "hawk" did not apply to scientists because while a physicist like Edward Teller was clearly a hawk, it was dovish, gentle Oppenheimer who oversaw one of "the most lethal innovations in modern warfare." Many of the Jasons who had promoted arms control were brilliant, creative physicists and caring and conscientious teachers. But while they might "publicly profess to be against the war, they continue to contribute their scientific talents to the military." One by one, SESPA rejected the kinds of justifications and excuses offered by the Jasons: that they weren't actually a powerful group and their advice was frequently ignored; that they were "boring from within" and working for the same change as outsiders like SESPA; that if good scientists didn't give advice, the Pentagon would look to mediocre scientists; that they hadn't known the true nature of the war early on; and, in what SESPA dubbed the "positive integral theory," that while Vietnam was a bad war, it was outweighed by other, more constructive U.S. military projects.[89]

SESPA countered that whatever the Pentagon reaction, the Jasons offered advice that was intended to be followed, with the goal of improving military effectiveness. The Jasons never disputed the objectives of military policy; they only evaluated the means: "Not whether to suppress guerrillas in Thailand, but only how." As for the claims of ignorance, SESPA noted that Jason members, like everyone else, had access to the writings of Jean LaCouture, Bernard Fall, and David Halberstam. (They didn't know that Fall himself had addressed the 1964 Jason summer session.) Instead, they proposed that the Jasons had fallen prey to their own vanity and ambition, succumbing to the allure of secrecy and power.[90]

But whatever the real reason "Why They Do It," the problem of accountability remained. The Jasons contended that they had indeed been concerned about the applications of their work and considered their decisions important matters of personal conscience. SESPA utterly rejected this standard. Matters left to the Jasons' "private conscience" affected the entire world, they argued, and had the power to destroy human life. "In such circumstances," the SESPA authors wrote, "a posture of 'I will decide what is best' is enormously arrogant." Personal conscience was no match for the military-industrial complex.[91]

* * *

Since the end of the war in Vietnam, public speculation about the Jasons has predictably ebbed but never disappeared entirely. The Jasons themselves still exist, preparing classified reports for Pentagon clients on traditional matters of weaponry and defense but also covering new fields such as cybersecurity, bioengineering, nanotechnology, data mining, quantum computing, urban surveillance, and climate change. In 2006 the science journalist Ann Finkbeiner published the first book-length treatment of the Jasons, based largely on recent interviews with current and former members, many of whom were still requesting anonymity even decades after their years of service had ended. Writing about the aftermath of the Jasons' work in Vietnam, Finkbeiner defended the group from most of the accusations of SESPA and other New Left critics. The Jasons had been naïve and made mistakes, she wrote, but they had not been immoral. "Perhaps at the conjunction of science and the military," Finkbeiner explained, "no universal moral law exists that says, this side of this line, good; this side, bad." Fundamentally, any decision about the proper location of the ethical line had to be left to the individual scientist. She agreed with Sidney Drell that "scientists must decide for themselves."[92]

In the academic world of the early 1970s, however, it was not the individualism of the Jasons but the sweeping ethical reckoning of SESPA— formulated on an institutional and global scale—that was ascendant. On campus after campus, radical critics were arguing that a scientist's personal hand-wringing meant little if his academic funding rested on multimillion-dollar Pentagon contracts and his professional societies welcomed visits from the president and his cabinet. Reports on technological inefficiencies did nothing to address the moral implications of U.S. foreign policy or the global death toll facilitated by the military-industrial complex. A new era of ethical calculation was dawning, and it rendered insufficient and archaic the old Los Alamos ideal of patriotism tempered by personal conscience.

SIX

Institutional Reckonings at MIT

"WAR IS INTERDISCIPLINARY," the radical historian Howard Zinn told a packed auditorium of students and faculty on March 4, 1969. Standing in Kresge Hall at the Massachusetts Institute of Technology (MIT), Zinn's words carried a powerful resonance and immediacy. A year earlier, the nascent MIT chapter of the Students for a Democratic Society had lashed out at the university's ties to the war in Vietnam. "MIT does not directly service the war," they wrote. "Nevertheless, MIT personnel advise and co-operate with the government, do studies on Viet Cong Defectors and Prisoners, and, most important, accept grants from the Defense Department to produce the technology capable of waging a war like the one in Vietnam . . . MIT must bear the responsibility for its actions." Now, on a national day of protest organized by graduate students in the physics department, MIT listened as Zinn and a host of other speakers scrutinized the university's institutional complicity in what they saw as an immoral war.[1]

It was, for MIT, an extraordinary but perhaps unsurprising turn of events. Since its nineteenth-century founding, MIT had cultivated and maintained strong ties to the nation's industrial and military establishment. It was one of the country's first engineering schools, an institute devoted explicitly to the practical fields of industrial science. The Reserve Officer Training Corps (ROTC) was born on the MIT campus in 1911, and faculty members served on the War Industries Board and the National Research Council during World War I. During the first postwar period, the school enjoyed substantial research support from key business and philanthropic entities, including the Rockefeller Foundation, General Electric, Du Pont, AT&T, and others and funneled its graduates into top and midlevel executive and research positions. Though the Great Depression devastated this system of corporate-sponsored university research, the massive mobilization of World War II sent funding levels skyrocketing. The bulk of military contracts went to industry, but academia also attracted vast new investments, with MIT and the California Institute of Technology (Caltech) leading the pack.[2]

During the war, former MIT vice president and Carnegie Institution head Vannevar Bush urged President Roosevelt to create the National Defense Research Committee, which soon became the influential Office of Scientific Research and Development, which Bush chaired. From this position, Bush steered funds to MIT researchers, and the school, with its undergraduate population reduced by nearly two-thirds due to the draft, transformed itself into a kind of national center for war technology, achieving extraordinary breakthroughs in electronics and radar. In creating the institute's Radiation Lab (Rad Lab), Bush drew on all of his academic and industrial ties. The Rad Lab employed thousands, spent more than a billion dollars, and yielded crucial radar technology that proved instrumental to Allied military success. Bush recruited top outside scientist-managers to run the lab, including I. I. Rabi, the future Nobelist, and Lee DuBridge, the future Caltech president and science advisor to Richard Nixon. The interdisciplinary nature of the Rad Lab, as well as its ties to government and industry, inspired MIT scientists and administrators to incorporate the lab more closely into the MIT campus and to create other facilities along similar models. The institute's sponsored research contracts increased over 3,400 percent over the span of the war, and MIT ranked first as the country's top "nonindustrial defense contractor," with Caltech and Harvard a distant second and third.[3]

Bush viewed this type of academic and defense coordination as only a temporary wartime necessity. The Rad Lab was officially dismantled in 1945, but with support from the newly created Office of Naval Research and other military funders, its resources shifted to a new entity, MIT's Research Laboratory of Electronics (RLE). Partly the brainchild of Norbert Wiener, the interdisciplinary RLE devoted itself to technologies of communication, from Weiner's invention of cybernetics to Noam Chomsky's study of linguistics. But it was also through the RLE during these early postwar years that Project Lincoln and other air-defense and missile-guidance projects got their start, and the first small defense-technology spin-off companies began to populate the areas adjoining what would become Route 128.[4]

The gentle dip in federal research and development (R&D) funding after the war lasted only a few short years until growing Cold War tensions elevated contract levels and, in one historian's estimation, "made the university, for the first time, a full partner in the military-industrial complex." The 1950s saw the creation of a spate of new special university labs, including MIT's Lincoln Lab, Berkeley's Livermore Lab, and Stanford's Applied Electronics Lab, all pumped full of defense money. By 1969 MIT was, in the words of *New York Times* columnist Walter Sullivan, "one of the pillars of American military research." The U.S. government paid for almost 90 percent of MIT's on-campus research, with the Defense Department (including the three service branches and the Advanced Research Projects Agency (ARPA)) accounting for more than a third of the total in 1967, including 40 percent of all engineering research. This support extended to the social sciences as well, epitomized by MIT's Central Intelligence Agency-funded incubator of modernization theory, the Center for International Studies. But by far, the largest magnets for government funding were the university's two "special labs": the Lincoln Laboratory, a pioneer in multiple weapons-related research fields located near Hanscom Air Force Base in bucolic Lexington, and the Instrumentation Lab (I-Lab) in Kendall Square, specializing in inertial guidance systems and run by Charles Stark Draper, the Montana-born MIT physicist and gyroscope pioneer.[5]

The origins of Lincoln Lab lay in the U.S. military response to Soviet nuclear testing in 1949. At that time, the air force's Scientific Advisory Board created an ad hoc committee on air defense, packed with MIT representatives. Within a year, the outbreak of the Korean War pushed air force top brass to request that MIT broaden its early air-defense study efforts to include full-scale research laboratory work. Though RLE researchers Jerome Wiesner and Jerrold Zacharias initially balked at the expansion of military projects, a succession of small summer studies and well-funded, carefully designed initiatives eased MIT faculty into the new work. Project Charles, conducted in the summer of 1951, paved the way for the official creation of Lincoln Laboratory out of RLE resources.[6]

The off-campus location at Hanscom Field facilitated its expansion, and Lincoln grew rapidly throughout the 1950s, particularly after the 1957 Sputnik funding boom. Lincoln also provided a convenient new home for the talented Rad Lab veterans whose expertise lay in the fields of radar technology, communications, and data processing. New Cold War projects such as the SAGE (Semi-Automatic Ground Environment) system and the Whirlwind computer required just such a specialized staff. Whirlwind led to major collaborations and spin-off projects with IBM, AT&T, and H. P. Robertson's System Development Corporation. SAGE provided the impetus for the creation of MITRE Corp., a portmanteau of "MIT Research," which removed responsibility for SAGE's "systems engineering" and technical support from Lincoln Lab. (At the time, MIT did not consider systems engineering a sufficiently academic line of research "for an educational institution.") In the meantime, however, Lincoln continued its research on weapons systems and detection techniques, including work on ARPA's VELA program.[7]

By the late 1960s, 98 percent of the lab's funding came from the Department of Defense, through a Federally Funded Research and Development Center contract. In addition to its weapons-related work, the lab took on projects in radio physics, radar observations of planets, and tests of relativity. Though the details of many Lincoln projects were classified, the lab published general descriptions of its major research areas; according to MIT's 1969 Directory of Current Research, Lincoln projects in 1969 included seismic data analysis for VELA Uniform as well as research on masers, thermal tracking, computer memory, radio astronomy, atmospheric studies, lunar surface and planetary properties, and satellite communications. The lab also ran ARPA's Project PRESS (Pacific Range Electromagnetic Signature Studies) program, part of a larger effort to study the physical traces left by missiles, decoys, and other intercontinental ballistic missile components. Such research was crucial for devising mechanisms for an antiballistic missile (ABM) system that could distinguish between incoming nuclear weapons and incoming decoy projectiles. Researchers studied the types of ionization, optical radiation, fluid-flow fields, and other phenomena produced by various objects traveling through the atmosphere in order to work out methods of reverse analysis—to see, in the words of one Lincoln associate division head—if given ionization and radiation data, for example, they could "infer the body." Related work in this area included studies of the efficacy of radar technology in decoy identification and general "defense system analysis and synthesis."[8]

Perhaps most controversially, the lab's radar work included the development of techniques for detecting underground tunnels and distinguishing the movement of people and vehicles through dense foliage. Newspaper reports indicated that Lincoln employees had even traveled to Vietnam to demonstrate the technology for military leaders. For MIT's antiwar contingent, it was these two categories of research—the ABM system and Vietnam-related technologies—that made the lab a major target of protest.[9]

✳ ✳ ✳

MIT's I-Lab followed a similar trajectory of expansion. Charles Stark Draper, a young MIT physicist and engineer from Montana, had founded the lab in 1932 with a substantial Depression-era investment from Sperry Gyroscope. Draper was primarily interested in "shipboard gunsights"—optical, gyroscopically stabilized devices to facilitate aiming a weapon from a moving ship. This early exploration into inertial frames of reference led to more complicated aiming, tracking, and guidance systems, drawing on improvements in gyroscopes, accelerometers, servomechanical-feedback systems, and radar. Draper's research shaped his teaching, and many of his students pursued Sperry-approved dissertation topics and were rewarded with corporate positions upon graduation. Among the lab's early successes were the Mark-14 gunsight used during World War II, the A-1 sight for fighter aircraft employed in the Korean War, and the "Black Warrior" automatic pilot system capable of loops and complex evasive maneuvers.[10]

World War II led to a broad expansion of laboratory research at MIT generally, but during the postwar boom it was Draper's I-Lab that was singled out by the Eisenhower administration for ballistic missile research, largely through the influence of key advisor and MIT president James Killian and several well-placed I-Lab alumni. Millions of dollars of navy and air force contracts rolled in throughout the 1950s. As had been true before the war, the curricular components of the "teaching laboratory" promoted by Draper heavily reflected the military orientation of the lab's funders. Officer-students, of which there were many, could take courses in "Weapons Systems Engineering," and navy and air force administrators worked closely with MIT faculty to design optimal graduate programs for officers. Summer courses and seminars attracted researchers from military labs and computing and aerospace corporations, many of whom might end up hired as "resident engineers" at the lab. Employment statistics document the work done by engineers from AC Spark Plug, Raytheon, Bendix, Honeywell, and other companies. According to the historian Stuart Leslie, Draper reportedly "kept salaries low to encourage turnover to industry." Throughout the late 1950s,

Draper's push factor met corporate pull factors—the establishment of new local research facilities in towns such as Wakefield, Burlington, and other suburban sites. Military contracts funded both the research results, which were funneled to industry for construction, and the researchers themselves, who were funneled to industry for further employment.[11]

By the 1960s the I-Lab had become a world leader in inertial guidance systems: technologies enabling "extremely precise guidance of a vehicle without magnetic compass or ground control." As one lab researcher would later explain, in the parlance of a professor of freshman physics: "This is what inertial navigation allows you to do. You go in a closet, you shut the door, you move it any place you want, it tells you where you are, how fast you are going, and what direction you are going." Lab researchers developed guidance systems for planes, ships, submarines, space vehicles, and missiles, most notably for the Apollo program and the Thor, Polaris, and Poseidon missile systems. They developed helicopter technology allowing vertical takeoffs, hovering, and backward flight.[12]

According to the lab's public literature, its guidance research in 1968 included the SABRE ballistic missile program and the controversial multiple independently-targetable reentry vehicle (MIRV) warheads intended for use with Poseidon submarines, allowing a single launch to hit multiple targets. On the Poseidon project, the I-Lab held a contract with the navy's Special Projects Office but subcontracted out various components and electronics to Bendix-Honeywell-Nortronics, Raytheon, Hughes, and General Electric. The lab's ties to private industry extended beyond collaborations through subcontracting. I-Lab staffers had consulting arrangements with an extensive list of defense firms: Boeing, Lockheed, Convair, Northrup, Rand, Hughes, Grumman, Westinghouse, Systems Development Corporation, General Motors, Nortronics, Itek, Control Data, Allied Research, Honeywell, General Electric, Sperry Rand, Watertown Arsenal, Raytheon, Avco, MITRE, IBM, Sperry Gyroscope, Univac, and General Dynamics, to name some of the most well known.[13]

Records from 1968 show that the I-Lab employed about nineteen hundred people, roughly a third of them "professional engineers and scientists," a third technicians and assistants, and a third administrative and support personnel. Of the technical staff, the largest group were electrical engineers, with mechanical engineers, aerospace engineers, physicists, and mathematicians also represented. Everyone on the lab payroll—including each student worker—was required to have security clearance. In 1967 the lab spent nearly $19 million on salaries and wages, $10.4 million on overhead, $11 million on materials and services, and $12.9 million on subcontracts. More than half the budget was supplied by the Department of Defense. As with Lincoln Lab, decisions on the acceptance of new contracts were made by the laboratory directors in consultation with Jack Ruina, the former ARPA head serving as MIT's vice president for Special Laboratories.[14]

✳ ✳ ✳

MIT's ties to government and industry had been debated and reaffirmed by administrators many times before the March 4 protest in 1969. In 1945 MIT president Karl Compton recalled that since the first moments of crisis during World War II, MIT had vowed *"never* to let the self-interest of the institution prevail over the interests of the nation." The wartime effort had demonstrated that the country's "educational institutions rank with our manufacturing industry and transportation system as the principal supporting lines of military power in time of war," a sentiment that would be reiterated by an MIT lawyer in 1966, who described the school as "a scientific arsenal of democracy." Compton's views were similar to many of the Manhattan Project scientists; he saw no differences between institutional and individual obligations in times of war. Universities, like the citizenry, must serve their nation but should take strict care they did not profit financially from their efforts. In MIT's case, "the prestige gained from our war research" should be reward enough.[15]

MIT administrators trod cautiously in the postwar period, dismantling the Rad Lab but continuing to host other defense-sponsored research. In the 1950s Louis Smullin, head of the RLE's microwave-tube lab, proposed that MIT pursue civilian goals with the same laboratory zeal that characterized the creation of Lincoln Lab, but his vision elicited little interest. Many other prominent MIT faculty members—including Victor Weisskopf, Philip Morrison, Martin Deutch, and Bernard Feld—held leadership positions in the Federation of American Scientists, which throughout the 1950s and 1960s issued warnings and public statements on the dangers of university militarization. By the late 1960s, the Federation of American Scientists' position was clear. "Except in times of national emergency," they wrote, "the university should not be a part of the military establishment and should not directly or indirectly take part in military operation or participation in the collection of military intelligence." They were skeptical of classified programs, viewing secrecy as "counter to the values . . . of a university" and opposed explicit weapons research or work on any project "designed to destroy human life."[16]

"Times of national emergency" was a phrase open to interpretation. In 1967 Provost Jerome Wiesner weighed in on "The Federal Presence at MIT" in the pages of *Technology Review*. Although he acknowledged that MIT's general policy was "to strive always for the maximum of free, open, uninhibited discussion—in other words, for unclassified research," he agreed with Compton that moments of crisis—such as World War II, the war in Korea, or key periods of the Cold War—justified drawing on MIT's unique technological expertise in the service of the national interest. Moreover, MIT's avoidance of key areas of classified research could mean falling behind other elite institutions more willing to take up controversial topics. Wiesner summed up bluntly: "Either we have enclaves of secret research on or near to the campus; or we bypass interesting and vital areas of modern technology." But Wiesner didn't confine his analysis just to MIT. He worried that government funding for academic research had become more restricted and more spread out in recent years so that more schools received smaller amounts of money. The long-term effect, in his view, would be the disappearance of the kind of "big centers" that had proved so innovative and productive in the past and the emergence of widespread, homogenized, mediocre research.[17]

But Wiesner's invocation of MIT's cutting-edge reputation could not mask the significant unpopularity of certain areas of research in the "big centers," especially work related to the development of an ABM system. By the late 1960s, even former MIT president James Killian acknowledged that the positive attitudes of the Sputnik boom years—the era of SAGE and Whirlwind and Project Charles—had soured; the days when academic enthusiasm for air-defense research had sometimes even exceeded military support were long gone. Killian nevertheless defended the heavy levels of defense funding at MIT, convinced that the university was still able to "preserve the policies and academic environment" of the institute. MIT was no Pentagon lapdog, he observed, joking that "somebody once in Washington made the comment that you didn't enter into a contract with MIT, you negotiated a treaty." But in the Vietnam years, all university contracts with the Pentagon were suspect, no matter how well negotiated. The young MIT Students for a Democratic Society (SDS) chapter set the tone for a new wave of campus activism with their 1967 pamphlet "MIT and the Warfare State," a condemnation of

MIT's institutional complicity in the war. "The man who designs the gun is just as guilty as the man who pulls the trigger," they wrote, "for without him, there would be no gun. . . . Scientists are not instruments of society, they are MEN with conscience. Technology is not neutral, it is directed."[18]

* * *

For much of the 1960s, MIT's campus had been relatively quiet.[19] Beginning in 1965, faculty-led efforts by Noam Chomsky and others laid the groundwork for the creation of RESIST, a war-tax resistance organization. But it was student activism, beginning in 1967, that culminated in the radicalism of the protests of March 4, 1969. Undergraduate activist groups included a small chapter of the Society for Social Responsibility in Science, the Rational Approach to Disarmament and Peace, and SDS. By the end of the decade, SDS, Resistance, Sanctuary, and other antiwar organizations had attracted substantial followings: in 1967 hundreds of MIT students protested against the campus recruitment efforts of Dow Chemical, and over a thousand attended a lunchtime panel discussion among prowar and antiwar professors, a Dow representative, and a medical expert testifying about the effects of napalm exposure. In the fall of 1968, hundreds of students occupied a portion of the student center in order to provide sanctuary to an AWOL soldier, and for two weeks the building hosted a steady stream of antiwar teach-ins and performances. A student poll from that year showed majority support for Eugene McCarthy's antiwar presidential bid. Faculty members formed a chapter of Scientists and Engineers for McCarthy in 1968 and participated in the Boston Area Faculty Group on Public Issues and Group Delta, a short-lived coalition of antiwar professors.[20]

Murray Eden, an electrical engineering professor, described the March 4 movement as originating from a dinnertime discussion among physics graduate students on the possible impact of an antiwar science-research strike. The students—Ira Rubenzahl, Joel Feigenbaum, and postdoctoral fellow Alan Chodos—quickly began reaching out to students and faculty in their own and other departments, eventually recruiting Bernard Feld, Salvador Luria, Bruno Rossi, Victor Weisskopf, and Noam Chomsky, among others, to join their cause, as well as Hans Bethe and George Kistiakowsky from outside the MIT community. Jonathan Kabat, a biology graduate student, signed on as another enthusiastic organizer. The result was a draft statement endorsed by prestigious faculty members on the "misuse of science" and the creation of a new group, the Science Action Coordinating Committee (SACC). A one-day research strike was planned for March 4, 1969.[21]

Initially, students comprised the major organizing force, drawing on faculty support in order to "provide prestige, enhance publicity, and help to persuade eminent men of science and politics to participate," in Eden's words. Many soon became frustrated by what they perceived as the "well-intentioned but timid liberals" of the MIT faculty. Tensions mounted, causing an organizational split: activist students remained within SACC, while faculty members, led by physicist Herman Feshbach, formed a new group, the Union of Concerned Scientists (UCS). Meanwhile, the MIT administration itself obviated the divisive characterization of the planned action as a "strike" by agreeing to cancel classes and devote the day to open discussions of the applications of scientific research to military causes.[22]

As scheduling efforts progressed, the planned day of events stretched into additional evening and weekend sessions. SACC and the UCS reached out to other schools, labs, and cities, coordinating simultaneous actions throughout much of the Ivy League and at major state research universities. Organizers recruited supporters at conferences and annual meetings, and the efforts yielded encouraging editorials in the *New York Times* and *Chemical and Engineering News*. Bell Labs agreed to send "a busload of participants." Both SACC and the UCS also circulated petitions and issued multiple public statements with demands both specific and sweeping. SACC proposed a series of MIT policy reforms, including an end to military-related co-op jobs, the elimination of Defense Department research funding, the eviction of ROTC, and the conversion of the university's special labs to nonmilitary projects. But the group also issued a more general call for *involvement:* an end to scientists' invocation of political neutrality. "When a scientist—or any other citizen—claims political detachment and non-involvement," they wrote, "his decision perpetuates militarism and waste, for that is the thrust of America's status quo." The ethical problem was not a matter of individual morality or personal culpability for the misuse of science but the *institutional framework* in which scientists operated. SACC explained:

> Science is misused by men, but men who function within an institutional framework. . . . The pressures upon the men who make policy flow from deep within the structure of this society. The limits on their actions are not solely the boundaries of pure reason. The men who rule have been screened again

and again, at each step of advancement, so that those who agree with the status quo are chosen, while those who might rebel are weeded out. Those who are chosen find that their way of life, their livelihood, their very lives come to depend on the maintenance of the established order. . . . Set against the forces of self interest, acculturation, pressure from peers and threats from higher up, we predict that reasonable arguments will not alone suffice. The question is one of power as well as reason.[23]

The UCS offered a muted variant of this analysis in their founding faculty statement, expressing deep disillusionment with government policies and hope that a silent but "concerned majority" of scientists would rise up. "Through its actions in Vietnam our government has shaken our confidence in its ability to make wise and humane decisions," they wrote. "The response of the scientific community . . . has been hopelessly fragmented. There is a small group that helps to conceive these policies, and a handful of eminent men who have tried but largely failed to stem the tide from within the government. The concerned majority has been on the sidelines and ineffective. We feel that it is no longer possible to remain uninvolved."[24]

The UCS's "concerned majority," however, did not materialize everywhere. Though by March 1969 the group included roughly a hundred members from MIT (a third of total membership), many Boston-area academic scientists opposed or ignored UCS outreach efforts. Counterpetitions circulated in MIT's engineering and political science departments. *Electronic News* reported that "there is little sign that the activist efforts of the physicists are extending into engineering or the electronics industry" and quoted Lee DuBridge's dismissal of the protesting scientists as "extremist elements," a phrase picked up by a negative editorial in *Industrial Research*. At the Argonne National Lab in Illinois, researchers formed a "Federation of Responsible Scientists" and announced a sixteen-hour "work-in" to coincide with the MIT strike.[25]

The reaction to the events of March 4 reflected this division of viewpoints. Professors and students packed Kresge Auditorium throughout the day, listening to speeches by Zinn, Victor Weisskopf, the radical linguist Noam Chomsky, and the charismatic Harvard Nobelist George Wald. But as the *Tech* reported, "Business was about normal in most laboratories": though some researchers took breaks to attend talks or other meetings, no major work stoppage occurred. Around the country, about fifty thousand students at over fifty universities expressed their discontent with the militarization of science: the University of Pennsylvania cancelled classes, and at Columbia, several hundred students and faculty—many referring to themselves as "Sputnik children" steered into defense-science careers—participated in a research strike. At Rockefeller Center in New York, thirty members of the Computer People for Peace distributed literature and voted not to "program death." And in the U.S. Senate, George McGovern symbolically introduced a bill proposing a national commission to oversee the country's "conversion to a peacetime economy."[26]

Back at the MIT campus, highlights of the many planned activities and speeches included historical reflections; assessments of the meaning of academic freedom and participatory democracy during wartime; a warning from economist Thomas Schelling on the dangers of moral crusades; speeches on protesting the draft; a debate about arms control and disarmament featuring Hans Bethe and historian Gar Alperovitz; and public statements from the organizers, SACC, and the UCS. The rhetoric of the day emphasized three broad themes: the ethical obligations of both individual scientists and their institutions, the agency of members of the MIT community, and the specific proposals for reform. As graduate student Jonathan Kabat reminded his audience, "You're MIT. You people are MIT." The MIT "system" was still composed of individuals capable of collective action, which could take many forms, including new contracting alternatives, research standards, and laboratory regulations. From the abstract and the philosophical to the concrete and the economic, the MIT community was debating the ethics of defense research on campus.[27]

Two key panels deserve closer analysis. The first, straightforwardly titled "The Responsibility of Intellectuals," featured short statements by SACC student organizer Joel Feigenbaum; linguistics professor Noam Chomsky; Physics Department Chair Victor Weisskopf; and UCLA chemist William McMillan, an advisor to Gen. William Westmoreland. The four men espoused very different ethical outlooks. Feigenbaum, the youngest, offered a clear articulation of SDS and SACC analyses. He opened with an impassioned condemnation of the failure of intellectuals to empathize fully with the plight of impoverished inner-city residents and victims of U.S. military actions. MIT intellectuals hid behind claims of impartiality, declaring themselves "apolitical" or "values-free," he complained, all the while building MIRV systems and publishing defenses of U.S. foreign policy in political science journals.

Weisskopf, representing the older Manhattan Project generation, offered a general reflection on the worth of scientific research: despite potentially dangerous applications to weapons science, in the long term, "it is good to know more." Nevertheless, he felt scientists bore special obligations due to their knowledge of potential risks. Scientists should not work on weapons projects *solely* out of curiosity and interest, he explained, and they ought to prevent the misuse of their research through active political engagement, either within the system, through government and military advising roles, or without, through conducting independent studies, organizing protests, or educating the public. Trying to play the role of peacemaker, Weisskopf warned that though both insider and outsider scientists would likely distrust each other, both groups were working toward the same goal: the responsible application of technology to world problems.[28]

Overshadowing both Feigenbaum and Weisskopf, however, were the presentations by Chomsky and McMillan. Chomsky spoke on "The Responsibility of Intellectuals," a subject on which he had published an influential essay two years earlier in the *New York Review of Books*. In that work, he had argued, like Weisskopf, that Western intellectuals bore a special burden because they were among the very few "in a position to expose the lies of governments, to analyze actions according to their causes and motives and often hidden intentions." American intellectuals had not taken up the challenge, however. They assumed the purity of government motives and so offered only technical critiques instead of more radical analyses. Chomsky targeted social and behavioral scientists in particular. Citing nuclear weapons theorist Herman Kahn as an example, Chomsky observed that much social science could not actually be tested empirically but that the trappings of empirical analysis allowed social scientists to promote themselves as rational, trained practitioners with "sophisticated" approaches, who could then dismiss the moral arguments of nonsocial scientists—arguments that lacked the prized technical veneer—as "emotional" and "irrational." The result was a class of entrenched intellectual experts whose own professional success created a vested interest in the maintenance of the status quo.[29]

Now, in front of the MIT crowd, Chomsky turned his attention to engineering and the natural sciences. He observed that as was true in the social sciences, engineering experts who weighed in with technical evaluations were considered "responsible" while those who criticized deeper motivations were dismissed by policymakers. But Chomsky reiterated his belief in the potential of human action. He told the student audience: "A good deal depends on our conscious choices. The scientists who are called upon to construct the ABM need not do so . . . They can organize and encourage others to join them in this refusal. They can also help to create the mass politics that provides the only real hope for restraining and ultimately dispelling the nightmare that they are now helping to create." The universities, in turn, ought to halt their acceptance of defense contracts and their complicity in the "war machine." Doing nothing would only preserve "the distribution of power as it exists."[30]

Chomsky's powerful words were countered by those of McMillan, fresh from his two-year stint in Vietnam as science advisor to General Westmoreland. He offered a spirited defense of his background and political positions, identifying himself as "a relatively independent member of the 'Establishment'" and listing his various governmental and military advisory roles as well as his academic training. McMillan took issue with several key aspects of the March 4 protest. First, he pinpointed SACC's complaints of "the present overemphasis on military technology" as a false target. Compared to the skewed military research agenda of World War II, MIT in the Vietnam era hardly emphasized military work, let alone overemphasized it. McMillan mourned the "literally hundreds" of possible contributions MIT scholars could make but refused to out of antiwar sentiment. "Where today is any organized effort on the part of the academic community to contribute?" he asked. "The answer is, there is no such organization. The university has been satisfied to sit back, accept the lopsided view of Vietnam presented by the news media, and on that unscientific basis decide to have nothing to do with Vietnam." McMillan then vigorously defended the existing university research on weapons systems, including the ABM, on the grounds that the aggressive stances of Soviet and Chinese leaders mandated cutting-edge defensive technologies.[31]

At the heart of McMillan's presentation, though, was a fervent defense of the good motives and responsible efforts of the nation's expert military scientists and planners. He chastised the MIT protesters for what he considered facile and ill-informed criticisms: "It is altogether too easy and too tempting to believe that the decision makers are stupid," he explained. "They are not. Whether military or civilian, they are among the most capable, thoughtful, dedicated, and responsible people we have. Moreover, by contrast with most lay intellectuals, they are exceptionally well-informed." The "lay intellectuals," in McMillan's view, refused to trust the expertise of the decision makers but failed to conduct sufficient research themselves—to acquire their own expertise. In a final damning analysis, he upheld—and defended—Chomsky's description of the implacability of the status quo, assuring his audience that "things are the way they are because it is exceedingly difficult for them to be otherwise."[32]

* * *

In contrast to the sweeping moral debates over the responsibility of intellectuals, the members of the panel dedicated to "reconversion" focused on concrete opportunities for policy reform, encouraging institutional efforts to shift funding sources and research emphases away from military projects and toward more socially useful areas. David S. Dayton of Technical Development Corporation, like many others had during the course of the day, emphasized scientists' "special responsibility" and implored his audience, "Don't sink into the system. Don't sink in. But don't drop out either. . . . If you can't join a big company, form one. . . . The creative process is here, in universities like this. It's in your hands." Los Alamos consultant Leonard W. Gruenberg echoed Weisskopf's endorsement of working "within the system," noting that a lab employee worried about losing his job would be unlikely to urge reform, but "if he can be educated, and if the administrators can be educated, into recognizing that by reorientation of their research they'll be in a less vulnerable position, I think that it's possible . . . that very great changes can be made with Defense-Department-supported research itself." His speech was a gentle call to action among weapons scientists themselves.[33]

Offering the most specific and detailed blueprint for how reconversion could work, Ronald Probstein discussed his experiences at the engineering department's Fluid Mechanics Laboratory. In 1966 the lab had held $300,000 in defense-sponsored research contracts, all focusing on research relevant to "missile re-entry and space exploration." Though this research was entirely unclassified, the lab's faculty and graduate staff, during months of soul-searching meetings, worried that the emphasis on military applications and money constituted an "imbalance." To compensate, they brainstormed how they could "reconvert" the lab's work to incorporate more socially palatable projects. Fluid mechanics specialists turned to new topics in water and air pollution, desalinization, and biomedical applications. By 1969 the lab's outside research contracts had doubled in size but drew only 35 percent of funds from defense sources. Instead, money flowed in from the National Institutes of Health, the Public Health Service, and Edison Electric. Probstein observed that although the new projects might seem "vastly different . . . from the types of problems encountered in nuclear explosions or missile re-entry," in fact "they all involve fluid mechanical and chemical kinetic concepts, so that the real efforts were in reconverting our own thinking from one area of research to another." Probstein acknowledged that many of the lab's grant proposals and applications for funds had been rejected, but he nevertheless pushed MIT to take a leadership position in defining and pursuing new areas of scientific research unconnected with defense objectives.[34]

Reactions to the reconversion talks ranged from enthusiastic to skeptical. During the closing discussion session, older faculty and audience members eagerly expressed interest in pursuing reconversion. But other commenters doubted the radical potential of trying to reform funding and research systems from within. As one questioner put it, "*You* built the atomic bomb, the H-bomb, the missile systems, and so on. How do you expect that, if you stay in a system, you're going to be able to do any different in the future than what they tell you?" To this, Gruenberg answered that "to stand on the outside and act morally superior and moralistic is also not of use . . . just standing on the corner and rabble-rousing is just as useless as going ahead and working on missiles." But the audience rejected his reply, questioning "the inference that one can subscribe to the system and do his job well enough to maintain it and yet not be influenced by the system that surrounds him every day." Concerns about "the system"—and the limits of individual agency within it—permeated nearly every question and comment posed during the discussion.[35]

✳ ✳ ✳

In the aftermath of March 4, SACC continued to exert pressure on both university researchers and the administration. They targeted researchers working on J. C. R. Licklider's Project CAM, a program to use the computer time-sharing system at MIT to develop a "behavior science date management system" linked to resources at RAND and ARPA. SACC members feared it would be used to enable studies of peasant groups and student radicals so as to be "more effective in suppressing popular movements around the world." They also helped plan another round of teach-ins, the "Agenda Days" to be held in May, featuring discussion groups and panel sessions addressing topics such as Pentagon research, Project CAM, the arms race, Poseidon and other weapons systems, the morality of defense research, MIT's place in the military-industrial complex, and insiders' views of Lincoln Lab. Participating speakers included Henry Kendall, J. C. R. Licklider, George Rathjens, and others, with portions to be broadcast on Boston public television.[36]

Most importantly, SACC demanded an end to the I-Lab's work on the Poseidon program and army helicopter research. In late April their efforts paid off. The university administration, acknowledging the widespread unease with campus weapons research, agreed to reevaluate the activities and funding of both of MIT's special laboratories. Headed by William Pounds, dean of the Sloan School of Management, a new panel was formed: the MIT Review Panel on Special Laboratories. The initial roster of members included professors of engineering, group leaders from the special labs, an undergraduate business student, graduate students from engineering and biology, physics professor Victor Weisskopf, former MIT president Julius Stratton, Yale historian Elting Morison, and Frank Press, professor of earth and planetary sciences and future science advisor to Jimmy Carter. Within a few weeks, two additional staffers from the special labs joined, as did Noam Chomsky, though his participation required biweekly flights from England, where he was lecturing at Oxford. While the panel met throughout that spring, MIT imposed a temporary ban on classified research. The administration had taken the protests of March 4 seriously.[37]

* * *

From the end of April through May of 1969, the committee met in both private and public sessions, mulling over the ethical issues at stake. They heard testimony from researchers, administrators, politicians, and activists, and convened to discuss among themselves the most appropriate recommendations to make. The committee's ethical debates and their evaluation of lab practices offer a snapshot into attitudes about science and morality at arguably the nation's premier technical research university.

In its early private sessions, committee members discussed the broader contours and history of university research. Why was it necessary for university labs to conduct military research? Peter Gray of the MIT Alumni Advisory Council observed to his fellow panelists that government research had traditionally taken place in three kinds of sites: government labs, university labs, and industrial labs. Recently, in-house Defense Department labs, such as those run by the army and the navy, had increasingly subcontracted out work to industry. Even MITRE, over the previous decade, had siphoned off much of its experimental work to focus on problems of management. In this analysis, university labs comprised the last, best haven for truly inventive, creative, independent people to work. Moreover, the intellectual atmosphere and prestige of university affiliation would draw top researchers. As Chomsky put it, "Whenever there is an opportunity, financial and otherwise, the genius will go to the universities." Nevertheless, Chomsky and other panelists objected to the work of the special labs on moral as well as logistical grounds—Chomsky worried about their size and the risk of "the tail wagging the dog," particularly in the context of Lincoln Lab's massive budget.[38]

The committee members also used their early meetings to articulate, with startling candor, their own "personal philosophies" on topics ranging from the proper role of student participation and campus democracy to the appropriateness of defense research at MIT. Marvin Sirbu, a graduate student in the electrical engineering department, opened the discussion by pondering the inherent difficulty of trying to identify universal contracting principles in a changing moral environment, noting that the original moral atmosphere of 1949, when the Lincoln Lab had been established, had changed dramatically by 1969, as had the lab's original function. He was unsure how to address this problem—he wondered, tentatively, if writing a university constitution that allowed for student representation in setting policy might be effective.[39]

Jonathan Kabat, an outspoken biology graduate student and SACC member, concurred with many of Sirbu's points but, surprisingly, downplayed the importance of participatory democracy, noting that SACC was not really "clamoring for student power." He doubted whether the MIT community possessed the "technical" or "intellectual competence" to understand what was at stake, suggesting that MIT students and staffers were too entrenched in their environments to be capable of diagnosing systemic problems and proposing viable alternatives. Unlike Sirbu, who had volunteered to support democratically enacted policies with which he disagreed personally, Kabat fervently advocated a policy of both individual moral action and institutional change. What if the MIT student body voted to build a gas chamber, for example? Kabat's views epitomized the New Left struggle to reconcile the ethical obligations of both individuals and institutions. Individual acts of resistance would be necessary to counteract dangerous institution-wide policies, even as Kabat's ultimate goal was a remaking of society such that its institutions promoted more benign ends.[40]

Like the graduate student Sirbu, most panel members expressed moral unease but uncertainty as to the proper policy course. Victor Weisskopf accepted defense research so long as attention was focused not just on the "gadgets" themselves but "the deeper use and consequences of use of these gadgets." Eugene Skolnikoff of the political science department echoed Sirbu's concern, worrying about "committing to making major decisions in changing the institution" in the context of "changing morality." But he wasn't sure how these concerns would translate into policy for the special labs. MIT Dean of Humanities and Social Science Robert Bishop found his Quaker beliefs hard to reconcile with his pride at the labs' accomplishments while Peter Gray wondered if reconversion was truly feasible. Though he would later take a more radical position, Sloan School undergraduate George Katsiaficas initially supported MIT's war-related research but also endorsed a review panel to ensure the school did not become a "yes man" to the government. David Hoag of the I-Lab's Apollo Group suggested that nuclear weapons might be like taking out the garbage—a dirty job that had to be done; Victor Weisskopf worried that talented MIT students who wanted to participate in large, cutting-edge projects had no choice but to head to the special labs; chemical engineering professor Edwin Gilliland advocated a more carefully controlled style of defense research; and graduate student Jerry Lerman lamented that the committee was only addressing the symptoms of living in a war-obsessed society, not the cause.[41]

Of the representatives from the labs themselves, computer specialist and Lincoln assistant director Gerald Dinneen wholeheartedly supported the strong ties among MIT, industry, and the Defense Department, noting proudly that he was a self-aware researcher who considered his work with the lab important and beneficial to the country. Irwin Lebow, a group leader at the Lincoln Lab, defended his work as both intellectually stimulating and worthwhile; he rejected arguments that "mission-oriented" work somehow violated university priorities. He observed: "I look upon the mission part of this work very much as the dissemination of knowledge, the dissemination of university-obtained knowledge toward the practical problems . . . of government." These government problems are "too important to leave to the generals," he argued. "It is too important for the university not to get involved."[42]

As the discussion wound down, Frank Press offered a clear synopsis of the problems identified by a large majority of the group, who simultaneously wanted MIT to distance itself from weapons research while exercising increased political influence. Press opposed classified research on campus but nevertheless believed the labs' technical work was important. The problem, as he saw it, was the war in Vietnam. "The emotionalism of a very bad war completely cut off American universities from rendering advice . . . in problems of defense," he observed. Vietnam was overshadowing the labs' other contributions. He was particularly impressed with MIT-generated data calling into question the feasibility of an ABM system. The fact that the ABM was even being debated was, in Press's view, "a revolution . . . the Senate Armed Forces Committee is being challenged for the first time on a major strategic system. The vote may go against the Department of Defense in Congress." In other words, the unpopularity of Vietnam had called into question MIT's ties to the Defense Department, even as its labs' technical input fueled unprecedented political resistance to an ABM system. Was there any way to distance the school from Vietnam while preserving its contributions to arms control? The pervasiveness of Defense Department funding across multiple fields and laboratory research programs suggested otherwise.[43]

* * *

One of the panel's first tasks was to visit the special labs to hear presentations on current projects, tour the facilities, and solicit statements from researchers and administrators. At both labs, the panel's questions and lab staffers' responses emphasized the same key issues: the ratio of basic to applied research, the autonomy of the labs, the importance of the labs' work to MIT's prestige, the inevitability of weapons development, and the morality of demonstrating the feasibility of specific weapons systems.

The panel began its investigation at the I-Lab, where Deputy Director Roger Woodbury offered his audience a brief history of Draper's work on gunsights, tracking technologies, and autopilot systems. He emphasized that the lab's work on inertial guidance stemmed from fundamental research questions, including the basic task of "establishing geometry and measuring quantities with respect to inertial space." Moreover, even the development of specific devices, such as new kinds of gyros and accelerometers, depended on basic advances in materials research and metallurgy. Woodbury and a stream of additional presenters hammered home this point—that basic research underlay much of the lab's work.[44]

Woodbury also argued that the I-Lab differed fundamentally from other industrial labs because it had no profit motive and because its military contracts supported versatile research that enabled advancements in nonmilitary fields. For example, I-Lab researchers had recently developed a blood viscometer from technologies developed for other projects. This portrayal elicited critical questions from the committee members until Woodbury conceded that the lab did not receive any funds for truly basic, "unrestricted research." Nevertheless, Woodbury focused on convincing panel members of the lab's autonomy from both military demands and industry competition. He noted proudly that "I think what our charter is, is to really maintain a technology that is sufficiently advanced that we are not competing with industry, we are providing information to lead the way." He chose his words carefully; when asked about a particular missile contract, he self-consciously explained: "We were instructed to work on an inertial, not instructed, we *agreed* to work on an inertial guidance system for Titan."[45]

Woodbury also distinguished between the proprietary restrictions of industrial research and the more open atmosphere of government-sponsored work. He explained that at industrial labs, researchers were often hamstrung by requirements that they draw on their own patented techniques, building outward from limited earlier technologies. He offered the example of Sperry Gyroscope, which lost its government contract for a civilian inertial navigation system due to an inefficient reliance on "proprietary information" rather than "available technology." In the case of the I-Lab, researchers had much more freedom to be creative and innovative. Woodbury described the development of one inertial guidance prototype: "We built a tremendous thing which was impractical, unsaleable, and we knew it, but it demonstrated the principle, the principle was feasible, and then we went on to develop the state of the art and get it down to size." He estimated that without the innovation of the I-Lab, private industry might have taken another two decades to make the same advances in inertial guidance.[46]

* * *

Other lab administrators echoed this sentiment while addressing specific controversial projects, including the powerful Poseidon missile. The men explained that the lab had been the "design agent" for Poseidon's guidance system, with industrial work farmed out to General Electric, Raytheon, Bendix, and Honeywell, among others. The lab's task was to construct a prototype "to show it can be built," as Program Manager Samuel Forter put it. Conveniently, the lab was simultaneously at work on guidance for MIRV, and although not part of the initial Poseidon missile design, flight tests for the new "MIRVed" missile began in 1968. Experts told the committee that even if MIT were to halt its Poseidon work immediately, the MIRV research would continue, albeit at a higher cost, at Raytheon and General Electric.[47]

Here again was the question that had haunted the Manhattan Project scientists—what were the ethical implications of "demonstrating the feasibility" of a new weapon system? The lab workers protested that if they didn't do it, other industry or military labs probably would—but not as quickly and not as creatively. And at stake was the reputation and prestige of MIT. As Edwin Porter Jr., associate director of the I-Lab, would argue in a later memo to the panel: so long as the United States existed in an imperfect world, there would be a need for defense research. The government could find other researchers besides those at MIT, but MIT's decision to abandon the labs would excommunicate it from the adult world of real responsibility: "MIT . . . would be like the little boy who took his ball and went home," he wrote. "MIT would no longer be a part of the defense effort, and thus would have little or no chance to influence it."[48]

Though several years would pass before the final decision to sever ties with the lab, the seeds of disunion had been planted in these early discussions between panel members and lab researchers. Woodbury's criticism of inefficient corporate research suggested the possibility that the I-Lab could continue as a nonprofit entity, though not necessarily one affiliated with MIT. When the charismatic Draper himself appeared before the committee the following week, he offered a rousing defense of the lab, but his joviality failed to placate the critical minds of his interviewers. Victor Weisskopf worried that research was a zero-sum game and that defense research crowded out other options. Most notably, Chomsky criticized lab requirements that staffers gain security clearance and argued that, in a sense, accepting U.S. national security decisions had become "a necessary condition for being associated with the Instrumentation Lab." Lab researchers might have chosen their particular interests and projects honestly, but their presence in the lab indicated acceptance of the "political ideology the labs represent." This acceptance was not neutral, Chomsky argued. It supported the status quo and thus fundamentally undermined any lab claims to meaningful autonomy.[49]

✳ ✳ ✳

Just as I-Lab researchers had, the scientists and engineers at the Lincoln Lab emphasized the freedom that came with academic—rather than military or industrial—affiliation. Their prize example was work connected to the proposed Sentinel ABM system. The lab could demonstrate feasibility of weapons systems, they acknowledged, but it could also demonstrate *unfeasibility*. This theme was emphasized during a presentation by Joel Resnick, an assistant group leader at Lincoln, who explained the lab's work on system analysis and the "technical evaluation" of missile programs. In his view, this work consisted of "guidance" and advice as to how best to improve maintenance techniques and further research and development without any specific recommendations concerning deployment. The Lincoln Lab was uniquely positioned for this task, he explained, because researchers were well versed in the technological details and data necessary but also had the benefit of MIT affiliation to give them "an objectivity and independence here which will allow us to do the analysis without regard to the sponsor's reaction to the conclusion"; to "let the chips fall where they may." In the case of the lab's review of the Sentinel program for the Defense Department and the President's Science Advisory Committee (PSAC), Lincoln analysts reported several potential threats to the effectiveness of the system. Resnick summarized, "When we finished the briefings there was no question about whether the Sentinel system would be effective . . . it was generally realized that it would not be." Resnick differentiated between objective technical evaluation and the politics of deployment decisions: "We believe it is important to have an understanding of the implications and limitations of ballistic missile defense. This is true whether the country will ever deploy it or not."[50]

But members of the panel did not immediately accept this justification. As he had during an earlier exchange with Woodbury of the I-Lab, Jonathan Kabat wondered whether any kind of technical research into weapons systems exacerbated the arms race simply by being conducted. After presentations by lab director Milton Clauser and William Lemnos, a Lincoln group leader working on developing decoys, jammers, and other "penetration aids" to enable American missiles to breach a potential Soviet ABM system, Kabat finally broke in:

Kabat: . . . sitting here and hearing this stuff over and over and over again, the offense and defense, and the penetration, and do any people feel after a while that it's all just completely insane? How do you feel about this stuff personally? I am addressing this to anyone who cares to answer it. It is depressing the life out of me.

Clauser: I can only refer you to the New York Times story about the Russian missile launches. Do you believe they are unreal?

Kabat: No. I am willing to consider their reality.

[Unidentified speaker]: I can answer that this way: when the Kwajalein [missile-testing] site was first set up I went down there and lived on Kwajalein for a year. When I saw my first re-entry it made me sick and the reason I am here, as far as I am concerned, is trying to prevent that from happening in reality. You may or may not agree this is the way to do this, but I believe that only a laboratory like this connected to MIT can allow us to make an input which is an independent and free input and that is what we have been doing as individuals and groups. Some of us have been sick for a long time about this, I am sure. It is a nasty world that we live in.

Kabat: The question is, what is the most effective way to change that?

Clauser: This is something we all have to think about. It is to be noted that a number of us operate privately. We haven't given up our citizenship. We operate privately and make our voices heard, but we don't confuse this role with the Laboratory role.[51]

In an open hearing the following day, Clauser and the panelists also clashed over the nature of Lincoln's Vietnam-related research. Under stiff questioning, Clauser acknowledged that the lab had done some work on "moving target" radar after he had received "a personal appeal from Johnny Foster . . . to help some of the people in Vietnam." A Lincoln-designed mobile radar unit designed to detect person-sized movements had been sent to protect a U.S. camp in Vietnam. Lincoln researchers were also at work on developing an antenna system to detect underground anomalies, such as the presence of tunnels. In other presentations, Lincoln staffers had taken pains to distance themselves from any Vietnam applications, suggesting that while most lab workers considered ABM work valuable, Vietnam perhaps presented more complex ethical challenges.[52]

Some of this ambivalence emerged in a joint statement to the panel submitted by staff members from Lincoln on May 15, agreeing that "research on defense problems is presently overemphasized at the Special Laboratories" but imploring the panel not to sever ties to the lab. The staffers carefully defended the lab's acceptance of classified research, noting that most of the classified material at the lab consisted only of "externally generated" background government documents used to "guide" specific projects, not actual results from ongoing lab work. (A "Fact Sheet on Lincoln Laboratory" produced by MIT's Office of Public Relations, however, reported that 40 percent of Lincoln's research was classified.) More importantly, the staffers argued, the lab's role as a source of candid advice—an "honest broker" who could "help in the settlement of disputed technical questions"—could not be replicated by any other group of experts, including "an Ad hoc review board . . . a JASON group, or . . . a PSAC subcommittee." These other well-intentioned but part-time groups simply could not bring the same amount of time and experience to the task. Who better to assess ABM-system efficacy than Lincoln scientists, after all, with their extensive firsthand understanding of the kinds of radar and optical data available to make discrimination decisions?[53]

As their colleagues at the I-Lab had, Lincoln staffers also invoked the potential blow to MIT prestige and influence. How could MIT *not* weigh in on these crucial decisions? Should the university cut its ties to the lab, they would be giving up their access to key technologies and their authority to offer meaningful, influential assessments. MIT's reputation "would be tarnished for having turned its back on an active, progressive laboratory. It would be said that instead of continuing to serve society through the lab on the large scale necessary for today's problems, MIT turned down the challenge."[54]

As the committee contemplated their policy options, they met with Jack Ruina, former ARPA head and MIT vice president for Special Labs since 1966, and Paul Cusick, the MIT comptroller. Ruina clarified aspects of the process by which lab contracts were approved. A small group of university administrators—including the president—oversaw lab contracting decisions. Their determinations took into account several key factors: ensuring that the labs pursued broad programs of national importance (as opposed to small, specialized applications); whether the work constituted a "public service" of some kind; whether the labs actually possessed the technical capability to do a good job; and campus attitudes. The labs did not accept contracts blindly and had previously refused certain projects, including chemical and biological weapons (CBW) research and intelligence work. Cusick estimated the possible economic consequences of detaching the labs from the university. He worried that although MIT was a nonprofit institution, losing the labs could potentially result in an extra $12 million in overhead expenditures. He also warned that the character of labor performed at the labs would be altered if they were severed from MIT due to changes in contracting. Whereas MIT maintained its own grant-like contracting system, the spun-off lab would be subject to more traditional fee arrangements, such as those at MITRE, where employees submitted time sheets and were monitored like industrial workers. Cusick called such a system the norm in industry, but "unacceptable" for a university lab.[55]

* * *

Among the politicians, top university administrators, and laboratory group leaders, few voices from the middle tiers of research staffers were heard. Only Paul Easton, a former Lincoln staff physicist from 1967–1968, addressed the panel at any length, and his convoluted narrative and obvious disenchantment irritated Pounds, who cut his testimony short. In the short time Easton spoke, however, he offered a detailed glimpse into a world that was nowhere as intellectually stimulating or ideology-free as others had portrayed it. Easton had applied for a plasma physics job at Lincoln after a graduate school stint working for the air force. As part of Lincoln's Project PRESS, Group 35, he analyzed data collected at Kwajelein in order to develop techniques for discriminating between reentry vehicles and decoys. The job paid well and his boss was laid-back, but he found the work itself to be easy and "low-level" despite the group's reputation for brilliance. He recalled his coworkers spending hours playing the stock market.[56]

Though opposed to the war in Vietnam, Easton initially justified his classified military research on the grounds that an ABM system was a defensive, deterrent technology. Over time, however, he began to wonder if other groups at the lab—those working on "penetration aids," for example—used his group's data for offensive purposes. Even then, the nature of his research didn't worry him excessively for the simple reason that he, like most of the other members of his group, did not believe the ABM system would work. The more high-profile politicians publicly supported the ABM, the more his colleagues laughed "at how dumb those congressmen were." The real problem, for Easton, was his anger about Vietnam and his slow realization that if he traced the chain of command high enough, he was working for Robert McNamara. He offered the panel a kind of psychological analysis of the lab culture that mirrored accounts of wartime troop cohesiveness:

> Essentially, what was motivating people to work there was not interest in the work, because it was not interesting work; it was not idealism because it was obviously not useful work, but it was the kind of team spirit . . . people like one another and they were working for the approval and respect of the other people on the team. And then, somehow . . . you have to have some sort of

positive feeling about the other people on the team, and you also have to have positive feelings about the guy your boss reports to and the guy he reports to, and eventually you report to Secretary McNamara or some guy who I would consider a maniac. . . . And the necessity to consider yourself a member of a group which includes Secretary McNamara and his friends in the administration . . . does something to one's thinking.

Easton closed by warning that the lab staff was not "a neutral and disinterested group" because the experience of working there promoted loyalty to government leaders. As he put it, in language that surely failed to convince the skeptical Pounds, working at the lab "does something to your head."[57]

* * *

Meanwhile, outside the committee room, public pressure mounted. The UCS demanded that Poseidon research and MIRV testing be "terminated as soon as possible," since successful MIRV flight tests would be "irreversible" and would "raise arms control to a new level of difficulty." To the UCS, the I-Lab housed the most objectionable classified research and therefore required immediate intervention. Lincoln, in contrast, could be more gradually converted to other projects. But equally urgent for many were the employment consequences of any such changes: Local 254 of the Building Service International Union, representing 860 lab employees, threatened to seek an injunction if MIT curtailed its research support. The group's business leader, Edward Sullivan, complained to *Electronic News* that "it's a group of pseudo-intellectual vagabonds who are stirring up the trouble at MIT. It's no longer a question of campus fun and games. It is the economic security of thousands of people which is at stake."[58]

Hundreds of letters poured in, from alumni, current students, concerned citizens, and MIT staff, overwhelmingly defending the labs. Two hundred and thirty-six I-Lab employees signed a petition to MIT president Howard Johnson supporting the Poseidon program and their contributions to its guidance system. Lab workers also submitted dozens of personal position papers to the panel. Some, like John Allen, an associate division head at Lincoln, emphasized their antiwar credentials while defending the nature of their research; he wrote that while he considered the war in Vietnam "one of the greatest mistakes in the history of the U.S.," he saw nothing "irreconcilable" in working to develop technologies to protect the lives of American soldiers while politically agitating for withdrawal. Others, such as Allen's colleague Thomas Casey, offered more philosophical assessments of the nature of scientists' obligations. Casey denounced claims that scientists bore special responsibility for the applications of their work. "I see no reason to believe that the man who designs the guidance system for a missile is necessarily more qualified to control its use than is a man who designs television sets," he wrote. A few paragraphs later, however, he acknowledged that MIT's enormous influence would be lessened if it gave up control over its labs. Therein lay the tension. As an individual, Casey wanted no extra moral burden due to his work at the I-Lab, but he recognized—and endorsed—MIT's institutional power: "When MIT speaks," he wrote proudly, "people listen."[59]

Charles Broxmeyer of the I-Lab further explored this tension between individual guilt and institutional responsibility in a deeply emotional statement read to the members of the committee. Broxmeyer, an electrical engineer, had joined the I-Lab in 1955 after receiving degrees from Drexel and Penn and serving a stint at the Naval Air Development Center. At the I-Lab he specialized in inertial navigation and guidance, literally writing the book on the subject in 1964, and spent two years on loan to Raytheon's Space and Information Systems Division. Now a "technical consultant" for the I-Lab's Deep Submergence Systems Group, he defended the work of the special laboratories to the Pounds Panel, complaining that due to capricious shifts in campus opinion, "magically the work which was always so important, useful, aesthetically satisfying and necessary has now become trivial, harmful, ugly and definitely superfluous." More powerfully, he accused the faculty members on the panel of acting out of guilt and cowardice, projecting their own regrets about Manhattan Project contributions and their frustration at their own inability to stop the war in Vietnam on Broxmeyer's innocent colleagues—an ill-conceived and dangerous act of scapegoating. In a particularly evocative section, he observed:

> The agony of the Vietnam War has induced intolerable feelings of anger, guilt, frustration and self-hatred in the American people. Further, we are even deprived of an object for our hatred. We can't hate the government because the government is reported to be trying to stop the war. We can't hate President Johnson because he isn't there anymore. We desperately need objects on which to deposit our hatred.

On the campus, the feelings of guilt for what is going on reside mainly in the faculty. What the students feel is fear. They are frightened, of being drafted, and being shipped to some jungle in Vietnam like livestock, and forced to kill, and then finally, slaughtered themselves for no reason. . . . The special laboratories, which every shred of evidence indicates have nothing whatever to do with the Vietnam War, have been selected to be destroyed. Only then will the collective guilt of the MIT community—that part of it which remains—be expiated. Only then will the students and professors feel cleansed and purified.

There were more than just shreds of evidence linking the labs' work to the war in Vietnam; testimony to the committee had established as much. But with startling precision, Broxmeyer had tapped into the guilt, disillusionment, and anxiety of many MIT scientists: guilt for choosing a field with applications to weapons and war, disillusionment with government policy, and anxiety that recourse for changing the state of affairs no longer existed.[60]

✳ ✳ ✳

Perhaps most influential for panel members were the statements and testimonies of the towering former presidential science advisors and PSAC members, veterans of the Sputnik push for greater scientific research funding, now called to account for the backlash. On May 8 Jerome Wiesner, then provost of the university, appeared before the committee. Though generally opposed to classified research on campus, he distilled the committee's problems to two key questions: first, did the labs add or detract from MIT's educational mission; and second, did they contribute to the national welfare? On the first question, Wiesner answered yes; the labs provided access to sophisticated technologies, ideas, and equipment. The value of weapons research itself was trickier. He observed that unilateral disarmament was impractical, and specific types of weapons evaluations were worthwhile as means to "prevent panic" about hostile foreign systems. Nevertheless, he distinguished between general research on weapons systems and actual development of operational technologies. He explained, "We should try not to be engaged in the development of things that kill people. We can be involved in understanding of fundamental technology that might be useful in the development of . . . military weapons, make a deep contribution to the defense field."[61]

Caltech president Lee DuBridge, soon to be appointed science advisor to Nixon, agreed. In an influential article in the *Bulletin of the Atomic Scientists,* he asserted that "it is not appropriate for secret military research to be carried on within university campuses . . . but I do not agree that universities should not accept any research support from the Department of Defense." He endorsed military-sponsored basic research and lauded the "wholesome influence" of university professors and other independent advisors on the military establishment.[62]

George Kistiakowsky also acknowledged that "scientists and engineers have a very major social responsibility to society." But by 1969, he had difficulty enumerating the exact policy obligations such responsibility entailed, beyond individual activism. He confessed that he could not "conceive as bad" the "acquisition of scientific knowledge" and that he did not believe that "the scientific technological community" was "competent" to judge the social value of applied projects. Nevertheless, the same community ought to be "urging and pushing people . . . into making better decisions." Without revealing the specific nature of his work with the Jasons, he told the committee of his disillusionment with the Pentagon: "When it became clear to me that the military had sabotaged—and I use now the word deliberately—the project . . . I resigned." But despite his disassociation from the Defense Department, Kistiakowsky still represented an older, more individualized ethical stance. In part, this view stemmed from his conviction that military-academic ties were nowhere as strong as protesters had suggested. Dismantling the I-Lab would not halt the arms race or strike a major blow to weapons research; likely it would result in somewhat "inferior guidance for Poseidon." The military was not utterly dependent on MIT, and MIT's decisions would not drastically alter the course of U.S. foreign policy.[63]

Kistiakowsky's focus on individual decision making influenced his views of appropriate MIT policy. Though he considered a blanket rejection of military funds to be "silly," he did not endorse accepting any offer "without any qualifications." Defense dollars were not "tainted," he observed, and MIT policy ought to take into account the different individual views of the researchers involved in order to find a "golden median." An MIT panel should review contracts and strive for imperfect moral decisions that would be "maximally satisfactory" to the MIT community. And in this process, Kistiakowsky saw little place for students. They could be consulted, of course, but in general he thought "students should be seen and not heard very much. By and large they are too emotional, they don't have time to look into issues in sufficient depth." Judgments on projects were best left to individual, thoughtful experts.[64]

∗ ∗ ∗

As the Pounds Panel members began to digest the hours of testimony and thousands of pages of documentation in order to draft an interim report, the issue of personal and institutional responsibilities loomed large. In a long, eloquent internal memo to Pounds, Noam Chomsky reviewed many of the philosophical and practical concerns facing the committee. In his view, the panel's work—and their eventual public report—could be enormously influential to both a national audience and MIT policymakers. For the national audience, he wanted a strong statement that government priorities "must be reordered, with war-related expenditures dropping sharply and national energies turned to other pressing social needs." In promoting this vision, he hoped the committee would "maximize" its impact by "being even more optimistic . . . than we may privately feel." In this context, Chomsky again emphasized the special position of scientists and the special position of MIT as their institutional home, arguing that a certain "form of politicization" would be welcome. He wrote: "Those who develop science and technology have in their hands a powerful weapon of destruction, and a major instrument for overcoming the problems of contemporary society. They must be aware of this fact, and conscious of their responsibilities in regard to the use of science and technology. To exercise this responsibility, they must, continually, make political and historical judgments. This is true of the work of an individual. It is far more important when the university makes an institutional commitment to the support of organization of research."[65]

Chomsky hoped these commitments would be subject to democratic evaluation by the entire university community and, ideally, would result in serious and "dispassionate deliberation." Though he wrote that all people in democratic societies ought to exercise these responsibilities, Chomsky felt it was "particularly important in the case of scientists and engineers because the consequences of their acts—their research, study, and teaching—are potentially so immense." This type of politicization, Chomsky wrote, ought to be encouraged and should take place in democratic committees composed of faculty, staff, and students.[66]

The report's second, related, goal, Chomsky wrote to Pounds, should be to provide realistic plans and policies for MIT's immediate future. Here he emphasized a different, darker politicization already taking place, a "consensus set elsewhere" that had warped MIT research priorities. This took hold through stipulations that lab employees agree to submit to security clearance and receive defense money as funding. Chomsky therefore urged that MIT not sever its ties to the labs. Instead, the university should set clear guidelines as to acceptable research, including bans on any work that "contributes to offensive military action," any "involvement in any form of counterinsurgency operations, whether in the hard or soft sciences," any "actual development of weapons systems," and, in general, any contributions to the "unilateral escalation of the arms race," though research into defense and deterrence might be acceptable. Chomsky estimated that these prohibitions would "rule out" CBW research, MIRV, and any "steps toward deployment" of ABM. He also called for an end to Project CAM, which he didn't think would "contribute to science in any serious sense" other than software development, but would "be used, primarily, for repression of popular movements and interference in the internal affairs of other nations, and perhaps for domestic repression as well."[67]

The better course of action, then, would be conversion. Chomsky envisioned the I-Lab as "an interdepartmental laboratory . . . sharing some of the characteristics of RLE or the Magnet Lab" with no classified research, no MIRV work, and the promotion of "socially useful technology" and basic research. Lincoln Lab presented a more complicated scenario. Chomsky expressed deep skepticism of lab administrators' claims that they provided "objective and independent evaluation of weapons systems." He noted that he "did not doubt the integrity of the laboratory staff, or their competence" but nevertheless felt that they represented "a limited range of opinion" working in a lab entirely funded by the Defense Department, where employment was subject to security clearance. MIT ought to apply the same standards to Lincoln as it did the I-Lab, but he acknowledged that if the restrictions proved too "sharp," the Pentagon might "simply take over the laboratory." Nevertheless, MIT should try for restriction and conversion, especially since it would send a strong national message. Even better might be a system in which Lincoln received funding from a more neutral government body—for example, Congress—and became a truly independent facility not subject to clearance requirements.[68]

Most of Chomsky's ideas found expression in the panel's interim and final reports, issued that spring and fall, accompanied by appendices of "Additional Statements by Members of the Panel." In the main reports, the panel endorsed retention of the laboratories coupled with explicit efforts to shift research to more civilian-oriented missions and provide better educational opportunities to undergraduates. The statement did not reject defense work entirely, nor did it propose permanently halting classified research, which had been temporarily banned while the panel met. In the strongest language of the report, panel members recommended an immediate end to Poseidon research, on the grounds that MIT labs should not produce prototypes and deployment-ready weapons systems. They also deeply criticized Lincoln's people-detector research, suggesting that the lab should have transferred this work to the military once it reached the testing phase. But in far weaker language elsewhere, the panel stated that these concerns were separate from any "collective judgments about military and strategic national policies"; judgments that the majority of the panel disavowed as inappropriate. To manage future contracting decisions, the panel recommended creating a "Standing Committee" with faculty, student, administrative, and lab representation to advise MIT's president on approvals.[69]

Chomsky attacked this weak language in a separate appended statement. He called for an end to all war-related research not strictly confined to defensive or deterrent work and affirmed, in powerful terms, the authority of MIT to make judgments on military strategy and policy. In his view, accepting defense contracts already constituted a form of political judgment. Scientists and engineers must not "remain blind to the question of how their contributions are likely to be put to use" or act as if phrases such as "national interest" or "public service" meant ideological neutrality, he argued. Chomsky concluded: "In an institution largely devoted to science and technology, we do not enjoy the luxury of refusing to take a stand on the essentially political question of how science and technology will be put to use, and we have a responsibility to take our stand with consideration and care. Those who find this burden intolerable are simply complaining of the difficulties of a civilized life."[70]

* * *

Issued after the end of the spring semester, after the departure of most undergraduates, reaction to the interim report garnered largely favorable reaction. A *New York Times* editorial lauded MIT's decision not to sever ties to the labs or "spin them off" in some way, noting that: "The sole effect of such a spinoff, in the absence of basic redirection of Government research activities, would be to force the armed services to set up more laboratory complexes of their own or to give huge new contracts to aerospace and other corporations. The result would be precisely opposite to the one desired by student and faculty dissenters—a strengthening of the military-industrial complex and a diminution in the capacity of university scientists to exert any useful influence in the shaping of public policy on military matters."[71]

MIT administrators took the Pounds Panel's reports seriously and convened a series of meetings of the Academic Council, which included every dean and vice president, to plan implementation of the recommendations. The body reiterated support for ending research related to "operational deployment of weapons systems" and remade Jack Ruina's old position as vice president for special laboratories into a vice presidency for research.. The Academic Council also approved the creation of the standing committee called for by the interim report, initially headed by chemistry professor John Sheehan and including two student representatives and two lab staffers among its ten members. The group was to meet in private sessions to evaluate potential lab contracts. In her 1971 account of the turmoil at MIT, sociologist Dorothy Nelkin summarized the nine key criteria used to judge contracts:

The potential for favorable interaction with MIT's educational operations

The uniqueness of the MIT contributions

The degree to which projects would evoke favorable MIT attitudes

The intellectual challenge

The national importance of the problem

The degree of basic research represented as opposed to production, field test, and deployment

The absence of immediate identification with a weapons system

The adequacy of existing national review of the problem

The potential for civil application

Nelkin observed that the function of this committee challenged notions of "academic freedom" and fostered resentment among lab staffers, some of whom reportedly referred to it as the "morals committee." Meanwhile, SACC complained about the lack of procedural transparency and small student representation. But by and large, student support for halting MIRV research and genuine efforts at reconversion remained strong.[72]

Far more unpopular than the panel's report and creation of the standing committee, however, was the subsequent administrative decision to replace

Charles Draper as head of the I-Lab. Charles Miller, chair of MIT's civil engineering department, took over, with the blow softened by officially renaming the lab after its founder. Draper angrily told reporters that he had been fired, and his bitterness captured national attention. Joseph Alsop churned out column after column that fall criticizing MIT's treatment of Draper, which he called "Oppenheimer in reverse." He consistently referred to the New Leftists and other lab critics as "storm troopers" and "neo-McCarthyites." Other industry publications echoed Alsop's language. *Air-Force/Space Digest* labeled the "self-appointed zealots" responsible for Draper's exit "latter-day Savonarolas" and quoted Draper's description of the Pounds Panel as "an inquisition." The same article referred to "professors still torn with guilt over Hiroshima and Nagasaki" and closed with extended quotations from Broxmeyer's emotional statement on the dangers of scapegoating. Nelkin reported the presence of "Doc Draper forever, SDS Never!" signs posted throughout the Draper lab.[73]

Clashes on the polarized MIT campus continued throughout the fall. In September MIT's Executive Committee of the Corporation announced that the university would no longer "incur new obligations in the design and development of systems that are intended for operational deployment as military weapons" though the status of Poseidon research was left unaffected. In October the Pounds Panels' final report appeared, with initial support from President Howard Johnson and a majority of faculty members. Even Draper eventually endorsed a trial period during which the special labs would be subject to the Pounds recommendations, with the final decision on divestment contingent on the success of the trial.[74]

But after several months of continued student protests (including the arrest of radicalized Pounds Panel member George Katsiaficas), escalating threats and rumors of violence, gridlocked standing committee meetings, and lab resistance, faculty support for total divestment began to swell. At a March 1970 faculty meeting, Ascher Shapiro of the Fluid Mechanics Lab, the model for reconversion, offered a surprising endorsement of divestment, arguing that national priorities and budgeting decisions lay at the heart of the militarization of university research and that attempts to convert MIT labs would only lead to fiercer competition for scant resources, not major shifts in defense contracting. After a well-guarded meeting of MIT's Executive Committee, the university announced in May 1970 that attempts to reconvert the Draper Lab had ended, and MIT would sever its administrative and financial ties to the facility. President Howard Johnson explained, "I do not believe that we have the right to hurt the capability of the laboratory by continuing to impose a restriction that neither the laboratory nor its contractors are willing to accept. Were we to force that situation, we would be wrong, and it would not work." Outside observers speculated that the administration had failed to locate sufficient funds for the nonmilitary projects necessary to keep the lab open in accordance with the Pounds Panel's recommendations.[75]

Though the MIT *Tech* published a supportive editorial and reported that faculty reaction was "difficult to gauge," Nelkin observed that the policy met with "ambivalent" reactions at the lab and general dissatisfaction among student activists. She wrote, "The administrative decision, intended to accommodate as many interests as possible, pleased few." Within a year, the unpopular Johnson had been replaced by Wiesner as MIT president, and a new interim Draper Lab board of directors was created with Draper, former Kennedy advisor Carl Kaysen, and IBM scientist Emanuel Piore among its members. In the summer of 1973, the Draper Lab officially remade itself as an independent nonprofit corporation, still housed in buildings adjacent to the MIT campus but no longer a part of the university.[76]

Lincoln Lab, in the meantime, had tried desperately to emphasize its relevance to arms control efforts and to make itself more available to students, including opening the Lincoln Laboratory Ballistic Range—a large chamber in which a light gas gun could fire spheres and cones at speeds approaching 26,000 feet per second—and providing free shuttle service from Lexington to the MIT campus. In the end two factors most likely saved Lincoln's relationship with MIT. The first was its status as a source of criticism for the ABM, rather than a producer of hugely unpopular weapons technology (as was the case with the I-Lab and MIRV). Second, the lab's off-campus location meant that MIT could consolidate and relocate any remaining classified research projects to the suburban lab site, thus easing some of the political tensions in Kenmore Square without fully severing its valuable contractual links to the Defense Department. For the duration of the Cold War, Lincoln would follow the Livermore model: a site for classified, cutting-edge defense research that was geographically—but not financially or administratively—separate from a top-tier undergraduate campus.[77]

* * *

In the short term, severing ties with the Draper Lab coincided with a difficult period of federal funding cuts and other economic woes facing universities in the 1970s, and MIT scrambled to secure funds for graduate students, researchers, and additional programs. After some initial turbulence and staff reductions, however, Draper Lab fared well. MIT graduate students still worked at the lab as "Draper Fellows," undergraduates attended seminars there, and the facility's resources brought together academic researchers, corporate scientists, and military personnel. During the 1970s Draper researchers continued work on the Trident and Minuteman missile technologies as well as the high-profile Apollo program, including assisting in the rescue of Apollo 13. Draper employees tackled the technology of the space shuttle in the 1980s and robotics and advanced guidance systems throughout the 1990s, including GPS. As historian Stuart Leslie observed, "In every way that mattered, nothing had changed except on paper."[78]

After a brief period during the 1970s, when its research program broadened to include topics related to health and energy, Lincoln Lab followed a similar path. Rather than allow for easier reconversion transitions, Lincoln's status as a Federally Funded Research and Development Center led to continued weapons-related work that was "tilted even further toward applied research and direct military applications." By the early 1980s, the lab was a major recipient of Strategic Defense Initiative funding. In 1986 a second MIT review panel investigated the work of the lab, determining that it had not followed the Pounds Panel stipulations to pursue nondefense research and educational goals. John Deutch, a chemist and MIT provost at the time, acknowledged that "the 1969 strategy to broaden nondefense work at Lincoln has just not worked out, and frankly, is not likely to work out in the near term given the priorities in Washington." No steps were taken toward divestment, however, and the lab became the centerpiece of the famed Route 128 Massachusetts defense industry, a ribbon of great and small contractors capitalizing on the Reagan defense boom. A decade later Deutch himself would take top positions at the Pentagon and the Central Intelligence Agency during the Clinton administration.[79]

The short-lived 1970s efforts to "civilianize" engineering and hard-science research ultimately lost out to the increased militarization of campuses in the 1980s, including the creation of the DOD-University Forum, heavy defense funding, and the rise of new military-sponsored university powerhouses: Georgia Tech, Carnegie-Mellon, Penn State, and others. The academic trend was matched by the late Cold War success of the Draper Lab and the many Route 128 spin-off defense contractors dependent on Pentagon funding.[80]

Despite these outcomes, the March 4 movement and the actions of the Pounds Panel at MIT were not simply weak temporary roadblocks in a larger long-term race to build a "Massachusetts Miracle" out of the growing defense industry. MIT's failure to reconvert its labs contributed to this process, as did the related trend of locating defense firms in suburban areas conveniently shielded from campus radicalism and urban activist groups. But the protests at MIT also pushed the ethical dilemmas of weapons research, university-military relations, and institutional responsibility into a broader national political discussion. Like Livermore employees in the 1980s, researchers at MIT's special labs were forced, often with hostility and anger, to confront the nature and applications of their research. This process was painful for some and irritating for many, but nevertheless constituted the kind of individualized intervention that many Manhattan Project veterans later wished had occurred during World War II, even if it would not have resulted in any change in their decisions to work on the bomb.[81]

From an institutional standpoint, MIT set a precedent for establishing some limits on university-managed defense research. As Nelkin wrote in 1971,"MIT, more than many other universities, lives with the recognition of its relationship to the 'real world'; indeed, it was the scope of its involvement outside the university that led activists to argue that the claim of political neutrality was a myth; that even the acceptance of given research priorities is an act in support of a particular political system." The protests and debates at MIT inspired similar actions at universities across the country and the globe, extending even to professional organizations with deep-rooted claims to neutrality, such as the American Physical Society and the American Association for the Advancement of Science. All of these institutions would face the difficulty of creating universal moral standards in a postmodern world of contentious politics and shifting values.[82]

Jerome Wiesner reflected on these trends, filtered through his personal experience, in his final commencement address to MIT students in the spring of 1980. "As science advisor to Presidents Kennedy and Johnson, and now as president of MIT, I have for years been under heavy pressure to defend science and technology," he told his audience. "At first, I did this with some of the same trepidation one would have in defending one's naughty child, but with each new challenge . . . I found that I was entering a new culture— or at least subculture—that saw the problem (and the world) though an intellectual filter quite different from mine." As he described his struggles to reconcile differing worldviews and to consider the balance of moral and technical concerns, Wiesner invoked the name of one of the leading scholars in the history of science, Princeton professor Thomas Kuhn and Kuhn's account of "the difficulty of changing scientific structures" and the inherent challenges of overturning deeply held, paradigmatic assumptions.[83]

Wiesner noted ruefully that what Kuhn had written about science applied equally to assumptions about social priorities, including the social priorities of science and technology. At the height of the Vietnam War, the most prestigious professional organizations and universities in the country were learning this difficult lesson. The American Physical Society epitomized this trend in the late 1960s, when an insurgent internal movement rejected the very notion that science could be "neutral." A similar debate soon echoed through the storied gothic halls of Princeton University as administrators and academics, including Kuhn himself, confronted the powerful intellectual challenges of the New Left.

SEVEN

The New Left Assault on Neutrality

FOUNDED IN 1899 at Columbia University, the American Physical Society (APS) by the late 1960s was arguably the most prestigious professional scientists' society in the United States. Its membership approached thirty thousand, ranging from the rising stars of graduate schools to Nobelists and grizzled Manhattan Project veterans, and it constituted one-seventh of the umbrella American Institute of Physics, which published sixteen scholarly journals, including the *Physical Review, Physical Review Letters,* and *Physics Today.* The purpose of the society, and its parent, AIP, was publicly affirmed in the opening pages of these journals: "the advancement and diffusion of the knowledge of physics and its applications to human welfare."

Behind the façade of scientific star power and technical prestige, however, lay deep political, economic, and generational divides. An early 1970s informal study of the makeup of the APS executive committee revealed a predictable homogeneity of background and position: of the eleven members, all but two were over forty years old; all held tenured university professorships or similarly secure laboratory positions; all but two had some association with Columbia, Cornell, Harvard, or MIT (Massachusetts Institute of Technology); and all but one had been educated in the northeastern United States. Observers referred to the organization as a "gerontocracy" and noted an "alarming pattern of conformity" among the executive committee members, whose professional ties skewed heavily toward "large institutions, PhD mills, with fat research contracts . . ." The comfort and placidity of the organization's leadership alienated many younger physicists, particularly those who faced precarious employment futures or who found military-funded university research opportunities ethically unpalatable in the context of the war in Vietnam.[1]

The presidents of the APS from 1966 through 1972—John Wheeler, Charles Townes, John Bardeen, Luis Alvarez, Edward Purcell, Robert Serber, and Philip Morse—were brilliant men who had all worked on military projects during World War II: on the atomic bomb, radar, mine detection, and even nascent operations research. But the implicit Manichean moral terms that had facilitated the total research mobilization of those days no longer applied. From the late 1960s into the early 1970s, the APS leadership confronted a youth-driven, politically radical reform movement determined to shift the priorities of scientific research away from military applications and toward ending U.S. involvement in Vietnam.

✳ ✳ ✳

Among the leaders of this effort was Berkeley physicist Charles Schwartz. Schwartz had been born in Brooklyn in the 1930s to parents with strong ties to Eastern European Jewry. His father, a Russian immigrant with little formal education, found enough work as a photographer and amateur inventor to support the family and eventually relocate to Connecticut. From there the younger Schwartz made his way to MIT, where he received first his BS and then his PhD in physics. As a graduate student, he worked in the field of nuclear structure with his advisor, Victor Weisskopf, whom he idolized (even going as far as to briefly affect a German accent). Despite MIT's strong military connections, he adopted the department's general disdain for applied work and, upon graduating, headed off to a short stint as an assistant professor at Stanford before landing at Berkeley in 1960.[2]

Professionally, Schwartz dutifully continued his physics research, gradually specializing in complicated problems of particle systems and field theory. He held an air force research grant that paid his summer salary and supported his graduate students, and it was through this grant that Kenneth Watson identified him as a promising young candidate to work with IDA's Jason group, inviting him to the 1962 summer session. Schwartz absorbed the glamour and "heady" sense of his potential influence as a government advisor, and even after IDA rejected him for future participation, he maintained a kind of Washington "world view" deeply rooted in principles of gradual escalation and a technical, problem-solving approach to world conflicts.[3]

In the summer of 1966, Schwartz was devastated when his brother was killed in an airplane crash. As he later recalled, the experience "provided a deep psychological emotional opening" through which new philosophical and political attitudes suddenly flooded. His mild criticism of the war hardened into opposition: he suddenly noticed and began signing the anti-Vietnam petitions circulating among his colleagues; he withheld taxes designated for war funding; he wrote letters, joined peace groups, and protested outside the Oakland induction center; and he immersed himself in grassroots political organizing.[4]

* * *

The following year, Schwartz submitted a letter to the editor of *Physics Today,* recruiting physicists to express opposition to the war. Citing its failure to address "physics as physics and physicists as physicists," the editors rejected the letter for publication. Schwartz interpreted the action as a kind of political censorship. Drawing on his newfound political skills, he responded by circulating a petition to amend the Physical Society's constitution so that with the support of at least 1 percent of the membership, any "matter of concern to the society" could be brought before the entire organization for a vote. The results of the vote, if publicized, would constitute an official organizational statement—potentially concerning matters of politics or U.S. foreign policy.[5]

Though Schwartz gathered only 248 signatures in an organization of over 24,000, his actions provoked a deluge of arguments and angry letters sent to the APS's top officers; a tempest in the APS teapot that drew outside criticism and even press coverage. Letters poured in, and the editors who had initially turned down Schwartz's Vietnam request found themselves devoting unprecedented time and space to debates over the political implications of the proposed amendment. Responses tended to fall into one of three categories. As the editors themselves summarized, most advocates of the reform believed that "physics is already deeply involved in the lives of nations, and its discoveries have broad technological and social implications in both peace and war. If physicists as a group can clarify a problem or suggest a solution not clear to others, they should have the means to do so." As Victor Paschkis and other nuclear critics had in earlier decades, these scientists sometimes invoked the specter of Nazi complicity: "German science remained pure and unpolitical during the 1930s. It has not yet recovered," wrote C. H. Blanchard of the University of Wisconsin.[6]

In other letters, however, opponents worried about "the danger of ineptitude and arrogance." The *Physics Today* editors summed up the views of these critics: "[They argue that] the essential nature of physics is that it deals with simplified models of simple systems (point masses, hydrogen atoms, the compound nucleus) and seeks solutions where such solutions are available. People trained to deal with such models may not have any special skill in dealing with more complex systems like the human mind and international politics. They might offer a lot of bad advice. Public assertion that physicists have special competence might be an arrogance that would bring discredit to a community in place of the respect it now enjoys." Rather than focus on the special *moral obligation* of physicists to address the social consequences of their work, these criticisms emphasized the potential *intellectual unsuitability* of physicists for just such a task.[7]

Finally, a third group worried about the administrative consequences of a policy shift: for example, would the publicity of physicists' political views affect federal funding decisions for major projects? Would politicization sabotage physicists' precious permission to travel and convene with counterparts in the Soviet Union and other nations? Could the APS lose its tax-exempt status? Would the addition of political issues detract from the organization's ability to promote scientific objectives? After distilling and summarizing these arguments, the editors sided with the third group and endorsed alternate modes for physicists' political expression: for example, through advocacy work with the Federation of American Scientists (FAS) or publication in the *Bulletin of the Atomic Scientists*. The editors warned that members who wanted to change the function of the APS "should look carefully at what they stand to lose while they are seeking routes to what they hope to gain."[8]

After this introductory synthesis, the editors opened the letters page of the journal for reactions, vowing to print all or part of every submission. From January to April 1968, diatribes and lamentations filled the section, from the most prominent physicists in the nation to concerned nonscientist readers. At a rate of more than four to one, they vehemently opposed the amendment. Eugene Wigner of Princeton wrote that changing the APS would be a "corruption of democracy." Frederick Seitz, former head of the National Academy of Sciences, worried that adding political discussion to the organization's journals would damage their "essentially professional character." Edward Teller responded to the problem of physicists' suitability for political pronouncements by arguing that the APS ought to maintain "a proper division of authority," preserving the relegation of "public issues" to "traditional channels." Teller acknowledged one exception, however: a case in which "a question of fact (rather than of opinion) on which all reputable physicists agree on a professional basis and concerning which there is a public misunderstanding." (Just such a scenario would later arise in the form of the Strategic Defense Initiative debate, although in that case, the great majority of "reputable physicists" would take positions opposite from Teller.) Teller himself had advised military and government leaders about weapons technology and policy, but he didn't believe that the APS as an organization ought to act in a comparable manner.[9]

Most opponents, however, followed Seitz's lead, warning of the damaging effects of politics on the purity and professionalism of their organization. They worried that change would "jeopardize the purely scientific nature of APS and the harmony between its members" and urged their peers not to "dilute our professional efforts by becoming a debating society" or "contaminate physics with politics." David McCall of Bell Labs summed up this criticism in a simple declarative sentence: "The APS is not a political instrument."[10]

Nearly two decades later, Charles Schwartz would dismiss the critics who called for the preservation of political neutrality by noting that "physicists know which side their bread is buttered on . . . even those who have . . . liberal views on these things understand we're all part of that arrangement, and you don't rock the boat. You don't do things that might offend the powers that be. So you claim to be neutral and apolitical and resist any attempts that might put you in a position where you might encourage the disfavor of important people."[11]

Not all of Schwartz's critics were Pentagon beneficiaries or timid liberals, however. Eugene Saletan, a young Northeastern University professor, took pains to note in his letter to the journal that he was as radical as Schwartz when it came to political activism, but he nevertheless saw the APS as a poor forum for outside issues. He wrote, "To establish my credentials, I am an adamant 'extremist' on the war in Vietnam. I have demonstrated against the administration's policy, have spoken at public meetings against it, have signed resist petitions, am faculty advisor to SDS, etc. etc. Nevertheless I believe that APS should remain pure. There should be an organization of physicists whose purpose involves only physics."[12]

Saletan's entreaty that the APS "should remain pure" assumed that the society's reluctance to issue political statements constituted a kind of neutrality, one that member-driven efforts to force votes on "matters of concern" would undermine. But was the APS "pure"? And what exactly were the "matters of concern" that might be raised? The amendment itself had yet to be passed and interpreted through practice, and the debate among physicists suggested a wide range of fears and assumptions. Jay Orear, the forty-two-year-old president of the FAS, considered the "matter of concern" clause to be a "safeguard," noting in a May 1968 interview that "if 1% of the membership proposed a resolution outright condemning the Vietnam War, the APS council would have to rule that out of order as inconsistent with the purpose of the society. I consider that a quite adequate safeguard." But Schwartz himself disagreed. He was convinced that the APS was *not* a neutral organization, and his concern that a double standard existed in its political activities had fueled his amendment drive. As an example, he cited the APS's invitation to Lyndon Johnson to attend an annual meeting and the subsequent publication of a photograph of the president with APS head Charles Townes. That, in his view, constituted "a legitimization of his Vietnam policies" by the APS membership. If such legitimization was allowed, then surely APS members should be able to register their opposition as well.[13]

Future Nobel Prize winner Martin Perl of Stanford offered a related analysis in the March issue, tackling the problem of Vietnam specifically: "While almost half the people in this country are opposed to the war, the expression of this opposition is comparatively small. The reasons for this problem are multiple. Certainly one of the reasons is that the almost automatic reaction of most *regular* and *established* methods of communication to issues like the war is to try to ignore these issues if possible. . . . The crucial thing about that silence is that it is *not* nonpartisan. No matter how one justifies or defends silence and no matter what sound reasons there are for silence, the result of silence is clear. Silence supports the war, and that is a sad silence."[14]

In the final tally, APS members voted against the amendment by a margin of nearly three to one, and such a sound defeat mooted the question of practical implementation. But perhaps more tellingly, out of an organization with over twenty-four thousand members, roughly half had taken part in the amendment vote, the highest participation numbers in APS history. The question posed by F. Jona of IBM's Watson Research Center was left unanswered: "Schwartz and his friends . . . would like to see APS officially oppose the Administration policy in Vietnam. Did it ever occur to Schwartz that APS might decide officially to *support* that policy?"[15]

✳ ✳ ✳

Several years after the vote, Carleton College physicist Barry Casper, chair of the newly created APS Forum on Physics and Society, reflected on the legacy of the failed effort: "The Schwartz amendment had an effect that endured long beyond its defeat. The debate over the amendment had raised broader questions about the responsibilities of the APS to its membership and to society." Schwartz's efforts had had dramatic short-term effects as well. At the 1969 APS annual meeting, an inspired coterie of young physicists called for the establishment of a new internal division devoted to the discussion of physics and society, an effort that would culminate in the creation, three years later, of Casper's Forum. Meanwhile, Schwartz and Martin Perl, stymied in their internal reform efforts, proposed the creation of a radical new organization, Scientists for Social and Political Action, wholly independent from the APS. (The organization would quickly open its ranks to biologists, chemists, mathematicians, and engineers, eventually changing its name to the more inclusive Scientists and Engineers for Social and Political Action (SESPA).) Schwartz's failed amendment thus sparked two significant political drives: one to create an internal mechanism for the airing of political issues within the APS, and one to establish an outside body devoted to explicit political activism.[16]

The impetus for the forum sprang from the minds of two young physicists from MIT's Francis Bitter Laboratory, Brian Schwartz and Emanuel Maxwell, who led the drive to create an APS division concerned with "the problems of physics and society." At the 1969 annual meeting, held in the shadow of the continuing war in Vietnam and Richard Nixon's inauguration, the pair found a ready audience of supporters. Conference attendees sported anti-ABM (antiballistic missile) pins while Kurt Vonnegut warned the American Association of Physics Teachers that "the virtuous physicist is one who does not work on weapons." The forum petition quickly accumulated over five hundred signatures, enough to alert the APS's Executive Council and force a response.[17]

Unlike the amendment proposed by Charles Schwartz, the new division would not require the APS to issue any formal political pronouncements or subject the entire membership to referenda. Rather, the internal group would not be political in any active sense, focusing instead on helping concerned physicists "clarify their own ideas." This moderate mission found grudging support in Edward Purcell and the APS Council, who eventually accepted the establishment of a Committee on Problems of Physics and Society, precursor to the APS Forum on Physics and Society.[18]

While the forum's approval process dragged on over the next three years, Charles Schwartz and Martin Perl charted a quicker and more activist course with SESPA. Announcing their inaugural meeting at the APS conference, they joined with Michael Goldhaber of Rockefeller University (son of Brookhaven head Maurice Goldhaber) and Marc Ross of the University of Michigan to publicize a new "independent body of socially aware scientists free from the inhibitions which abound in the established institutions." They charged that the APS, and scientists' professional societies in general, had "deliberately remained aloof from the desperate problems facing mankind" by promoting the simplistic mantra that "research means progress and progress is good." In SESPA's view, such an attitude encouraged young scientists to pursue weapons research without contemplating the moral and social consequences of their work. The time was ripe for a new organization.[19]

In response to the call, three hundred scientists showed up for a meeting held in a Hilton hotel room, "a broad coalition," in Casper's words, "that included arms controllers and environmentalists, liberals and radicals." An elated Martin Perl proclaimed that the youthful new organization would shake up the elderly APS Executive Council and would force the old guard "to come out of their ivory towers and bomb shelters." But exactly how, and to what end, remained unclear. As Charles Schwartz later recalled, "Marty Perl [gave] the first speech in which he made it very clear that this was not going to be a radical organization. And then I gave the second speech in which I said in my opinion this was going to be a radical organization. . . . It was designed as very much an unorganized organization." The first meeting covered a wide range of topics, from opposition to the antiballistic missile system to the possibility of a research strike at MIT.[20]

Though critics such as Lee DuBridge, Nixon's science advisor, and John Bardeen, two-time Nobelist and APS president, dismissed the protesters as extremists with little influence, the SESPA faction managed to attract coverage in a broad spectrum of print media, from *Electronic News* to the *New York Times*. Members led the charge against Jason scientists employed on their campuses; campaigned to prevent future APS meetings from taking place in Chicago, home of the violent 1968 Democratic Convention; lobbied against the antiballistic missile system; called for a boycott of Los Alamos and Livermore scientists; organized a nonparticipation pledge for war research; promoted a Hippocratic oath for scientists; and oversaw another unsuccessful amendment attempt, this time to include stronger moral language in the APS constitution. Specifically, they demanded that "the object of the Society shall be the advancement and diffusion of the knowledge of physics in order to increase man's understanding of nature and to contribute to the enhancement of the quality of life for all people. The Society shall assist its members in the pursuit of these humane goals and it shall shun those activities which are judged to contribute harmfully to the welfare of mankind."[21]

With the narrow defeat of this measure, Charles Schwartz and his supporters had once again demanded an accounting of the social consequences of research and failed. And with their failure, the broad coalition within SESPA began to fracture. In Boston the local SESPA chapter volunteered to publish the organization's newsletter, previously the province of Martin Perl. As Charles Schwartz observed, the Boston group was "quite active," and the change in the newsletter "signified a transformation to a much more radical perspective and posture." The SESPA newsletter became the radical magazine *Science for the People,* whose very logo—a red clenched fist and an Erlenmeyer flask—proved sufficiently radical to alienate Perl and other liberal group members. Early editions published exposés of Polaroid's involvement in South Africa, the work of Jason members at Columbia University, and cases of discrimination at the University of Massachusetts. Typical article titles included "Engineers in the Working Class," "Fighting the Police Computer System," "The Social Impact of Modern Biology," and "People's Science Projects for Vietnam," which called for medical and agronomic assistance to heal the ravages wrought by the U.S. military.[22]

* * *

Frustrated by the radical turn of SESPA, some liberal APS members, such as Martin Perl and Brian Schwartz, turned their attention to the nascent forum. Though Charles Schwartz and others complained that the forum amounted to "cooptation" and "a way to get all those people who are concerned . . . under the control of the establishment," disgruntled moderates nevertheless found it a more appealing alternative to SESPA. By 1972 the forum had roughly one thousand members and an annual budget of $4,200.[23]

The aims of the forum were indeed more limited than SESPA's mass boycotts and efforts at deep institutional reform. Forum-sponsored panels and discussions nevertheless brought the problems of antiballistic missile systems, energy policy, international conditions for science, and even the work of the Jasons to a wider APS audience. The forum also sponsored a congressional fellowship program, allowing two physicists a year to work on a congressional staff. But throughout, the executive committee of the APS exercised subtle oversight, applying behind-the-scenes pressures and limits on forum activity.[24]

One notable example of the tensions between the executive committee and the forum occurred during preparations for a forum symposium on Physicists and Public Affairs, to be held in April 1972 and moderated by Jay Orear, president of the FAS. Planned speakers included Cornell high-energy physicist Raphael Littauer, discussing the Cornell Air War Study and physicists' potential contributions to political science; congressional aide Leonard Rodberg, describing his work with Sen. Mike Gravel in the publication of the Pentagon Papers; Stanford physicist Pierre Noyes, detailing ways to use the legal system to oppose war; and, most controversially, William Davidon, a pacifist mathematician and physicist from Haverford College, speaking on the social responsibility of scientists.[25]

Davidon had begun his career as the director of research at the Nuclear Instrument and Chemical Corporation in Chicago after World War II before enrolling in the graduate physics program at the University of Chicago, where he was a star student. After a short stint at the Argonne National Laboratory, he settled into a long career at Haverford College in 1961. The Quaker-influenced school was a near-perfect fit. Davidon had long been active in peace-seeking scientists' organizations: he supported the Pugwash Conference, the FAS, and the Society for Social Responsibility in Science. During the 1960s, however, his opposition to the war in Vietnam spurred him to more drastic acts of protest and disobedience, particularly after an eye-opening trip to South Vietnam in 1966, sponsored by the Committee for Nonviolent Action. Described by one ally as "someone with a knowledge of the scene, a keen sense for tactic & detail & little fear of risk for himself," he refused to pay taxes earmarked for war funding, joined students in destroying records at the Georgetown draft board, and, in 1971, assisted Daniel and Philip Berrigan and a small cadre of other activists "in a conspiracy to blow up heating systems in Federal buildings and kidnap [Presidential Advisor] Henry Kissinger." Davidon ultimately escaped prosecution for this last act as an "unindicted co-conspirator," and the remaining defendants in the "Harrisburg" group were never convicted. (In an unusual meeting during the spring of 1971, however, Davidon and two fellow co-conspirators, antiwar activist Thomas Davidson and Notre Dame Sister Beverly Bell, actually spoke with Kissinger in the Situation Room of the White House.)[26]

A year after the conspiracy, Davidon's scheduled forum talk focused on the logic and ethics that had propelled his initial commitment to sabotage and property destruction. In Davidon's view, destroying draft records had been a means both to protest the war and "protect and build respect for life." In the abstract of his talk submitted to the APS *Bulletin*, Davidon argued that U.S. reliance on "technologically advanced weapons systems" in Southeast Asia meant that "scientific and technical workers are replacing the foot soldier." This left "scientists, engineers, technicians, and those who teach them" with the obligation to prevent the misuse of technology and, most crucially, to resist and impede the war effort. This resistance might take the form of publicizing the effects of deploying new weapons systems or, more radically, "inactivating equipment intended for killing or harming people."[27]

Davidon's call for sabotage deeply distressed the executive council of the APS, who refused to print the abstract of the talk in the APS *Bulletin*, as was customary. American University's Earl Callen, president of the forum, warned that suppressing the abstract would prove more harmful to the APS's reputation than publishing it. As he emphasized, Davidon was only *advocating* sabotage, after all, not *inciting* it. Callen believed that Davidon should be allowed to present his talk, and APS members could let the "Darwinian process of natural selection—of ideas, of speakers, of session chairmen, and APS Divisions" operate. Jay Orear agreed. Though the panel itself hardly reflected a broad spectrum of viewpoints, the council's action still constituted "a clear case of censorship": the forum had an obligation to present Davidon's view of the misuse of science, even if advocates of the opposing viewpoint (such as John Foster) had turned down invitations to participate in the forum panel in the first place.[28]

The decision to withhold an abstract from the APS *Bulletin* might seem a trivial affair in the grand scheme of militarized science, the war in Vietnam, and campus protests, but, like Charles Schwartz's amendment attempt, it required another APS self-evaluation and offered another opportunity for critics to voice their dissatisfaction. In the case of Davidon's abstract, the most vociferous critic of APS policy was Stanford physicist and fellow panel member Pierre Noyes, who personally took it upon himself to photocopy scores of Davidon's abstracts to distribute at the meeting and who further declared his opposition in a press release the day of the conference.

Noyes himself had traveled a great ideological distance to reach the day's panel discussion: after receiving physics degrees from Harvard and Berkeley he had spent much of the Eisenhower years as a group leader at the Lawrence Livermore Laboratory, as a consultant for Project Orion studies of nuclear propulsion, and as an AVCO visiting professor at Cornell. But he had been outraged at the conduct of the war in Southeast Asia. In 1970 he had burned his bridges to military science and brought a class-action lawsuit against the Nixon administration, citing the use of taxpayer dollars for actions violating international war crimes agreements and the U.S. Constitution. Now, in his talk and his press conference, Noyes drew heavily on the example of Nazi collaboration and the lessons of the Nuremburg trials. In his panel presentation, he warned his audience:

> Nazi judges who enforced racial laws legally passed within the German system were convicted at Nuremberg. Those of us who receive rather than give orders are still under the legal obligation not to carry out criminal acts. A soldier with a gun held to his head who obeys an order to kill or torture a prisoner is still guilty of a criminal act. The fact that he had no reasonable moral choice but to obey may be properly argued in mitigation of sentence, but cannot alter the fact of his guilt. In contrast, the extent of the guilt of the citizen in a war-related job who fails to search diligently for alternative employment is not an easy *legal* question to answer, but the moral imperative of the Nuremberg precedents is unambiguous."[29]

Noyes built on this analysis in his press conference, now including the APS and its claims of neutrality in his distributed text. "We perpetuate the myth that science and technology are neutral, and only people evil," he asserted, "rather than training our students to humane traditions. We welcome to our professional meetings representatives of military laboratories to discuss the unclassified spinoff from their labors, and to recruit for their illicit activities. We have come down heavily on the wrong side, and it is past time to redress the balance."[30]

A year earlier, Noyes had offered a similarly themed speech at the annual banquet of the American Physical Society in Washington, DC, an event covered by the *Washington Post*. Warning that "the moral imperative is unambiguous," he had demanded the public resignation of Edward David, Nixon's science advisor. The *Post* noted that the "conservatively-dressed" Noyes spoke "with cold heat" and intellectual rigor, in a "style [that] goes far past the shouting, slogan-rich radical style of the late 1960s." Unlike the public disdain heaped upon an angry young microbiologist who had earlier interrupted the proceedings, Noyes "was listened to by all and was applauded by many." Noyes's status as a "shirt-and-tie" radical, with impeccable physics credentials, warded off trivialization and easy dismissal.[31]

* * *

Physicists were not the only scientists grappling with problems of professional neutrality. New organizations and accompanying publications were springing up everywhere among science and technology professionals. The Computer People for Peace exhorted fellow programmers to "Join with other workers to make computers serve the people!" and, in addition to their journal *Interrupt,* published pamphlets on data banks and repression, computer technology and warfare, and other controversial topics. Alan Mc-Connell, a University of Illinois math professor, founded *MAG,* a periodical designed to "make the American Mathematical Society, and math community in general, more socially aware and responsible." Even the scientists at Brookhaven Laboratory set up an underground "free press" and radical newsletter.[32]

In New York, area engineers from academia and industry joined together to form the Committee for Social Responsibility in Engineering, proclaiming that "thousands of engineers feel that their engineering talents are misused in both civilian and military projects, and believe that the constant development of weapons technology spells ultimate disaster for mankind." Committee for Social Responsibility in Engineering members marched on Washington in April 1971 (under an "Engineers for Peace" banner) and held a "counter-conference" during the 1971 national convention of the Institute of Electrical and Electronics Engineers in New York City, where they protested the invitation of Assistant Secretary of Defense David Packard, whom they called "Mr. Military Industrial Complex." The counterconference itself drew high-profile attendees and speakers. Ed Koch, then a New York congressman, attended the opening press conference; Seymour Melman lectured on conversion to a peacetime economy; Jeremy Stone of the FAS discussed the importance of expert testimony before Congress; and Victor Paschkis offered an environmentalism-tinged lecture on predicting "secondary effects" of new technologies. Haverford's controversial William Davidon also made an appearance.[33]

From 1969 through 1972, protests and confrontations disrupted the national meetings of nearly all the major scientific professional societies, taking a wide variety of forms: from impassioned interruptions of speeches to carefully calculated invocations of Robert's Rules of Order. In some cases concerns about job security, dire employment conditions, and other labor problems comprised the major motivations for reform efforts, as with the American Chemical Society, but mostly, it was anger at U.S. foreign policy and the perceived complicity of scientists and engineers that fueled scientists' protests. As had been true of the discontented physicists in the APS, manifestations of that anger ranged from moderate calls for reform in traditional venues to choreographed disruptions and street theater.[34]

＊ ＊ ＊

Many of these tensions came to a head during the notorious 1970 national meeting of the American Association for the Advancement of Science (AAAS), the largest professional society for scientists and the publisher of *Science*. The most detailed account of the meeting's proceedings appeared in the January 8, 1971, issue of *Nature*, in the tellingly titled "Dissent Blooms at AAAS Circus" under the byline "by our Washington Correspondent." The author described the meeting as a combination of responsible, methodical science activism, embodied by the efforts of Matthew Meselson and Stewart Udall, and "noisy," "theatrical" stunts choreographed by SESPA and other groups. In the category of responsible activism, he was particularly impressed that mere days before Meselson's AAAS group was to present its final report on the consequences of herbicide use in Vietnam, Nixon's Defense Department had agreed to stop the spraying. The "Washington Correspondent" compared Meselson to Ralph Nader, noting that he, too, had "forced a giant corporation to change its policy."[35]

Also praised was the heartfelt speech on the political and moral obligations of the scientific community delivered by Udall, the former secretary of the Interior. Udall's words were harsh. He deeply criticized the attitude of organizations such as the APS, which sought only to offer technical but not political advice. To Udall, this made scientists into "political eunuchs— mere technicians detached from a value system and its attendant 'political' judgments." The National Academy of Sciences was no better: "While courageous individuals in the scientific community were raising the alarm about the lethal threat of chemical and biological warfare, what was the academy doing? It was working under contract to the Defense Department to select bright young scientists to work in the Defense Department's Chemical and Biological Weapons Centre."[36]

Nature characterized Udall's speech as more serious, and more seriously delivered, than the boisterous SESPA theatrics, which came largely in re-action to a volatile session titled "Is There a Generation Gap in Science?" featuring presentations from "two venerable Hungarians": physicist Edward Teller and physiologist Albert Szent-Gyorgyi. Moderated by Margaret Mead, the panel also included Richard Novick of the Public Health Research In-stitute, *New York Times* science writer Nancy Hicks, and Harvard freshman Stuart Newman. To a packed room filled with students, activists, and tele-vision crews, Gyorgyi opened by acknowledging that the application of sci-entific research to weapons development justified popular antipathy for science: "Because science is used for war . . . there is a revulsion against scientists." He referred to the recent bombing of the Army Math Research Center at the University of Wisconsin by antiwar radicals, an act that had killed physicist Robert Fassnacht. "The bombs dropped in Vietnam make bombs go off in Wisconsin," he observed.[37]

Gyorgi's critical assessment was easily overshadowed by the appearance of Edward Teller. Teller's mere presence seemed to offer an unapologetic defense of weapons science, and he provoked outrage by arriving flanked by a bevy of bodyguards. As SESPA members waved signs labeling him a war criminal, Teller denounced the protesters as thoughtless and unreasonable, likening them to the Nazi thugs who had persecuted him in his youth. Ac-cording to *Nature,* Teller's invocation of Hitler temporarily quieted much of the audience. Nevertheless, disruptions began anew when Richard Novick, the third speaker, offered Teller the "Dr. Strangelove Award," a Nazi-evoking trophy displaying a police officer in the act of shooting, cap-tioned "I am just following orders." Novick sided with Gyorgi and called for scientists to create a new form of labor union in order to exert control over the applications of their research. He explained, "By such actions it may still be possible without a bloody revolution to radically reorganize this society so that it serves the people and is therefore unable to misuse our science."[38]

Of the coverage of the conference in the mainstream press, *Nature* of-fered by far the most sympathetic account, painting SESPA as a cheerfully disorganized group whose outbursts may have alienated many conference attendees but who succeeded in making the general point that "science is political." In *Science,* for example, Philip Abelson complained that the SESPA radicals had succeeded in disrupting enough meeting activities "to tarnish the image of the AAAS" but had failed to win over the vast majority of actual attendees. Abelson, like Teller, invoked the specter of Nazi storm troopers, quoting a similarly themed *Washington Post* complaint.[39]

SESPA itself offered some critical self-reflections in a postmortem published two months later in *Science for the People.* In retrospect the heckling of Teller had been "in good fun" and effective as "ridicule," but "the moralistic tone of the Strangelove award helps us not at all to understand Teller as a product of society, as an exaggerated example of what so many of us and our colleagues are in part or might be. It provides no basis for scientists to immunize themselves against the appeal of Teller's attractive personality or his obvious capability as a physicist."[40]

SESPA concluded that the attack on Teller, like Charles Schwartz's attempt to promote a Hippocratic oath for scientists at Berkeley, did not address the potential of researchers to become unknowing or even unwilling collaborators in military projects. Charles Schwartz himself had related a cautionary tale to the *Washington Post* in 1970 in the aftermath of the implementation of the Mansfield Amendment, which required that all military-sponsored research have clear defense applications. Schwartz, whose work had been supported by the air force, inquired as to the justification for funding his "entirely theoretical, non-secret work on the structure of the atom." He had been told, he reported to the *Post,* that he could "rest assured" that his work was "vital to the aerospace mission," but his funders could not be more specific for security reasons. He promptly terminated his contract. SESPA's analysis—that it was insufficient to judge an individual scientist's personal actions instead of the larger system in which he operated—echoed the arguments taking place on university campuses across the country. Students chastised professors for working on war-related scientific research and administrations for accepting millions of dollars of funding from the Defense Department. As the leaders of the APS had been made to do, moderate and liberal faculties and administrations were forced to confront the meaning of neutrality, academic freedom, and autonomy in the face of deep antiwar anger.[41]

✳ ✳ ✳

On a Thursday night in late April 1970, President Nixon announced on television what pundits and reporters had long been speculating—that the United States would expand the war into Cambodia and in fact would be sending in combat troops within the week. Clandestine aerial bombing of the region had been U.S. policy for several years, culminating in the famed "Operation Menu" secret bombings begun in the spring of 1969, but this night offered the first official announcement of U.S. incursions into Cambodia. Nixon openly acknowledged the imminent attacks designed "to clear out major enemy sanctuaries on the Cambodian-Vietnam border" but assured the viewing public that this was "not an invasion" and was part of his plan to de-escalate and conclude the war.[42]

Reaction on the nation's campuses was swift and angry, most tragically at Kent State, where the Ohio National Guard opened fire and killed four protesters, and Jackson State in Mississippi, where two student activists were shot and killed by local police. Among science students in particular, response varied greatly. *Chemical and Engineering News* reported that chemistry classes had gone on as usual on the campuses of Colorado State and the University of Nevada while students in other departments demonstrated and engaged in property destruction. California was a different matter—Berkeley was closed by order of Ronald Reagan, a quarter of chemistry students at the University of Southern California boycotted classes, and at Stanford, the overwhelming majority of the chemistry department members—professors, researchers, and students alike—signed a resolution opposing Nixon's actions. The chair of Stanford's chemistry department, Harry Mosher, noted that such political activity among his peers had "never happened before." And at the California Institute of Technology (Caltech), science faculty and students publicly wrote antiwar letters to politicians and took out newspaper advertisements. Caltech chemist George Hammond, who vehemently condemned the Cambodian invasion at a campus rally, saw his recent appointment as the new deputy director of the NSF revoked by the Nixon administration.[43]

At Princeton, thousands of students and faculty members reacted to the speech by gathering at the Princeton chapel for a "hastily called protest meeting" and, late in the night, voting to boycott classes. During the following week, Students for a Democratic Society and other groups demanded an end to what they perceived as Princeton's complicity in the war effort, emphasizing weapons research on campus and school ties to the Institute for Defense Analyses (IDA), the parent institution of the Jasons, which was housed on Princeton property. As one campus leaflet summarized: "As an institution Princeton aids and abets the ever-expanding war in Southeast Asia by allowing ROTC, IDA, military recruiting, and war-related research to operate on campus. Therefore to remain consistent with the demand that Princeton University take a stand as an institution against the war in Southeast Asia the resolution that was passed Thursday night also demanded that the University sever its ties with the Department of Defense."[44]

Responding to student criticisms and their own concerns, on May 6 the Princeton faculty voted in favor of a systemic reckoning of war-related research at Princeton. Within a week the Council of the Princeton University Community had taken up the faculty's resolution and formally called for a "special committee" to report back on Princeton's relationship with the Defense Department and the school's sponsored research policies. In particular, the council set out two proposals for the new committee to evaluate: "1) That the University refuse to accept any outside funds for research on campus which is directly and specifically related to weapons and weapons systems. 2) That Congress be asked to channel all funds in support of research to universities through civilian departments and organizations such as H.E.W. and the N.S.F. rather than the Department of Defense." To accomplish this evaluation, the committee was to assess the cost, financial and otherwise, of purging Defense Department contracts; find ways to offset these losses; and reach out to other universities in order to promote new pathways of funding.[45]

As plans for the new committee worked their way through the university's governing bodies in late May, the political climate at Princeton eased from tense to pragmatic. Students organized themselves into bands of political volunteers, most notably for the antiwar congressional candidate Nicholas Lamont in Philadelphia, thus opting to work within, rather than challenge the legitimacy of, existing electoral channels. Noting the trend, the *New York Times* reported that for Princeton, 1970 was not to be "a year of revolution as 1969 had been Harvard's and 1968 Columbia's." The school had been "politicized by the Cambodian actions, not radicalized."[46]

The makeup of the committee reflected this critical moderation. Members included undergraduate and graduate students in both the sciences and the humanities, faculty members from multiple departments, and voteless representatives from the university administration. But the committee's most prominent public face was that of its chairman, Thomas Kuhn, the well-regarded professor of philosophy and the history of science. Kuhn had earned a PhD in theoretical physics from Harvard in 1949, then authored the influential *Structure of Scientific Revolutions* in 1962. He embodied the committee's mix of scientific expertise and social criticism.

As the committee began meeting in late May, it confronted the enormous complexity of its mission, which reached beyond mere financial analysis to deeper questions of ethics and Princeton's institutional philosophy. Should the committee worry only about research sponsored by the Defense Department or any research funded by an outside source that might have military applications? What about basic research that might be useful in a particular field but also spawn destructive new technologies? (Their debates echoed the question posed by hydrogen-bomb theorist Stanislaw Ulam: "Without the invention of the infinitesimal calculus most of our technology would have been impossible. Should we say therefore that calculus is bad?"). More fundamentally, was it absolutely wrong for weapons-related research to take place on campus? Did restrictions on weapons research restrict academic freedom? What was the purpose of a university in 1970s America?[47]

Kuhn himself worried deeply about this last question. In June he wrote a long personal letter to Charles Hitch, president of UC-Berkeley, requesting information about Berkeley's policies and Hitch's general advice. He described the committee's work as "an attempt to make certain that the University has not, as an institution, involved itself so deeply with the federal government as to surrender an essential part of its traditional independence and assume functions inappropriate to its nature." To acting Caltech president R. F. Christy, he reiterated this fear, wondering if "current modes of federal funding" might cause "distortions of academic research." Such questions assumed an established role for a university: an impartial, independent institution providing fertile grounds for the advancement of knowledge, free from outside agendas and pressures.[48]

Not everyone agreed that was a valid assumption. The critical view was probably best articulated by Martin Summerfield, a professor in the Aerospace and Mechanical Engineering Department, who wrote to committee members after an open meeting during the following fall. Irritated by assertions that Princeton's "institutional neutrality" precluded certain faculty activities, such as weapons-related research, government advising, and the maintenance and use of the classified library, he invoked the old concept of "Princeton in the Nation's Service." Was such a slogan still meaningful, he asked, and if so, did it not "imply the availability of [Princeton's] resources and talents" for government work? Speaking for the engineering departments, he contended that "most of us, as loyal citizens, will continue to work on defense problems." He reminded the committee of Princeton engineers' contributions during World War II, for which the department had received "the gratitude of our government and the public."[49]

Summerfield also complained to committee members that research models touted by scholars in "contemplative" fields, such as English or math, lacked relevance for engineering, which was an "activist field." An engineer, he argued, is "like a physician—he responds to visible needs," often in areas deemed important by government agencies and the private sector. "In your field of contemplative scholarship," he wrote to Kuhn, "you are privileged as individuals to set your own private goals. An engineer who tries to secure such privileges for himself is a misplaced individual; if he persists he will find himself on public welfare." This passive view of engineers—as heavily influenced by the priorities of outside funders—was echoed within the committee by Robert Jahn, a NASA-funded professor of aerospace science, who informed his colleagues that "the selection of research topics is influenced by the sponsor" in "any good engineering school."[50]

As Kuhn and his counterparts at other universities weighed these deeper issues, the committee's research staff spent their first summer collecting information on every funded research project on campus, giving special attention to projects flagged by critical students and those conducted by research groups in the Department of Aerospace and Mechanical Sciences, which seemed at first glance to have the most questionable projects under way. The staff also evaluated funding statistics, solicited information from dozens of other universities, arranged for an economic study of the financial consequences of severing Princeton's ties to the Defense Department, and reviewed dozens of statements from faculty members about their own research and their views on appropriate research policy.

The results nearly paralyzed the committee. Setting aside the issue of IDA's presence in a Princeton-owned building and Princeton's maintenance of a classified library, the remaining problem of sponsored research proved far more ambiguous and complex than expected. For example, Professor William P. Jacobs's botany research had been noisily targeted by campus critics, who had earlier distributed notices arguing that "the concept of 'neutral research' is a fallacy. One must question, for example, what future applications the Army has in mind when it allocates some $26,000 to

study the 'Physiological Mechanism of Leaf Abscission and Senescence.'"
To critics, such a project could mean only one thing: defoliant research applicable to the war in Southeast Asia. Jacobs gamely defended himself in the *Daily Princetonian,* acknowledging army support of his work but characterizing it as "basic research on the relations of hormones to ageing in leaves." The same projects had earlier been funded by the National Science Foundation (NSF), he argued, and "research support from the Army did not change the topics or course of the research in any way. . . . None of it was concerned with the practical problems of defoliation."[51]

The issue was not quite so straightforward, however. Earlier in the spring, Martin Summerfield had been informed by one of his own sponsors that Jacobs's work carried a "NOFORN restriction"—it was not to be released to foreign nationals. The irritated Summerfield accused Jacobs of damaging the fate of Defense Department research at Princeton and demanded to know: "How can you defend to the Princeton community the motives of your research when you are willing to limit its distribution. . . . Is this not self-evident proof that your research is not really as open as you recently claimed in the Daily Prince?" His concern was not with Jacobs's "motives" but rather the risk of "jeopardizing your defense in the recent public argument over your contract." Summerfield, himself a recipient of defense dollars, was "concerned about my own position as it might be affected by any defeat you suffer." The military-funded Princeton researchers had to prove that their sponsored research was innocent, and Jacobs was sabotaging their case. A week later the beleaguered Jacobs sent copies of his correspondence with Summerfield to the Kuhn committee, noting simply that he hoped "that our D[efense Department] contracts cease, Congress be urged to give corresponding money to NSF, and that the University adapt to the much lower level of funding that will probably result."[52]

Princeton biologist A. J. Levine finally clarified the nature of Jacobs's research at a fall committee meeting. Sponsored most recently by the army's Fort Detrick, Levine explained, Jacobs had merely been pursuing his fifteen-year interest in leaf senescence, which naturally included an analysis of the defoliating compound 2-4-D, which happened to have military applications. Levine assured the committee that Jacobs's intrinsic interest had come first—and given that interest, the use of 2-4-D was a scientifically appropriate area of exploration, no matter what the army's priorities were. Besides, the compound was a "tool" rather than the "subject" of Levine's work. This characterization added yet another difficult scenario for the committee members searching for an ethical policy: in one instance, they had engineers who argued that their field, by definition, relied on outside influences to set agendas; and on the other, they had the apparent case of a conveniently co-incident agenda addressing one researcher's intrinsic interest and the needs of the U.S. Army. Apparently even the category of "basic research" contained gray areas. As Robert Frosch, assistant secretary of the navy for research and development, had argued in a 1960 speech read by committee members, "pure" research and "basic" research were not indistinguishable. Pure research was defined *psychologically* as work that was just intrinsically interesting while basic research implied work that advanced large areas of knowledge. More critically, Frosch approached the ethical dimensions of the research controversy by opining that whether work was "pure" or "impure" depended solely on the motivation of the researcher, and the only party responsible for immoral applications was the applier himself. For Frosch, ethics was a matter solely for individuals. But the Kuhn Committee was searching for ethics on an institutional level.[53]

As the committee members continued to review specific projects, they found that most of the projects flagged as possibly weapons related were just that—*possibly* weapons related. There was no explicit weapons research at Princeton and almost no classified research. No one was perfecting a neutron bomb or tinkering with weaponized napalm. Instead, plausibly basic research projects with multiple applications abounded: from Jacobs's leaf senescence work to engineering studies of air flows with convenient relevance to the design of bomb-dropping airplanes. Faced with such ambiguity, committee researchers tried to sort projects by sponsor but once again found that Jacobs's experience—receiving grants from both the Defense Department and the NSF—described many other campus researchers as well. The NSF, despite its reputation for funding only basic research, sometimes also funded research with military applications, while the U.S. military routinely sponsored basic research. As Physics Department Chair Marvin Goldberger told the *Daily Princetonian*, "The [Department of Defense] finances a lot of basic research because they think they're going to get another bomb out of us, and there is nothing we can do to dissuade them from this delusion. We have an Air Force contract on fundamental concepts in theoretical physics."[54]

The committee compiled data on the key outside funders for Princeton's science and engineering departments. NASA and the NSF were major funders of the departments of astrophysics and aerospace and mechanical science but so was the Department of Defense (DOD). The Physics Department received 73 percent of its outside money from the Atomic Energy Commission (AEC), with the Pentagon at over 18 percent and the NSF at 8 percent. The results reflected the difficulties of funding overlaps and raised additional questions about another set of sponsors—those that were neither the DOD nor the NSF. What was the committee to make of NASA, for example, or the AEC? Comparing projects in the Aerospace and Mechanical Engineering Department, the committee noted that NASA-sponsored research looked "no different" from Defense-sponsored projects. As the committee's winter deadline approached, fewer and fewer policy recommendations emerged as adequate solutions to the complex problem of war-related sponsored research.[55]

✳ ✳ ✳

At a loss as to how to resolve these issues, the committee finally settled on a stronger reiteration of the school's anticlassified research policy and formal opposition to the existence of a classified library (a recommendation that would be overturned by the Princeton faculty). The proposal to sever ties with the Defense Department completely had long since been quietly jettisoned. These recommendations failed to satisfy many committee members, however, most notably Kuhn himself. The committee could use the classified barrier to deter explicit, exclusively war-related research on campus, but were there ways to detect and prevent subtler changes in Princeton research caused by the influx of so much outside money? French literature professor and committee member English Showalter articulated this fear when he noted in a meeting that "it is striking how unaware researchers are of the extent to which they are influenced by outside agencies." Did their choice of research topics, course curricula, and graduate students betray the influence of defense contracting?[56]

Kuhn called the phenomenon "drift," and he had alluded to it in his early fears of "distortions of academic research." In his standard letter soliciting information about sponsored research at other universities, Kuhn had written to Cornell administrators of his concern about the dangers of heavy funding from "mission-oriented" sources such as the DOD and NASA. He wrote, "If, as many members of the Committee are currently inclined to presuppose, that mode of funding does somehow *distort the normal evolution of academic research as a whole,* then we must recommend ways in which

Princeton, hopefully in conjunction with other institutions, can help to effect a substantial change."[57]

Belief in the existence of a "normal" or pure path of research was odd for Kuhn, given his reputation as an advocate for the theory of paradigm shifts in science, which held that scientific theories do not, in fact, follow an independent, linear progression. His correspondents were quick to point out his error. W. D. Cooke, vice president for research at Cornell, admonished Kuhn for his naïvete. Every source of funding influences the course of university research, he wrote, unless it is completely unrestricted—a rarity. Why single out the DOD, NASA, the AEC, and others as potentially distorting and not the Ford Foundation or the NSF? He noted wryly that the "NSF is often accused of protecting the entrenched scientific establishment which generally dominates their policy boards." Berkeley president Charles Hitch, responding to Kuhn's request for advice, noted that Berkeley had been undergoing a similar review of sponsored research, though its mission as a public university perhaps differed from that of Princeton. He criticized Kuhn's "pejorative" category of "mission-oriented" agencies, noting that "the university research they sponsor is probably no more mission-oriented than, say, that of [Housing, Education, and Welfare] and [the NSF]." Moreover, this relatively decentralized "multiplicity of government agencies," originally envisioned by Kuhn's mentor, James Conant, probably worked better than the kind of "large-scale, government-run research institutions outside of universities" that operated in other countries.[58]

Preparing the final report, committee members went through draft after draft of recommendations regarding the problem of drift. In fact, they could not even declare that drift was authoritatively a problem. As Kuhn wrote in one draft, "One may categorically deny 'co-option' and still recognize the possibility that the differential availability of external funds for different university activities may gradually and subtly alter the direction of institutional development, producing over time a quite decisive transformation." The trend might be difficult to detect, and its effects might be negative or positive. Thus, there was no policy prescription other than "continuing vigilance."[59]

∗ ∗ ∗

Despite the endlessly qualifying language, Kuhn never quite relinquished his belief in the existence of an alternate, counterfactual, 'pure' path—the progressive journey that physics and other sciences would take in the absence of military funding. In the reality of the Cold War United States, however, no such path existed. The hopes of Kuhn and others—that the increased sponsorship of basic research by the NSF could herald a return to the pure path—had been thoughtfully criticized by Kuhn's correspondents Cooke and Hitch.

Even if one were to combine the views of Cooke and Kuhn and assume the NSF-sponsored path to be more benign, from the perspective of a weapons-science critic, in this scenario scientists still would not be *in control*. That is, they would not be able to pursue any topic they chose, relying still on institutional powers to grant funding for and access to expensive equipment (what a Marxist might label the means of production). Kuhn's analysis suggested that Big Science had diminished the autonomy and agency of researchers, even as they followed what they believed was a natural path of inquiry.[60]

What then, was a university to do? MIT had opted to sever official university ties with one of its war-related labs while Princeton had established stronger barriers to classified research and made arrangements to phase out its lease to IDA. But the same research continued at the Draper Lab, the Jasons continued to study weapons efficacy and advise the Defense Department, and classified research that might have taken place in a Princeton laboratory now surely occurred elsewhere, in a government facility or, more likely, at a private defense science firm. If not a Princeton scientist, then who? If not a laboratory with university oversight, then where? Had the academic critics of weapons science now pushed war research farther away from their means to restrain it?

✳ ✳ ✳

MIT and Princeton were not isolated cases—at campuses across the country, students, faculty, and administrators reviewed contracts and research projects and debated their schools' missions and ethical obligations. Cornell University sold its Aeronautical Laboratory, Caltech opened access and promoted more basic research at its Jet Propulsion Lab, and Stanford toyed with drafting mortgage documents that would sever ties to the Stanford Research Institute while imposing research restrictions as loan conditions. (Stanford eventually abandoned this plan, but severed ties to the Stanford Research Institute nonetheless.) Michigan's Elderfield Committee, tasked with setting the school's classified research policy, ultimately chose to ban research whose "specific purpose" was "to destroy human life or incapacitate human beings," but to allow other classified research to continue. At the other extreme, protesters occupied buildings across the country, including MIT's special labs, and in some cases committed acts of sabotage and domestic terrorism, as at the University of Wisconsin's Army Math Research Center.[61]

The problems of "drift," the authenticity of scientific curiosity, and university-mandated research codes revived the specter of academic freedom, interpreted in widely divergent ways by campus opponents. Michigan's Elderfield Committee argued that the preservation of academic freedom was a necessary barrier to the imposition of moral codes, noting that "one of the main values which distinguishes a university from other communities is its respect for and devotion to the principle of individual diversity—academic, political and moral. To restrict arbitrarily the activities of its faculty members on the basis of some concept of what a university 'ought to do' is to do violence to one of the main principles which a university should uphold." Princeton botanist William Jacobs and his defenders had espoused similar views, simultaneously lauding the importance of individual curiosity guiding one's research while also accepting the existing system of defense contracting and classified projects. Wolfgang Panofsky took a more skeptical but similar position, telling Stanford students in 1969 that "the initiative for wishing to do a given piece of research work should come from the professor and his students. It should not come from the sponsors who pay for it if special money is required." But to a panel of Trustees he qualified his views, observing that classified research was not inherently "immoral," and, in the absence of "unilateral disarmament," it was necessary to some degree.[62]

To Chomsky, academic freedom meant just the opposite: a constant monitoring of the inevitable politicization that accompanied government and military outside contracts. As he had warned the MIT community in the Pounds report, "The idea that a university preserves its neutrality and remains 'value free' when it simply responds to requests that originate from without is an absurdity." Requiring employees at the special labs to gain security clearance or, at the very least, accept Defense Department funding, constituted a kind of institutional politicization that undermined any claims of individual academic freedom and academic neutrality. Moreover, many research contracts at MIT were not simply requests from outside funders but were the products of ongoing negotiations between the university and defense agencies.[63]

But two key questions remained: was any kind of politicization—or any kind of Defense-sponsored research—acceptable? Would weapons work have proceeded without disruption had the war in Vietnam been less unpopular? At the University of Michigan, scientists debated these very questions: whether campus discontent was rooted in commitment to universal moral truths or frustration and anger at the war in Vietnam. No doubt the two were related; the Union of Concerned Scientists had emphasized their disillusionment in their founding statement, writing that "through its actions in Vietnam our government has shaken our confidence in its ability to make wise and humane decisions." This same problem had motivated Charles Broxmeyer's emotional invective before the Pounds Panel, as he accused its members of translating their frustration at Vietnam and guilt about the atomic bomb into inappropriately expansive moral language.[64]

But in this regard, scientists were no different from the rest of the country and the rest of the world. Across the globe the language of a new generation of theorists and analysts, including Kuhn and Chomsky, was gaining ascendancy. The military-industrial complex, American imperialism, cultural hegemony, the dehumanizing effects of weaponry and automation—all were popular targets that could be addressed in the sweeping terms of theory and illustrated through the horrors of the U.S. war in Vietnam, French involvement in Algeria, or a host of other brutal world conflicts. As anger mounted, scientists whose World War II-era work had made them heroes now found themselves condemned for their monetary connections to tainted government agencies. For physicists and engineers especially, there were few ways out of this predicament; as the example of MIT showed, research conversion was difficult and funding rejections risky. In a sense the possibility for individual choice was collapsing, but opportunities for significant institutional reform remained severely limited.[65]

Back on the local level, few college administrations translated these radical institutional criticisms into significant policy reforms, but the analyses of Chomsky and others influenced unexpected audiences outside their own elite campuses, namely the entrepreneurs and managers of new spin-off defense firms and the administrators of large second-tier universities hoping to cash in on whatever defense moneys might be made available. The intellectual debates among elite academic scientists were not the major factors contributing to changes in the defense industry in the coming decades. Nevertheless, the perceived risks of investing research funds in radicalized campus facilities or locating labs in hotbeds of student unrest contributed to new developments in the social geography of defense research: the movement of research funds to southern and midwestern institutions and the proliferation of suburban "spin-off" companies devoted to computer command-and-control technologies.

EIGHT

Collapse of the Sputnik Order

By the dawn of the 1970s, the power and prestige of the Manhattan Project generation was rapidly eroding, as elite science advisors clashed with key policymakers and campus radicals attacked the Cold War architecture of defense contracting and classified research. Meanwhile, the prolonged "hot war" in Vietnam demanded a constant diet of new supplies and technological innovation, and a new, technocratically minded cohort emerged to take on the mission. The ranks of in-house military and industrial weapons workers—including scientists, engineers, and technicians—were expanding. These transitions occurred just as the political influence of scientists in general was itself being tempered, first by the accession of Lyndon Johnson, a president less involved in science affairs than his two predecessors, and then by Richard Nixon, whose disregard for the views of scientists pushed some advisors into open rebellion. The Nixon years remade the character of presidential science advising completely. The President's Science Advisory Committee (PSAC), born out of the Sputnik boom, perished at the tail end of the Vietnam bust, its lifespan stretching from the idealism of 1957 to the disillusionment of 1973.

* * *

During the war in Vietnam, the PSAC, with its overlapping membership and function, had predictably followed a similar trajectory to that of the Jasons, increasingly tackling problems of counterinsurgency and warfare in Southeast Asia. Overseeing, or at least participating in this shift, was Johnson's special assistant for science and technology, Donald Hornig, appointed in 1964. Hornig's professional trajectory was typical of elite scientists of his generation. He was born in Milwaukee in 1920, and after receiving degrees in chemistry from Harvard, where he studied with E. Bright Wilson, Hornig joined the Manhattan Project during a two-year stint at Los Alamos from 1944 through 1946. He taught at Brown and served as dean of the graduate school, then left for Princeton in 1957. From there he joined his colleagues in entering government and military service; his memberships included the Advisory Committee of the air force's Office of Scientific Research, the National Academy of Science's Space Science Board, and the PSAC under both Eisenhower and Kennedy.

As head of both the PSAC and the Office of Science and Technology (OST), Hornig and his key staffers, Spurgeon Keeny, Donald Steininger, and Vincent McRae, guided the development and allocation of PSAC panels and other expert resources to key issues of national security. Defense-related PSAC panels included those devoted to antiballistic missiles, antisubmarine warfare, tactical naval warfare, biological and chemical warfare, ground warfare, and, eventually, the Ad Hoc Panel on Vietnam and the Special Sub-Panel of PSAC on Vietnam. A later report on Hornig's tenure described the delicate process by which the OST struggled to influence Pentagon programs, relying largely on "persuasion" exercised through "informal meetings" such as regular lunches with McNamara and other key officials and conferences hosted by PSAC panels. Not surprisingly, Hornig often found that "strenuous efforts" were required to secure implementation of even partial recommendations.[1]

A recently declassified report on the OST during this period notes that from 1963 through 1965, top-level science advisors had little to do with the conflict in Southeast Asia, with the exception of the PSAC military aircraft panel and "some limited activity of the BW-CW Panel." The PSAC panel on chemical and biological weapons, for example, weighed in on the use of chemical and biological weapons in Vietnam, arguing against the use of DM ("vomit gas") and recommending against unspecified "proposed extensive biological test programs." The military aircraft panel, led by Vincent McRae and Richard Garwin, advised the air force on effective radar systems and "electronic warfare."[2]

As the war escalated, however, so did OST and PSAC involvement. By 1966 PSAC meetings were addressing Vietnam in greater depth while Pentagon officials more frequently sought expert science advice. As the PSAC "became increasingly concerned and questioned the way in which research and technology were being used in support of the military operations there," additional action seemed necessary. In the spring of 1966, the OST report states, "Certain members and ex-members of the President's Science Advisory Committee (PSAC) (acting in an individual capacity) urged a broad study of technical problems associated with Vietnam." Hornig agreed, but he worried that the PSAC and the OST were losing influence with the Pentagon. He suggested that the study be undertaken elsewhere, such that it would be guaranteed to "receive very high level attention and implementation of its findings would be much easier." The result, of course, was the Jasons' Dana Hall summer study of 1966.[3]

＊ ＊ ＊

Hornig's suggestion was not an attempt to remove Vietnam-related topics from the purview of the PSAC or the OST, however. According to the OST's own administrative history, by the end of 1966 "almost one-fourth of the PSAC discussions . . . were on topics directly related to Vietnam." The establishment of an "Ad Hoc Group on Vietnam" quickly followed, which included IBM's Richard Garwin and the heads of the PSAC's four military panels devoted to aircraft, naval warfare, ground warfare, and strategy. The scientists recommended the allocation of greater scientific resources to the war effort, including an expanded technical staff assigned to McMillan, as well as greater flexibility in funding structures.[4]

Hornig himself was a tireless champion of close relations between scientists and the Johnson administration. Late in 1965, when informed of a small budget transfer away from the PSAC, Hornig warned Johnson that the "PSAC would be extremely sensitive to any thought that their distance from you is being increased." A year and a half later, Hornig complained to Johnson about the lack of scientists included in an "intellectuals' lunch" hosted by the president. (Hornig was upset by his own lack of an invitation, which could harm "my relations with the academic community, particularly the scientists, who also consider themselves intellectuals.") But more importantly, the president had missed an opportunity to reach out to scientists, whose political power he had clearly underestimated. "It would be a serious error to discount the interest and influence of the scientific community, either in Viet Nam problems or in social progress," Hornig wrote. "They are among the most worried and hard to deal with in connection with Vietnam and we continue to need their support."[5]

Indeed, much of Hornig's workweek consisted of speeches to and correspondence with other scientists, including those who wrote to complain to him about the progress of the war. Whatever Hornig's personal views, he was tasked with defending administration policies. As he wrote to one such acquaintance, a Welsh chemist, "I would rather we were corresponding now about infrared spectroscopy or hydrogen bonding than Vietnam." Hornig acknowledged "the strong opposition of many U.K. and European intellectuals to our operations in Vietnam," but he characterized much of it as "a highly emotional response based on incorrect information." Instead, he defended the war as an attempt to repel "direct aggression by North Vietnam," and invoked the wartime example most compelling to scientists: "We all learned very painfully in World War II the consequences of postponing the evil day when one would have to face the campaign of indirect and direct aggression by Nazi Germany."[6]

With this stated commitment, Hornig oversaw the assignment of Vietnam-related work to the military PSAC panels. By the winter of 1967, the Ground Warfare Panel was addressing a host of relevant topics: hamlet evaluation systems, search-and-destroy and base-defense operations, and military advisor training, for example. The creation of the PSAC's Naval Warfare Panel further typifies the transition away from the previous emphasis on conventional aspects of the Cold War arms race. In 1965 a PSAC panel devoted to antisubmarine warfare was deeply enmeshed in establishing an Underseas Warfare Technical Center and bickering with navy leaders and other top advisors about helicopter carriers, equipment performance, and the status of Soviet submarine capabilities. In 1967, however, Hornig oversaw the creation of a new PSAC Naval Warfare Panel. The new group initially took on a similar complement of projects connected to Soviet naval activity "until requested by [Hornig] to devote all of its attention to a pressing problem relating to the war in Vietnam," namely, "a technical review of US military capability for reducing the quantity of material imported by sea into North Vietnam." This included studying the potential effectiveness of blockades and of mining or bombing the harbor at Haiphong.[7]

This review came as part of a more sweeping PSAC decision, made at the August 16, 1967, meeting, to assign the four military panels to undertake "an intensive study of those aspects of the Vietnam problem which fall within their areas of interest." Each panel was also explicitly instructed to address a pressing issue: how technology could contribute to the effort "to secure and hold areas and hamlets." That winter the full PSAC met with

President Johnson to discuss their potential contribution to the war effort. The meeting eventually culminated, in the spring of 1968, in a widely distributed report on the effectiveness of previous bombing campaigns in North Vietnam and predictions about the likely outcomes of various potential future campaigns. The context was President Johnson's bombing halt in March 1968, and like the Jasons a year and a half earlier, the PSAC report authors presented detailed analyses to support the pause and additional de-escalatory efforts.[8]

Indeed, two of the report's contributors—Marvin Goldberger and Gordon MacDonald—had been authors of the Jasons' barrier study and contributors to the Jasons' four-volume bombing study in 1967. As with those previous reports, the PSAC members qualified their conclusions with the acknowledgment that substantial data remained unavailable: "Much of the information relating to this problem is subject to large uncertainties or simply non-existent." Nevertheless, the pages of careful analysis culminated in a set of key conclusions relating to bombing plans and an accompanying series of technical recommendations. The conclusions stated that bombing had improved morale in South Vietnam but had not weakened "resolve" in North Vietnam; bombing had not prevented the transfer of supplies, either from North Vietnam or through Laos; and though additional bombing in Laos aided by new technologies might yield improved results, further bombing campaigns in North Vietnam—including the mining of Haiphong—would at best only "marginally" affect operations. Overall, the scientists wrote, "None of the studies we have seen of possible expanded campaigns makes a convincing case."[9]

These conclusions were accompanied by a series of technical recommendations, several of which drew on Defense Communications Planning Group (DCPG) and barrier-related work as potential means to improve the accuracy and effectiveness of existing operations. The main body of the report assessed the widespread physical damage wrought by previous bombing campaigns in North Vietnam, including the high civilian death toll and destruction of infrastructure, but still concluded that the costs of rebuilding and resupplying had successfully been passed on to the Soviet Union and that North Vietnam could likely withstand "even greater costs." North Vietnamese transportation routes were rapidly repaired so that communications and transfers of supplies were actually improving, despite the bombing. In this context the effectiveness of the barrier should be reviewed to ensure "its most effective use as an integral part of future interdiction campaigns." The PSAC urged military planners to identify and procure new equipment, in-

cluding sensors, aerial photographic reconnaissance systems, electronic surveillance, laser-guided bombs, and small land mines. Like the Jasons, they
hoped for technological solutions.[10]

* * *

Drafts of the report were distributed to top Pentagon and military leaders.
As with the Jason bombing reports, reactions predictably split along the
military-civilian divide, with consensus coming only on suggestions for technical improvements. McNamara's successor, Secretary of Defense Clark Clifford, wrote to President Johnson that he thought the report's "general conclusion is probably valid" and "is consistent with other bombing studies which
have been made from time to time." He was particularly impressed by the
technical recommendations, however, promising "to ensure that full use is
made of the Committee's Report in our continuing efforts to improve the
effectiveness of our interdiction programs." He confirmed that much of this
work was already underway. Earle Wheeler, chairman of the Joint Chiefs
of Staff, agreed on this point: He informed the president that he was "in
accord" with the recommendations, many of which were already being addressed. "Laser, Electro-Optical (EO), and Infrared (IR) guided bombs"
were being tested, he reported. "The Laser Guided bomb is under combat
evaluation in Southeast Asia."[11]

But Wheeler also noted pointedly, "I do not agree with the rationale and
conclusions of the study." When the scientists discussed matters of technology, he accepted their advice, but in their analysis of strategy, bombing
assessments, and implications for morale and politics, he simply dismissed
their views. Walt Rostow evinced a similar attitude in his correspondence
with President Johnson. "In disposing of Dr. Hornig's report on bombing,"
he wrote casually, "I focused my questions around the 6 technical recommendations which were not essentially controversial." In this regard he likely
took his cues from the president himself. After reading Hornig's initial memo
introducing the report, Johnson had referred the entire matter to Rostow,
along with the dismissive observation that "I basically do not regard bombing
as a matter of science." As McNamara had the previous year, the president
and his top advisors easily drew out the desired technical advice from the
nation's elite scientists while quietly disposing of their political and strategic
recommendations.[12]

* * *

The PSAC members working on Vietnam-related problems did not appear to suffer the same ethical crises as many of the Jason participants in the DCPG steering committee, but they did engage in self-reflection. In January 1968 Sidney Drell, the Stanford physicist who was working with both the Jasons and the PSAC, wrote a long letter to Hornig describing some of his views on the war and related "political-philosophical" remarks. Drell was no pacifist; as head of the Ground Warfare Panel he offered Hornig a range of "long-range problems" for the PSAC to address, including how technology could be used in missions crossing into Cambodia and Laos, how expanded "search and destroy operations" in the Mekong Delta could change the character of the war, and how vocational and agricultural education programs and other forms of development aid could be useful tools.[13]

If Drell had reservations about current war policies, they were rooted more in domestic political concerns than his assessments of Vietnamese society. He urged a halt to bombing in the "far north" of North Vietnam, though not at the demilitarized zone, in order to win back the trust of the alienated youth in the United States, who were disillusioned and skeptical that their leaders were seriously pursuing peace. Drell similarly opposed draft exemptions for science and engineering graduate students "for reasons of improving the empathy with the youth in college." When Hornig forwarded Drell's letter to Rostow, he took care to portray Drell as responsible and moderate. Drell "continues to be a positive force," Hornig wrote, "and although he has some worries, he has not let them get in the way of trying to produce technological gains which could contribute to the solution of our military and political problems in Vietnam."[14]

Given the military reactions to most of the war-related PSAC and Jason reports, Hornig's words encapsulated the administration's definition of a useful science advisor in the age of Vietnam: someone who provided the necessary technical expertise without excessive political complaint. It was a far cry from the idealistic enthusiasm of the Eisenhower days and heralded worse conditions ahead. Less than five years later, the fragile relations between a president and the PSAC would finally disintegrate.

✳ ✳ ✳

The experiences of the Jason and the PSAC scientists during the Vietnam War reveal some of the ways in which elite academic scientists contributed to the war effort while still maintaining a kind of "outsider" status, even while they remained on government and military payrolls. They enjoyed a measure of autonomy in their work and felt free to criticize particular programs and strategies. But theirs was not the only kind of scientific contribution. The military engineers and researchers employed at the limited-war labs and in war-related industries conducted important studies to ensure the effectiveness of aircraft, instrumentation, incendiaries, bombs, and chemicals used in Vietnam. Some of this work consisted of retrofitting older weapons to new applications, as in the case of napalm and defoliants, but these engineers also developed cutting-edge new technologies rooted in laser research and advances in computing.

Assessing the political and ethical views among these nonacademic researchers is difficult, as unlike the high-profile Manhattan Project physicists, few recorded their thoughts in memoirs and interviews, left voluminous personal papers in university archives, or were surveyed by political scientists. An unscientific poll of Los Alamos employees conducted by Citizens for Peace in Vietnam in 1967 revealed strong minorities both supporting and opposing the war, with a plurality of respondents refusing to state their political position. In recent years a handful of social scientists have conducted anthropological studies of the laboratory environment, but their work has largely focused on areas of nuclear research or the particular methodology and epistemology of science, rather than the specifics of Vietnam-related work.[15]

In 1968 and 1969, however, doctoral researcher Jeffrey Schevitz conducted a series of thirty-four interviews among "weaponsmakers" in the Santa Clara valley, a region in which 60 percent of the working population was employed in defense and aerospace fields. Schevitz focused his attention on the nineteen respondents who expressed antiwar views and at least some moral qualms about their work. He set aside the remaining fifteen subjects, categorizing them as "flag wavers" and "do-nothing patriots" who assumed their work was moral and worthwhile. (Schevitz clearly assumed the opposite.) He attributed their professional decisions to the corporate pay and benefits of the defense industry, the opportunity for advancement, and the intellectual challenge of their work. Although Schevitz's conclusions strongly reflected his own antiwar outlook, his interviews and employee profiles revealed a wide range of attitudes toward the war and the ethical implications of weapons-related research. His subjects were aware of the context of their work and either embraced its applications (as in the case of the fourteen respondents who supported the war effort), rejected them and either changed jobs or tried to organize for change within their place of employment, or, in Schevitz's analysis, stuck by their jobs and "rationalized" their moral situation.

Though not a large and scientific survey, Schevitz's study nevertheless offers a glimpse of the prevailing attitudes among Vietnam-era defense workers, which mirrored, to some extent, national views. The majority of his respondents either supported the war or had qualms too minor to outweigh the benefits of their employment. An energetic minority expressed open opposition, resigning in protest or working from within to try to change policy. Over time, of course, self-selection would fill these lost positions with additional prowar or apolitical candidates; by the 1980s scientists employed by defense firms tended to have more favorable views of various military programs than academic and other nondefense scientists.[16]

In a prescient observation of this trend, George Kistiakowsky observed in the spring of 1968 that academic scientists were being pushed out of their Pentagon advisory roles, which were instead "largely taken over by professional military scientists and those in the aerospace industry and think-tanks." This shift was visible in the influx of MITRE engineers contracted to work on DCPG projects and in the makeup of the air force's Scientific Advisory Board (SAB), whose membership in the 1960s increasingly drew on experts from industry and the nation's military labs. Even as a vocal contingent of young scientists took strong stands against the war, many of their peers and mentors held contrasting political views and enthusiastically entered the defense world with a realistic understanding of their role within it. Two key military advisors, William McMillan and John Baldeschwieler, illustrate this development.[17]

Born in California at the end of World War I, McMillan studied chemistry at UCLA and chemical physics at Columbia, where he received his PhD and contributed to Manhattan Project materials research. After a brief stint working for Edward Teller at the University of Chicago, he landed at UCLA, where he was conveniently located for consulting work at Livermore and RAND during the boom years of the 1950s. He also spent time as a member of the AFSAB, chair of the Weapons Advisory Group, and chair of the Divisional Advisory Group at Eglin Air Force Base.[18]

McMillan later explained his advisory work at RAND as "driven by patriotism" and encouraged by friends and colleagues from UCLA. In his early years there, he researched the effects of ionizing radiation, contributed to a study devoted to reducing the size of thermonuclear warheads, and in 1961 cowrote the influential RAND report "Arguments in Support of the Proposed Atmospheric Nuclear Effects Tests," which offered justifications for the temporary resumption of atmospheric testing in 1962.[19]

In an oral history conducted nearly thirty years later, McMillan reflected on the character of science advising during the 1960s. In some ways, the decade was a kind of golden age "when we could [get] things done rapidly"—for instance, when work on ionizing radiation suggested potential problems for guidance computers, RAND scientists gained quick access to top military commanders and millions in further research funding. But in other ways, the 1960s seemed a step backward. McMillan mourned the retreat of academic scientists back to their university labs after the extraordinary cooperation of World War II. Working with the top Pentagon scientist Harold Brown, McMillan helped found the Defense Science Seminar, a Jason-like effort to recruit a younger generation of scientists as military advisors. The seminars, held over the summers on the UCLA campus, included presentations from prominent speakers, field trips and demonstrations, and other efforts to "pump them up so they would know what some of the issues and problems are."[20]

In 1966 John Foster, whom McMillan had known from a summer spent at Livermore, invited McMillan to serve as a science advisor to General Westmoreland. McMillan agreed and ended up spending two years in Vietnam, working for both Westmoreland and his successor, Gen. Creighton Abrams. In this capacity McMillan enjoyed widespread respect and unique access to military top brass. McMillan shared a house with Westmoreland and so had the ear of the general and a front-row seat for military planning decisions. Though a staunch defender of Westmoreland, McMillan was not uncritical—he was shocked at the lack of genuine debate and discussion at military meetings, for example, and convinced Westmoreland to take a cue from academia and encourage honest dissent. As a scientist—a military outsider—McMillan was sometimes more easily able to take critical positions without disrupting the hierarchical deference prevalent in military circles. On one occasion he gently chastised Westmoreland's unfounded fixation on the likelihood of a post-Tet coup by citing a *Peanuts* comic strip in which Lucy, informed that what she thought was an insect queen was actually a jelly bean, wonders aloud how a jelly bean ever got to be queen. Westmoreland laughed and acknowledged the message that "we mustn't become enamored with our theory." Indeed, Westmoreland was quite taken with McMillan's ability to offer brilliant, blunt, "practical and pragmatic" advice, which could quickly "disabuse the staff of any self-satisfaction and complacency." He particularly valued McMillan's role as a kind of "devil's advocate," noting that if someone of McMillan's intellectual stature "questioned why we did things a certain way, it compelled us to re-examine our methods."[21]

In addition to this rhetorical role, McMillan also contributed to hundreds of small efforts to use science to improve U.S. military efforts. He helped pioneer the use of Fragmacord linear mines, which would be set along drainage ditches on trails prone to enemy infiltration. U.S. soldiers would wait along one side of the trail and open fire on infiltrators, who would then dive into the opposite side ditch, lined with the Fragmacord. He also worked with the Advanced Research Projects Agency (ARPA) on a failed effort to use blimps as silent surveillance devices, on the DELILAH system targeting the radar of enemy surface-to-air missiles, on studies of the effectiveness of the bombing campaign, and on tunnel detection. McMillan also continued his recruiting work, inviting "scientists, engineers, and social scientists . . . to join the Office of the Science Advisor in Vietnam as consultants for periods ranging from one month to over one year. These individuals were drawn from government laboratories, nonprofit institutes and—where possible— from the academic community. At its maximum this group numbered about 40, with a total of about 100 during my tenure." To create this program, Mc-Millan worked closely with John Foster, Generals Westmoreland and Abrams, and other key Pentagon personnel. He described the positive effects: "The stimulation of seeing the Vietnam problems firsthand and of being able to grasp the opportunity provided by technology to help out in a constructive and substantive way gave rise to an almost unparalleled esprit among these consultants."[22]

In addition to collaboration with ARPA and army engineers, McMillan also worked with Los Alamos director Harold Agnew. In 1967, as plans for the creation of the DCPG barrier and other more traditional barrier constructions took shape, McMillan reached out to Agnew and the research scientists at his disposal. The problem was how to build a barrier in a particular area where the North Vietnamese had set up artillery and were constantly disrupting construction. As McMillan later recounted, Westmoreland's first response was to instruct the air force to attack enemy territory, to "bomb them out of existence." But indiscriminate bombing rarely destroyed artillery pieces that had been carefully shielded—only pinpoint accuracy would be successful. Researchers at Los Alamos, however, worked out an acoustic locator system and built several prototypes that McMillan tried desperately to have mass produced and distributed. He had control over the creation of prototypes, but mass production and adoption depended on frustratingly slow bureaucratic processes and quibbling over funding sources. Gen. Creighton Abrams eventually endorsed the acoustic system, but only a handful were ever distributed. McMillan came to deplore bureaucratic impediments to what he considered obvious technical improvements, inefficient decision-making processes, and, in the case of Robert McNamara, a lack of basic scientific knowledge (he recalled McNamara once asking a team of radar specialists what a decibel was). Despite McMillan's best efforts, the glory days of coordinated war research and development seemed hopelessly lost.[23]

Unlike Kistiakowsky, however, McMillan's frustrations led him to embrace the opportunities for weapons work available in the private sector. He returned from Vietnam more interested in tactical weaponry and started his own business, McMillan Science Associates, specializing in "operations analysis" and working with major contractors on a variety of projects, including the development of Scud missiles. He later became a champion of President Reagan's buildup of high-tech weapons and their successful use during the first Gulf War.[24]

John Baldeschwieler was one of the "students" in McMillan's 1966 Defense Science Seminar. Baldeschwieler's youth and his scientific background—a mix of elite academic, military, and industrial experience—seemed ideal for McMillan's program. Baldeschwieler had attended Cornell on a Standard Oil scholarship, where he studied chemical engineering and held a summer job at DuPont and Los Alamos, joining the Reserve Officer Training Corps (ROTC) during the Korean War. He was, as he explained in a 2001 oral history, interested in "the military and the weapons side of things." Specializing in infrared spectroscopy but known for his talent with instrumentation, he performed his graduate work at Berkeley, completed his military service in the form of "neutron permeability" research at Aberdeen, spent six years at Harvard (where he frequently taught Kistiakowsky's classes when the advisor was off on government work), and landed at Stanford in 1965.[25]

At Stanford, Baldeschwieler was first chosen for the UCLA summer seminar, for which he was a "natural." An invitation from the Army Scientific Advisory Panel quickly followed, which, to Baldeschwieler, made "good sense." He described the work as "interesting and fun" and was proud to have the opportunity to do "hopefully useful things." As for the war itself, Baldeschwieler recalled (albeit thirty years later), "I was very loyal to the US government, and I remain very loyal. My attitude at the time was that the war itself didn't make a lot of sense to me, but the young servicemen who were exposed were my age and my colleagues, and I felt . . . I would do the best I could to give them tools that would help save their lives."[26]

During a month-long visit to Vietnam, Baldeschwieler worked with a team from General Electric on a potential "people sniffer"—a portable sensor that could detect human odors and other exuded chemicals, to be used in dark jungle conditions. The technology proved less successful than hoped, but Baldeschwieler enjoyed the experience. He also became involved in the implementation of the DCPG barrier and the sensor technologies used at Khe Sanh and along the Ho Chi Minh Trail. Like McMillan, he found himself frequently challenging conventional wisdom and providing much-needed doses of scientific skepticism. He recalled one such instance, when junior high school math skills allowed a particularly harsh criticism of military policy:

I was given a briefing by a young lieutenant colonel in a nicely starched uniform who talked about the hamlet pacification program. You'll remember that McNamara at that time had stressed the quantitative evaluation of everything. So there was a hamlet pacification index. This was calculated by having a team go in and interview the village father and count the number of foxholes around and the other defensive measures. All kinds of things. You'd check this off, and it went into a computer. The colonel showed me a plot of the hamlet pacification index as a function of time. It was a straight line. It started in the mid-sixties and went on up through the time that I was there. . . . Well, I looked at the curve and I asked, "When were US forces introduced into Southeast Asia in large numbers?" I then said that what the curve showed was that the introduction of US forces into Vietnam made no difference at all. And the room was utterly silent. Do you remember when the presidential candidate from Michigan—George Romney—went to Vietnam? He came back and said he had been brainwashed, and that ultimately led to his withdrawing from the Republican presidential primaries and his loss of credibility. Well, I'm sure I heard the same briefings, the same stuff. It really didn't make much sense. I can tell you many, many such stories.[27]

Baldeschwieler's competence and candor attracted the attention of Donald Hornig, who invited him to join the relevant PSAC ground warfare subpanel. Eventually, Baldeschwieler became a full-fledged regular PSAC member, at the moment of transition from President Johnson to President Nixon and Donald Hornig to Lee DuBridge as chief science advisor. Baldeschwieler recalled feeling a bit out of place: "It was an interesting assignment, because . . . most of the people involved had backgrounds in either nuclear physics or were from the MIT Radiation Lab . . . I was the token chemist." He remained active in science advising, however, even taking a leave from Stanford at the start of President Nixon's second term to join the Office of Science and Technology, home to the remaining government science advisors after the dismantling of the PSAC.[28]

To the generals and presidents he served, Baldeschwieler was a brilliant applied mind. To an increasingly radical subset of the antiwar population, however, he embodied a kind of technocratic false consciousness. Even during his 2001 oral history interview, Baldeschwieler's casual reflections could have been the stuff of New Left antiscience pamphlets decrying the supposed neutrality of science. Describing the dangers of his work in Vietnam, Baldeschwieler observed that "the fighting was everywhere, and I was shot at a number of times. I recall wondering, 'What are they shooting at me for?' I hadn't done anything—I was just trying to get the technology straight."[29]

✳ ✳ ✳

While Baldeschwieler's cohort waited in the wings, the election of Richard Nixon further destabilized the old Sputnik order. Nixon's decision to invade Cambodia radicalized the antiwar movement and alienated many scientists while his controversial endorsement of an antiballistic missile system (ABM) and supersonic transport program (SST) pitted the president against key members of his PSAC. In 1974 physicist-turned-activist Frank von Hippel reflected that the war in Vietnam had revealed the "bitter lessons about governmental limitations and fallibility," which eventually emboldened scientists and other experts to take strong public positions against projects they felt were inappropriate or wasteful. The debates over the ABM and the SST exemplify this trend.[30]

The problem of the ABM had been the subject of a long-simmering debate that reached a full boil at the end of the Johnson administration, coinciding with the frustration of the Jasons and the resignation of George Kistiakowsky. At the center of the controversy was Richard Garwin, PSAC member and advisor to the Pentagon, Central Intelligence Agency, and the Atomic Energy Commission, as well as star scientist at IBM. In March 1968 Garwin and Hans Bethe had published an enormously influential article in *Scientific American* on the problems facing the Johnson administration's proposed ABM system. The plan, proposed by McNamara and known as Sentinel, consisted of a combination of radar-based detection technologies and interception via short- and long-range antimissile missiles. Garwin and Bethe argued that in addition to threatening the relatively stable deterrence of mutually assured destruction, rooted in the assumption that no matter which side initiated a nuclear war, both the Soviet Union and the United States would end up destroyed, the Sentinel system could be easily overwhelmed by the use of decoy missiles and other inexpensive tactics.[31]

Garwin's criticism didn't jeopardize his strong relationship with the Johnson administration, however. In January 1969 science advisor Donald Hornig lauded Garwin's extensive contributions to national security matters in a letter to the outgoing president, noting that "the tragedy is that most of his labor has been so highly classified that public recognition of the usual sort has been out of the question." To compensate, Hornig drafted a personal thank-you letter for LBJ to sign, referring to Garwin's "immense personal effort" on behalf of the government and lauding him as an "unsung hero."[32]

But with the transition to the Nixon administration, the writing was on the wall. Political calculations drove Nixon to endorse a new ABM system known as Safeguard, but his lack of consultation with his science advisors deeply antagonized the PSAC. Within a year, Garwin was engaging in the unthinkable. He continued his highly visible, public anti-ABM stance, personally appealing to members of the U.S. Senate to vote against the project. By 1970 he had publicly taken up a second issue, the SST, which he opposed on the grounds that it was wasteful and a potential environmental risk. Garwin provided Congress with scathing testimony and publicly revealed that the PSAC had reviewed the project and reached conclusions at odds with the Nixon administration's current supportive stance. (The ensuing lawsuits by various groups attempting to gain access to the PSAC SST report would eventually contribute to the expansion of the Freedom of Information Act.)[33]

James Killian, the first presidential science advisor during the Eisenhower administration, wrote in his memoirs that he believed that the job of PSAC members was to serve the president. If they disagreed with his policies, they ought to resign. He could not accept Garwin's behavior, writing in 1982 that while "I respected his conclusions and his right to them, I still cannot defend this act." Lee DuBridge, former California Institute of Technology (Caltech) president and Nixon's top science advisor, would offer a similar view of the PSAC's role, at one point telling Congress that, unlike Garwin, he considered himself "a soldier" who was bound "to support the President's decision" on the SST.[34]

In truth, the realities of the PSAC system had altered considerably from the days of Eisenhower to those of Nixon. Whereas Killian, Kistiakowsky, and Wiesner had been outspoken, respected advisors who were intimately involved in major national security debates (and who maintained excellent personal relationships with the presidents they served), DuBridge and his successor, Edward David of Bell Labs, were milquetoast science advisors, rarely consulted, lingering on the fringes of key decisions. A 1970 *New York Times* profile of DuBridge described him as a "gentle, grandfatherly science adviser" with apparently "little influence" and "no direct responsibility." He was followed by Edward David, whose access to the president was limited and who resigned after two and a half years.[35]

John Baldeschwieler, the chemist who had worked on the barrier and other Vietnam projects and who was, briefly, deputy director of Nixon's Office of Science and Technology, described some of the internal politics of science advising in a 2003 oral history for the Chemical Heritage Foundation. He recalled that under President Johnson, the PSAC would typically produce reports with several approaches to a particular issue—one option to do nothing, one to spend an exorbitant amount, and one somewhere in the middle that seemed most reasonable, which would usually be referred to the president. Under Nixon, however, the process changed: "[Nixon and Kissinger] requested a presidential-decision memoranda, employing a specific format to present a set of options, of which they got to choose one. Our recommendations were sent back to us for reworking if what the president wanted . . . wasn't on the list." Though Baldeschwieler credited the PSAC with a few notable instances of influence—for example, pushing Nixon to create the Environmental Protection Agency—he acknowledged that the bitter acrimony that developed between Nixon officials and Garwin constituted "a revelation that the argument was usually cast to fit the objective, rather than to fit the facts."[36]

It was in this political atmosphere that scientists who opposed Nixon's preferred policies suffered a swift backlash. Nixon revoked the appointment of Cornell chemist and ABM critic Franklin Long to head the National Science Foundation (NSF), a public humiliation that aroused an immediate outcry among scientists. As Mary McGrory noted in the *New York Times* at the time, "It touched the brotherhood to the quick." The Federation of American Scientists warned: "An aspect of the affair that is of even greater concern . . . is that the choice of scientists for important and influential positions in government, even those which are not directly related to the military, is being made on political grounds rather than on the basis of scientific and administrative competence. If this continues, the Nixon Administration will have so-called scientific advisors who are trained seals rather than intelligent men of independent scientific judgment."[37]

Nixon's move actually solidified scientific opposition to the ABM, McGrory argued, noting that "scientists have a strong sense of fraternity and an equally powerful sense of guilt for their silence about the moral consequences of splitting the atom. Now, goaded by the President they are probably in the anti-ABM fight to stay." Whether Nixon was to blame or not, scientists overwhelmingly opposed the ABM. A Scientists and Engineers for Social and Political Action poll conducted at the 1969 annual meeting of the American Physical Society revealed that 76 percent of the 1216 physicists who voted opposed Nixon's proposed Safeguard ABM program. During congressional hearings in 1970, all four former top science advisors (Killian, Kistiakowsky, Wiesner, Hornig) testified against Safeguard, as did Hans Bethe, Herbert York, and Jason members Marvin Goldberger and Sidney Drell. It was the unity and largely technical arguments of the arms-control scientists that bolstered congressional opponents to the program and set the stage for Nixon's eventual endorsement of the moderate limitations imposed by the Anti-Ballistic Missile Treaty, signed by the United States and the Soviet Union in 1972.[38]

Of course, it wasn't only the ABM and the SST arguments that deepened the antagonism between scientists and the Nixon administration. The war in Vietnam and the actions of President Johnson and his advisors had already created a dangerous gulf, and Nixon's invasion of Cambodia further fueled the disillusionment and opposition of many scientists. Though scientists' public pressure eventually forced Nixon to reoffer the NSF position to Long, he promptly refused it the second time around. A year later the president would pull a nearly identical stunt, revoking the appointment of Caltech's George Hammond to head the NSF. Hammond had participated in a Caltech protest against the invasion of Cambodia, part of the same wave of protests that led to the shootings at Kent State and the formation of the Kuhn Committee at Princeton.[39]

It was not surprising, then, that in 1973, Nixon opted to dismantle the entire presidential science advising system, eliminating both the PSAC and the Office of Science and Technology in favor of a much weaker White House Science and Technology Policy Office. Joel Primack and Frank von Hippel assessed the decision along the same lines as Baldeschwieler and McGrory, noting that "the abuses of the advising system arise out of its political exploitation, and the White House appears to have abolished PSAC precisely because it was not exploitable enough." They also concluded that "frustration with science advisors like Dr. Garwin" was at least "partly responsible" for the decision, especially considering that the SST and the ABM debates had constituted the first times that Congress had "really challenged the administration on a major high-technology program." If, as von Hippel would later argue, the Johnson and Nixon administrations had increasingly only drawn on science advisors "to give political decisions the appearance of objectivity," and the science advisors—or at least an influential handful—had now refused to play along, it made sense that Nixon would no longer see a need to maintain the existing system. Jerome Wiesner bluntly summed up this analysis in a 1973 statement to *Time Magazine,* observing that the dismantling of the PSAC was, in the end, a more honest result: "The reorganization simply recognizes the situation as it has existed throughout the Nixon Administration."[40]

Federal science advising would nevertheless limp along through the 1970s. In 1975 President Gerald Ford oversaw the establishment of short-term advisory panels, designed partly as a means to reach out to the science community and repair some of the Nixon-era damage and partly to investigate actual topics in science and technology. A year later Congress would create the Office of Science and Technology Policy, which included a presidential advisory committee. President Jimmy Carter dismantled this committee at the start of his administration, but the Office of Science and Technology Policy remained. Much of the science advising during the

Carter administration took place at the departmental level, however, particularly in the departments of agriculture, energy, and defense.[41]

* * *

The new advisory systems of the 1970s enabled and expanded the trend that had been in place since the mid-1960s: the elevation of a new class of experts to replace the aging Manhattan Project veterans and the theoretical physicists borrowed from the elite institutions of academia. The new class had much stronger roots in military research and defense engineering. As James Katz observed in *Presidential Politics and Science Policy,* during the Johnson administration, "PSAC . . . began recruiting its members from industrial, engineering, and even social-science backgrounds." The air force's SAB had already shifted in that direction by 1964, and Air Force Secretary Thomas Reed later described a concerted effort in the 1970s to elevate the status of "R&D program managers, information systems people," and other technical experts. ARPA in the 1970s likewise shifted its focus from the free-ranging promotion of cutting-edge new technologies to short-term projects with clear defense applications; in 1972 it was renamed the Defense Advanced Research Projects Agency, or DARPA.[42]

A similar transformation was occurring in the competition among universities for increasingly precious government research dollars. Whereas federal contracts had surged during the Sputnik boom years, they began to level off and decline in the early 1970s, during the Vietnam bust. In Massachusetts, where the labs of the Massachusetts Institute of Technology (MIT) and Harvard had "spun off" dozens of small entrepreneurial defense firms in the 1960s, the dip in defense contracting—to both universities and their spinoffs—coincided with the dramatic student protests demanding an end to weapons-related campus research and the beginning of a wave of layoffs at both major defense firms and smaller technology startups. Rather than spending precious financial resources on retaining the top scientists with PhDs from elite universities, many companies axed their more expensive employees and recruited younger, more specialized engineers who could be lured at a cheaper price. In 1970 Berkeley Rice published a depressing portrait in the *New York Times* of out-of-work scientists and engineers laid off from their cushy jobs along the "Golden Horseshoe" of Route 128, the "East Coast center of the electronics industry." PhDs from Harvard and MIT complained that they had been priced out of their own companies, which were downsizing and scaling back research as federal funding for defense and space-related technology dried up. Deborah Shapley reported a similar trend on the pages of *Science* in June 1971, conducting dozens of interviews in the Boston area and noting the common complaint of "de facto age discrimination," in which labs and research firms discarded older experts in favor of "fresh graduates" to "promot[e] economy in the company payroll."[43]

Both Rice and Shapley noted widespread bitterness toward the Nixon administration among the unemployed scientists and engineers. Shapley reported that in the thirty interviews conducted by *Science,* every subject had "stated adamantly that they wouldn't vote for Nixon in 1972," though the reasons cited had little to do with foreign policy and everything to do with cutbacks in aerospace and other defense contracts. Rice cited several engineers who theorized that Nixon's cuts were a form of retaliation against the Massachusetts senators who had opposed the ABM. At the other end of the political spectrum, in 1970 Science for the People complained of the "planned unemployment" orchestrated by the Nixon administration, designed to force recalcitrant engineers to go "begging for DOD [Department of Defense] and Space make-work."[44]

While Nixon's antagonism toward the scientific community is well established, the conspiracy theories of the Route 128 engineers and the Science for the People radicals are difficult to prove. But since the mid-1960s, the government had made efforts to redistribute research funding to promote geographical diversity; in other words, to redirect funding away from the established academic centers in California, Massachusetts, and New York and toward new regions and institutions. During the Johnson administration, a proposal by Donald Hornig to the Federal Council for Science and Technology was expanded by the president in late 1965 into a national policy to help "additional institutions to become more effective centers for teaching and research."[45]

One outgrowth was the Pentagon's Project THEMIS, which began as a $25 million a year program "deliberately seeking out departments in universities which have not been engaged significantly in defense related research, and supporting research in those departments which appear to have the capacity to contribute to the solution of problems of interest to DOD." During the three years of its existence, the program distributed $88 million to universities in forty-two states. Though little has been written about the politics behind the launch of Project THEMIS and the results of its implementation, at least one scholar, James Katz, has concluded that campus activism at key research centers was an important precipitating factor: "Of major concern to the DOD was that the traditional performers of university research (e.g., Harvard, Massachusetts Institute of Technology, Princeton, Berkeley) were increasingly shying away from military-related R&D because of pressure from the rising anti-war movement." The trend continued beyond the relatively narrow THEMIS program, with the South and southern universities in particular soaking up defense funds being redirected from more politically volatile areas.[46]

These trends—rising antiwar radicalism at elite research universities; private sector preferences for younger, cheaper labor; and federal contracts in need of new recipients—encouraged the expansion of existing second-tier research universities into powerhouses of defense research. The rapid growth of Northeastern University in Boston during this period exemplifies this process. In her comparative study of the technology firms of Route 128 in Massachusetts and Silicon Valley in California, AnnaLee Saxenian notes that in the 1960s, "The community and state colleges in Massachusetts were small, underfunded, and lacking in status, particularly compared with California's community and state college system." She cites the perception of some local technology employers that despite key spin-off genealogies, Harvard and MIT seemed standoffish and distant from the growing electronics and computing sector when compared to Stanford or Berkeley in California. As a consequence, some of New England's large defense technology firms began to offer employees training and education themselves. They also reached out to Northeastern, a large urban commuter-school with an ambitious president, Asa Knowles, and an innovative co-op program that blended classroom instruction and local work experience.[47]

Through the 1960s and early 1970s, Knowles set out to remake Northeastern into a prominent research university. During his tenure, the school's undergraduate enrollment more than doubled, and a host of new graduate programs offered professional degrees. Knowles was himself a canny surveyor of the political landscape. In the early 1970s, he pushed hard to bring defense contracts and research money to his school and to expand the co-op program as a feeder to big defense companies. Though Northeastern students were hardly docile, Knowles explicitly promoted his university as a safer alternative for weapons research than the more radicalized elite campuses of MIT and Harvard. In 1970 Knowles addressed an audience of representatives from some of the nation's most prominent defense contractors, including Westinghouse, Monsanto, Dow Chemical, and Du Pont. Hoping to secure new institutional ties, Knowles advertised the typical Northeastern co-op student as likely to be "more business and industry oriented and . . . likely to be a more understanding sympathizer of the establishment." He promised that companies hiring Northeastern co-op students could "avoid many of the extremist attitudes and difficult problems which may arise." Northeastern students were not apathetic, he explained, but their working-class backgrounds and interest in industry experience precluded antiwar radicalism. Along with Knowles's similar efforts to attract classified and military research to campus—the types of research that were so vilified elsewhere in the early 1970s—Northeastern succeeded in dramatically expanding its research and development contracts, as well as its student co-op program, during the 1970s.[48]

※ ※ ※

At the height of the Vietnam War, the Sputnik consensus collapsed. A younger generation of scientist-activists pushed for reform based on a moral and political reckoning focused on institutions, rather than individuals. In their minds the ethical obligations of scientists did not consist of mere personal decisions to embark on particular projects but included rigorous analyses of funding sources, laboratory affiliations, and military consequences, accompanied by obligatory political action. As institutions themselves became targets, the new terms of debate alienated career lab staffers and inspired students to revolt against their university administrations. Many sympathetic Manhattan Project veterans found themselves caught awkwardly in the crossfire between their radical students and their hawkish colleagues, seeking out a middle ground unlikely to please either side.

But the controversy over university defense contracting and the problem of neutrality yielded ambiguous results. University policies and funding structures changed, though not always in the ways hoped for and certainly not on the scale imagined by protesters. The failure to "convert" weapons laboratories to civilian purposes and their eventual purging from elite campuses contributed to the rise of defense research at second-tier schools and in private suburban research firms. At the same time, high-profile protests brought debates about the ethics of weapons work back into public view on a scale not seen since the days of the Manhattan Project, pitting scientists against each other and forcing public justifications of research projects and funding sources. This new discourse, often heavily critical of science because of its perceived links to weapons development and war, contributed to the general antiscience attitudes of the late 1960s and early 1970s.[49]

Combined with Richard Nixon's attacks on scientists and his dismantling of the PSAC system in the early 1970s, these campus and professional debates thus contributed to two contradictory outcomes. On one hand, the 1970s saw the expansion and elevation of a technocratic class of weapons workers and advisors, even as public disputes over science policy tended to erode popular faith in scientific expertise. At the same time, other scientists developed a stronger self-awareness of their own political value and the ways in which their work might be manipulated by others, facilitating a new kind of independence that would prove influential during the Star Wars debates of the 1980s.[50]

NINE

A United Front against Star Wars

In June 1985 Massachusetts Institute of Technology (MIT) president Paul Gray delivered the university's annual commencement address to graduating seniors, in which he discussed science and the politics of research. He focused specifically on the current debates over the Strategic Defense Initiative (SDI), known popularly as "Star Wars," a new kind of antiballistic missile (ABM) system proposed by President Ronald Reagan. Gray warned that the Pentagon was using lucrative government contracts with universities to create an "implicit institutional endorsement for SDI." It was a "manipulative effort," Gray observed, given "the controversial nature and the unresolved public policy aspects of SDI." In stark terms he announced that "this university will not be so used."[1]

Four years later, at an MIT teach-in held on the twentieth anniversary of the March 4, 1969, protests, engineering professor and self-described conservative James Melcher recalled that in his youth, he had actually felt alienated from the radical student movements of the 1960s. The antiwar protests had left him unmoved. In his case, he explained, his rebellion had come a decade and a half later, with Reagan's Star Wars speech in 1983.[2]

SDI created another turning point in the history of scientists' activism, in large part because so many scientists—including many who worked on nuclear weapons research—came to oppose it or at the very least deeply and publicly doubt its chances for success. Star Wars reopened the old wounds of the lopsided arms-control debate, once again pitting the elderly veterans of the Manhattan Project and the majority of academic physicists against the indefatigable Edward Teller and a handful of other SDI cheerleaders.

It led thousands of scientists to sign statements refusing SDI contracts and resulted in the resignations of key government advisors. Unlike the agonizing debates during the war in Vietnam, the SDI program elicited clear public condemnations from professional organizations, university presidents, and even Livermore employees. The variety and inclusiveness of anti-SDI activism reflected a tentative generational reconciliation, uniting former New Left radicals, Manhattan Project liberals, and even nuclear hawks concerned that SDI undermined the theory of mutually assured destruction.[3]

The scientific and political landscape of the Star Wars debates was not the same as that of the test-ban treaty and Vietnam controversies. Circumstances had changed by the 1980s, in large measure as a result of the earlier moments of upheaval. In the 1970s, in the aftermath of Vietnam protests and in the face of a renewed debate over the development of an ABM system, federal advisory mechanisms were dismantled and reconstituted in much weaker forms, scientists' compunctions about political activity shifted, and the social geography of weapons research itself stretched and expanded to include new regions, neighborhoods, and institutions. These changes contributed to the particular form of the Star Wars debate as well as the character and strategies of its participants. A broad-based resurgent arms-control movement kept the debate over nuclear weapons in the public spotlight. High-profile elite scientists, left with fewer channels to advise federal policy directly after the dissolution of the President's Science Advisory Committee (PSAC) in 1973, sought alternate means of influence over policy. The American Physical Society (APS), which during the Vietnam years had witnessed bitter internal struggles over the meaning of neutrality and the appropriateness of political activity, now undertook a massive study aimed at discrediting Star Wars on technical grounds. And, in a movement that bridged individual and institutional activism, a coalition of thousands of academic scientists pledged to boycott SDI research.

But supporters of SDI—including a small coterie of elite conservative physicists—had also learned lessons from the tumult of the Vietnam years. The New Left had challenged notions of objectivity and neutrality, and now the New Right launched a parallel attack, questioning supposed scientific consensus on a variety of issues, from pesticides and pollution to the feasibility of a space-based missile defense system. Critics might denigrate Livermore scientists as tools of the military-industrial complex, but conservative physicists such as Frederick Seitz and Edward Teller levied opposite charges upon the APS itself, identifying its leaders as longtime arms-control activists willing to misinterpret data to support a political agenda. The highly charged Star Wars debates of the 1980s thus simultaneously united activist generations of radical and liberal scientists, healing the generational wounds of the Vietnam years, but they also revealed the power of conservative physicists who increasingly had the ear of politicians in Reagan's Republican Party.

✳ ✳ ✳

By the time of President Ronald Reagan's 1983 Star Wars speech introducing the SDI program, the Massachusetts defense and technology industries were on the rise once again as key centers for computing and software research. But also ascendant was the nuclear-freeze movement, an outgrowth of the arms-control efforts of the 1950s now bolstered by the growth of environmentalism; the partial meltdown of the Three Mile Island nuclear reactor; and significant attention from popular media through news reports, films, and literature. The election of Reagan in 1980 further fueled nuclear fears and consequent political organization, as did the publication of a spate of books by politicians, activists, and scientists warning of the terrifying risks of nuclear disaster. The new wave of nuclear activists called for an immediate and complete halt to weapons development and production, an echo of Jerome Wiesner's proposals during the test-ban negotiations of 1963. Graduate students at MIT publicly supported the freeze in 1980, and two years later a New York City antinuclear protest drew a million participants. Even the sixty-five-thousand-member International Chemical Workers Union, which included DuPont and Monsanto employees, voted in favor of a nuclear-freeze resolution at their 1982 convention.[4]

While a full history of the nuclear-freeze movement is beyond the scope of this book, two aspects of its development are important to note. First, the movement incorporated many of the scientist-activist organizations that had flowered in opposition to the war in Vietnam and included members of the Physicians for Social Responsibility, the Union of Concerned Scientists (UCS), and many other groups. Second, the movement was particularly strong in Massachusetts, where Senator Ted Kennedy pushed for a nuclear freeze in Congress, and in California, where protesters held daily vigils at the gates of the Livermore lab, confronting the scientists and engineers who worked within and demanding justifications for weapons contributions. Hugh Gusterson, an anthropologist who studied nuclear-weapons scientists at

Livermore in the 1980s and 1990s for his 1996 book, *Nuclear Rites,* questioned his subjects about their reactions to the protesters outside. Many had been opposed to the Vietnam War and sympathized with the environmental movement but believed strongly that their own nuclear-weapons work promoted peace through deterrence. Gusterson concluded that, for many, the protests actually strengthened in Livermore employees a kind of individualized ethics rooted in "the central axiom that, given the nature of the international system, nuclear weapons offer the best hope of preventing war and saving lives." In many ways the ethical assessments of the Livermore employees of the 1980s resembled those of the Manhattan Project scientists during the early years of World War II: their weapons work was a necessary contribution supporting national security goals with which they agreed.[5]

Reagan's proposed SDI, offering a revamped ABM system with space-based interceptors, exactly undermined deterrence based on mutually assured destruction, a strategy held dear by Livermore scientists. He called for defensive measures that could, in theory, allow for a new kind of nuclear war that was both survivable and winnable. The origins of this shift lay in the resurgent conservatism that had swept Reagan to power and in the elevation of Edward Teller to a level of advisory influence he had not held since the first Eisenhower administration. Since the late 1960s, Teller had been lobbying then-governor Reagan in favor of missile defense, and immediately after Reagan's inauguration in 1981, Teller was hard at work with colleagues from the Heritage Foundation developing strategic defense recommendations and meeting repeatedly with top administration officials to present the new technologies being developed at Livermore that could contribute to such a system. In particular, he believed in the potential of "a third generation of nuclear weapons"—space-based lasers that could destroy incoming Soviet missiles with the directed energy of a nuclear explosion. A 1983 article on nuclear policy in *Scientific American* further dissected this shift, noting that whereas McNamara had reportedly advised both Kennedy and Johnson never to use nuclear weapons, Caspar Weinberger, Reagan's secretary of defense and a former Bechtel vice president, believed in the possibility of "waging a wide range of 'limited' or 'protracted' nuclear wars and the concomitant undesirability of a 'no first use' policy."[6]

Teller's influence trumped the weak science advisory mechanisms in place; Reagan neglected to consult his own administration's White House Science Council in the period preceding the announcement of the new policy. John

Bardeen, inventor of the transistor and a double Nobel laureate in physics, resigned from the council shortly after Reagan's speech, complaining that "President Reagan prepared his speech with no prior consultation either with technical experts in the Pentagon . . . or with his own science advisor." From his perspective, "there was no point in being on a committee which is supposed to give advice when such a truly major scientific and technical decision is made without consulting the body." (Or as Herbert York complained, Teller had exploited his "privileged" advisory position to sell "exceedingly expensive technological exuberance" to "an uninformed leadership" while knowledgeable science advisors had been left out of the loop.) Teller's preferential treatment was made possible by the absence of independent, influential science advisory channels and by the compatibility of his hawkish nuclear views and the strategic priorities of Reagan's cabinet, which represented a dramatic break from the arms-control liberalism of Robert McNamara and the détente of Nixon and Kissinger.[7]

In a televised address delivered on March 23, 1983, the president laid out his vision for an alternate approach to the Cold War arms race. He asked the nation to imagine a world in which defensive capability, not offensive might, provided American security; a world in which "we could intercept and destroy strategic ballistic missiles before they reached our own soil or that of our allies." Invoking the ethical obligations of American physicists, he called on "those who gave us nuclear weapons" to "turn their great talents now to the cause of mankind and world peace, to give us the means of rendering these nuclear weapons impotent and obsolete." Achieving this objective would require a massive research and development program, for which Reagan was willing to supply immense financial resources. He opened his speech with a withering attack on critics who wished to cut defense spending, likening them to appeasement-minded Europeans in the 1930s, unwittingly inviting "the tragedy of World War II." He closed his message with the announcement of a long-term, "comprehensive and intensive effort" toward the "ultimate goal of eliminating the threat posed by strategic nuclear missiles."[8]

∗ ∗ ∗

Reagan's genuine passion for a defensive system and Teller's technological optimism formed the ideological basis for what would become the SDI program. In the aftermath of the president's speech, his administration launched two influential studies. The first would explore the ramifications of strategic defense for arms control and deterrence, and the second would assess the status of the technology required for deployment. The latter, known formally as the Defensive Technology Study Team and informally as the Fletcher Panel, named for chair James Fletcher of NASA, was staffed by representatives from industry, academia, and the national laboratories and reported to the undersecretary of defense for research and engineering, Richard De-Lauer. Sandia scientist and panelist Gerold Yonas characterized the Fletcher group as composed of people with a variety of viewpoints and "strong convictions" and "intellectual skepticism." They were "duty-bound to suggest a myriad of possible methods to defeat any defense." Even with this rigorous skepticism, Yonas saw grounds for optimism. As he wrote later in an article for *Daedelus,* he understood SDI as a creative research program rather than a surefire short-term defense system, with a long-term goal of eliminating the threat of ballistic nuclear weapons: "Most of us who became involved in the technologies study did not feel that the president was urging us to pursue specific, short-term weapons development programs. Our understanding was rather that we were to initiate . . . approaches that, although perhaps unavailable today or even in the near future, could someday provide entirely new and effective technological options that would 'pave the way for arms control measures to eliminate the weapons themselves.'"9

The Fletcher panel ultimately proposed a five-year research program with a projected $26 billion budget, and in 1984 the Reagan administration formally created the Strategic Defense Initiative Organization (SDIO) to oversee the new program, with Yonas appointed as chief scientist. From his new position, Yonas continued to promote a long-term research program to devise an effective defensive system targeting all four phases of incoming ballistic missile flight: boost, post-boost, mid-course, and terminal. He encouraged the development of several key areas, including kinetic-energy weaponry, directed-energy weaponry, and advanced command-and-control technologies. Kinetic-energy weapons would rely on physical impact to destroy incoming missiles, whereas directed-energy weapons, when developed, would harness the power of nuclear energy to generate destructive beams of energy (for example, in the form of an X-ray laser). To integrate detection and defensive weaponry, complex and reliable new computerized command systems would be needed. The SDIO's Innovative Sciences and Technology Office, headed by James Ionson, oversaw funding for both basic and applied research that could contribute to long-term goals.10

* * *

For a scientific community that had been divided over the war in Vietnam, the ethics of government service, and the legitimacy of New Left criticisms, unified opposition to SDI began to heal some old wounds. A 1986 poll conducted by the National Academy of Sciences (NAS) found that of over 450 members who were employed in fields "relevant to the Strategic Defense Initiative," more than 80 percent opposed it. But without an effective, relatively independent PSAC, scientists had to find other ways to influence policy. As had been true in the 1970s, elite physicists continued to reach out as individuals in an effort to influence policy. In testimony to the U.S. Senate Armed Services Committee and reiterated in a 1985 article in the *Journal of International Affairs,* Richard Garwin carefully laid out what he considered the technical flaws and political risks of ABM defense. Garwin acknowledged the optimistic assessments of the Fletcher report, but he warned that planning a defense against existing threats was not enough—Soviet researchers would undoubtedly adapt to whatever defensive system the United States developed, and therefore anticipating these adaptations would be key for an effective defense.[11]

Garwin explicitly cited three early studies critical of SDI, all released in 1984: a widely publicized and influential report by the UCS, an analysis by Ashton Carter for the Office of Technology Assessment, and a Stanford report coauthored by Sidney Drell. All three of these reports challenged the assumption that a perfect defensive system could be created. As Garwin summarized the common conclusion: "There is no significant prospect for reducing to a tolerable level the destruction to society which would result from the launching of a large fraction of the ten thousand strategic warheads now in the possession of the Soviet Union." In particular, Garwin challenged the notion of "layered defense"—the SDI plan to defend against each phase of an incoming missile, from launch to atmospheric reentry. Land-based lasers, he argued, would be insufficient to "see" and destroy a Soviet missile during its short boost phase, and even satellite-based lasers would be unlikely to intercept more than a "small number" of incoming weapons, if the lasers themselves could survive the likely development of antisatellite weapons. Even more worrisome to Garwin was the persistence of the decoy problem so critical during the ABM debates of the 1960s and 1970s. Soviet decoys could easily overwhelm attempts at "midcourse phase" defense. Should enemy warheads survive these flawed defensive systems and reach the terminal phase, Garwin warned, a thermonuclear weapon could still destroy a city even if it were detonated 30 kilometers above it in the atmosphere. Finally, Garwin observed that even if all the proposed defensive elements for each phase were to work perfectly, an enormously sophisticated computerized command-and-control system would have to function flawlessly as well, with zero software errors or other liabilities.[12]

Garwin married this technical critique to a powerful condemnation of the political fallout of even a limited Star Wars defense. "Even if effective defensive means should be found," he wrote, "the resulting deployment of defenses on both sides could create a totally *unstable* environment." Both sides might choose to attack each other's defensive capabilities, for example. And in the much more likely scenario, in which both the United States and the Soviet Union developed imperfect defensive systems, there would suddenly be "far more advantage in striking first." A far more responsible policy, in Garwin's view, would be to accept the role of nuclear weapons for deterrence, try to reduce any accidental onset of war, work to prevent "non-national groups and irrational leaders" from developing nuclear weapons, and negotiate with the Soviets to ban antisatellite technologies and space warfare.[13]

Garwin's warnings represented a continuity with the type of argument and activism typical of scientists' arms-control efforts during previous decades. But anti-SDI efforts extended far beyond earlier anti-ABM strategies. Two key examples, the 1987 APS report on directed-energy weapons and the massive SDI boycott effort, reveal new approaches to activism that owed at least a small debt to the controversies of the Vietnam years.

* * *

The report of the APS was not the first, nor the only study to attack SDI on technical grounds. The earlier SDI evaluation by Ashton Carter of the Office of Technology Assessment had concluded that "the prospect that emerging 'Star Wars' technologies, when further developed, will prove a perfect or near-perfect defense system, literally removing from the hands of the Soviet Union the ability to do socially mortal damage to the United States with nuclear weapons, is so remote that it should not serve as the basis of public expectation or national policy about ballistic missile defense." In the summer of 1987, a report from the Defense Science Board and a study from Livermore itself criticized SDI technology as deeply vulnerable.[14]

The significance of the APS report, however, lay in the prestige of the organization and the evolution of its approach to Cold War politics. The APS of the 1980s was no longer wholly dominated by the old guard of Frederick Seitz and other objectors to political activity. The New Left generation had come of age, and many older physicists who had rejected anti-Vietnam activism had softened their views. Years later, hydrogen-bomb pioneer Kenneth Ford recalled, "The late'60s and early 70s was the period of student turmoil. The Vietnam era. . . . Within the American Physical Society, the ramifications were a little bit delayed. They came in early, up to the mid, and even the late 1970s. They changed the nature of the organization from a pure research society, just publishing scientific journals and holding scientific meetings, into a much broader based society with committees on minorities, committees on women, committees on education, and a forum on physics in society." Throughout the 1970s the Forum on Physics and Society organized panel discussions, prepared accessible materials, connected issues of nuclear energy and arms control, and explored society-wide activities in the "public interest." In January 1973 the forum had even hosted "an impromptu hour-long session on the JASON question" consisting of an "informal discussion" between Marvin Goldberger, Scientists and Engineers for Social and Political Action representatives, and an audience of APS members.[15]

Over time the ethical commitments of the forum had extended to the larger APS body itself, particularly on the issue of nuclear weaponry. As an internal APS report explained, the organization had "been active for many years in educating physicists and the general public about the physics of arms control," but these "efforts have been deliberately low key to avoid association with any political side in the arms control debate." By the early 1980s, however, APS members had become increasingly worried about the growing arms buildup. "These concerns came to a head" in 1982, resulting in an official APS endorsement of the NAS's resolution on nuclear war and arms control. Then, at the height of the nuclear-freeze movement, acting according to "the concerns of the Society," the APS Council itself issued a strong, separate public statement in favor of arms control. Noting that "thirty years of vigorous research development have produced no serious prospect of effective defense against nuclear attack," and given that "the afteraffects of general nuclear war . . . could destroy civilization," the council urged the United States and the Soviet Union to pursue arms-control measures "without preconditions and with a sense of urgency." That fall, candidates for APS leadership positions were polled as to their thoughts on the council statement. As Sidney Drell put it, in a reflection of gradually changing views, the statement was appropriate because it was general; a more pointed public position on "specific political issues" should still be avoided.[16]

The evolution of political commitment within the APS, combined with the historic link between physicists and support for arms control, had thus positioned the organization to take a stronger stance in the SDI debate than it had during the Vietnam War. In the aftermath of Reagan's March 1983 speech—and a provocative arms-control panel convened at a fortieth- anniversary celebration of Los Alamos the following month—APS President Robert Marshak first proposed the idea of an unclassified APS arms-control study in order "to raise the level of public debate." The key topic would be directed-energy weapons, the space-based technologies that would harness the power of a nuclear explosion to target and destroy incoming nuclear warheads. The APS Executive Committee quickly convened a prestigious group of consultants to evaluate the feasibility and utility of such a study. Advisors included representatives from academia and industry; among those consulted, formally and informally, were Hans Bethe, Charles Townes, and Sidney Drell. All agreed that such a study, focusing on the "physical basis" and "technological feasibility" of directed-energy weapons and their applications to SDI, would be an appropriate and valuable contribution by the APS.[17]

Throughout the following year, APS organizers finalized plans for a Directed Energy Weapons (DEW) study group and sent funding requests to a range of government sources and private foundations. In order to approximate a kind of nonanonymous peer-review process, the report drafted by the study group would be sent to a six-member "review committee" composed of "unusually distinguished senior scientists" who had neither conducted explicit SDI research themselves nor belonged to organizations publicly opposed to SDI. By the spring of 1984, two laser experts, Harvard's Nicolaas Bloembergen and Bell Labs' Chandra Kumar, had been selected as cochairmen of the study group, and the Carnegie Corporation and MacArthur Foundation had pledged hundreds of thousands of dollars toward the cost of the study.[18]

Choosing the remaining group members proved trickier. While physicists employed by the government labs at Livermore, Sandia, and Los Alamos were likely to possess necessary expertise in cutting-edge weapons technologies, the APS leadership worried about stacking the group with too many insiders. Internal correspondence reveals calls for "good physicists who have taken no public position on the space weapons issue but are recognized as sympathetic with the loyal opposition" and, to offset the heavy weapons-lab representation, a "member who is viewed publicly as a responsible critic of SDI." Too much public opposition could be problematic, however. Chemist George Pimentel, who had unsuccessfully recruited George Kistiakowsky to join his antiwar activism during the Vietnam years, resigned his membership in the study group in order to oppose SDI more openly. The final roster of study group and review panel members consisted of a blend of academic and government scientists, heavily weighted toward older scientists with a long history of experience with Cold War physics. Of the twenty physicists, only one, Bruce Miller of Sandia National Laboratories, had been born after 1945. Review panelists included such luminaries as Charles Townes, William Panofsky, and Herbert York.[19]

After a seven-month delay due to difficulties ensuring declassification of the report, the APS officially released the DEW study at their annual meeting in the spring of 1987. Printed in *Reviews in Modern Physics,* the April 1987 "Science and Technology of Directed Energy Weapons: Report of the American Physical Society Study Group" offered a damning assessment of the current state of DEW technologies. At every phase of interception and for nearly every technology discussed, the study group found dangerous gaps between what existed and what was needed, with significant uncertainty as to whether such holes could be closed with additional research and development. Despite "substantial progress" in many areas, the study group wrote in a powerful executive summary, there was "insufficient information" to evaluate whether the necessary advancements "can or cannot be achieved." Rather, the physicists estimated that it could take "a decade or more of intensive research" just to determine whether directed-energy weapons even had the potential to form a reliable component of a defensive missile system.[20]

The APS DEW study articulated a deep skepticism of the feasibility of Reagan's vision, but as promised, the physicists kept their analysis restricted to technical evaluations stripped of explicitly political language. Nevertheless, the policy implications of their study were obvious. The APS conclusions jeopardized congressional funding for further research while painting the optimism of men such as Gerald Yonas as unwarranted and Pollyannaish. The message was not lost on Star Wars advocates. An SDIO comment on the draft report acknowledged that it was "an impressive compilation of unclassified source material," for which the APS had "astutely applied physical principles in their analysis," but SDIO reviewers objected to what they considered to be "unduly pessimistic" and "subjective" conclusions. Shortly after the study's release, a *Wall Street Journal* editorial acidly referred to it as "The Naysayers' Report."[21]

If SDI supporters immediately worried about the political ramifications of the DEW study, the actions of the APS Council made the implicit arms-control message explicit. In December 1985, as the DEW study was underway, Cornell physicist Jay Orear first proposed that the APS Council issue a public statement opposing SDI. Orear, a former chairman of the Federation of American Scientists, had had been a vocal supporter of Charles Schwartz's ill-fated 1968 campaign to encourage political statements by the APS membership. Now, Orear argued in a letter to council members, the American public had been presented with a "scientific hoax" in the form of Reagan's Star Wars program, and it was "important for [the President] to see that the organization which represents all the American physicists is strongly opposed to SDI-1 on technical grounds." He distinguished between two phases of Star Wars: the first, SDI-1, consisted of population defense that would render nuclear weapons obsolete, and the second, SDI-2, was essentially the resurrection of a 1970s-style ABM system. In his proposed text for the statement, Orear asserted that "no amount of effort and cost could provide a nuclear weapon defense of population so efficient and reliable that it would make nuclear war obsolete."[22]

Orear's proposal sparked another internal debate within the APS, but one far more muted and less vitriolic than its Vietnam-era counterpart. Orear's colleagues questioned the wording and timing of his proposed statement, voicing concerns that he had oversimplified matters and that issuing a political statement before the release of the DEW study might jeopardize its impact and suggest that "the Physical Society had made up its mind before it undertook the study." Even the APS Forum's executive committee expressed skepticism. As hydrogen-bomb veteran Kenneth Ford acknowledged, "Most in this small group probably share my view that SDI is ill-conceived, dangerous, and wasteful." But to Ford, "actions by individuals (guest editorials, articles, letters to the editor, briefings) are the preferred course, rather than resolutions by societies."[23]

Orear declared himself open to suggestions regarding the timing and text but defended his proposed statement from deeper critiques by characterizing it as technical rather than political. "I am uncomfortable with the argument that the Council should not act just because some people may label a non-political statement by the Council as political," he wrote to Robert Park. "In my mind it is a scientific fact that SDI-1 cannot make nuclear weapons obsolete." But he also intimated that the DEW study itself had been borne out of explicit arms-control hopes. "I am not as optimistic as you seem to be that the APS DEW study will help kill SDI-2," he wrote. His words suggested an unspoken agenda behind the study as well as his own belief that a lengthy technical report might not be enough to halt SDI; in his view, a strong statement by the APS Council would hammer home the technical case against the program.[24]

Despite some reservations within the APS, Orear successfully persuaded the council to consider the issuance of a statement. By April 1986 the council had appointed a drafting committee, headed by Case Western Reserve physicist Thomas Moss. Two proposed versions emphasized deep skepticism about the feasibility of a reliable "perfect missile defense system" and warned of the cost. One draft added a call for greater arms reduction: "What the American Physical Society can strongly support is the Administration's policy of seeking to negotiate deep reductions in nuclear weapons. That is clearly a timely and technologically valid direction in which to seek security from the nuclear threat." APS physicists also debated holding a special videoconference to explain the conclusions of the DEW study to all physicist members, who could then spread its message with maximal impact. Members of the study group ultimately rejected this last proposal, protesting that it would be "unseemly" and that verbal explanations and presentations were preferable.[25]

On April 24 the APS Council followed through on Jay Orear's proposal and approved a public statement on SDI. The vote was not unanimous. Council members jettisoned earlier draft language concerning arms control and opted to refer explicitly to the conclusions of the DEW study, citing the ten years of additional research needed to evaluate whether directed-energy weapons might even be useful for strategic defense. Given the high stakes of nuclear warfare ("human suffering and death far beyond that ever before seen on this planet") and the enormous monetary costs necessary for just the preliminary research, the United States ought not to commit to the deployment of SDI, they wrote. Their assessment was consistent with decades of scientists' arms-control rhetoric: it was a technical argument drawing on moral concerns. In an effort to head off criticism, the council defended their issuance of the public statement by asserting that "the Council of the American Physical Society believes that it has a public responsibility to express concerns about the Strategic Defense Initiative that go beyond issues of DEW covered in the Study."[26]

By attaching the executive summary of the DEW report to their public statement, however, the council inextricably linked their explicit policy proscriptions to the more limited technical assessments of the study itself. The move rankled some APS physicists, including DEW study cochair Nicolaas Bloembergen and six other study group members, who wrote an open letter to President Val Fitch to protest "being included in the council's statements on matters neither we nor they studied." West Point physicist and study group member Thomas Johnson complained about the lack of organization-wide input, noting pointedly: "If the APS wishes to enter the arena of public policy formation, it should find out what the membership actually wants to say. If APS Council members, themselves distinguished scientists and citizens, wish to make statements on policy matters, they are certainly free to do so over their own signatures, individually or collectively."[27]

In his response, printed in *Physics Today,* Fitch acknowledged that "it might have been better" not to restate the DEW conclusions along with the council statement, and the timing might have "made it easier" for people "intent on discrediting a report for their own purposes." But he downplayed the impact of the statement, arguing that it had been largely ignored by the press. It was the DEW study report that had sparked debate and attention, not the council's call for a halt on SDI deployment.[28]

But to the report's main critics, the issuance of the council statement was valuable ammunition in the battle to discredit the APS and characterize the physics community as politicized and divided, even over the issue of technical feasibility. The month after the report's release, former APS and NAS president Frederick Seitz and Livermore physicist Lowell Wood met with Republican members of Congress to pick apart the APS findings. Both men were members of a tiny coterie of elite scientists who reacted to the tumult of the Vietnam years by embracing the conservatism of Ronald Reagan. Wood, an astrophysicist and pioneer in the applications of laser and computer technology to missile defense, had been profiled two years earlier in a book-length portrait of Livermore employees written by *New York Times* science journalist William Broad. Broad described Wood's shift to the political right during the 1960s as stemming in part from disgust at Berkeley activists' attacks on his nuclear-weapons research. But by the early 1980s, Wood himself reflected a kind of mirror image of New Left critiques of science, evincing a skepticism of objectivity and neutrality that could be used to attack other scientists when their claims proved politically unpalatable. "I would be very surprised if very many scientific endeavors, maybe even minor ones, happen because a disinterested scientist coolly and dispassionately grinds away in his lab," he told Broad. Scientists were motivated by "competition, peer esteem, his wife and family, prizes and recognition . . . Pushing back the frontiers of knowledge and advancing truth are distinctly secondary considerations."[29]

Even more than Lowell Wood, Frederick Seitz embodied this new, conservative attack on scientific objectivity. Seitz was a Manhattan Project veteran and former PSAC advisor who found his staunch anticommunism and hawkish views on Vietnam increasingly out of place among his liberal arms-control colleagues. In 1979 he retired from academic research and spent the next decades defending RJ Reynolds from tobacco regulation, advocating expanded nuclear-weapons development, and attacking environmental regulation. In 1984 he joined with former NASA astrophysicist Robert Jastrow and former Jason physicist William Nierenberg to found the George C. Marshall Institute, a think tank devoted to providing conservative causes with scientific backing. In the late 1990s, he would publish a controversial attack on the legitimacy of climate change that resulted in explicit condemnation from the NAS, of which he had formerly been president. Science historians Naomi Oreskes and Erik Conway later labeled Seitz a "merchant of doubt," arguing that Seitz had learned from his experience at RJ Reynolds that the introduction of opposing scientific viewpoints—no matter how lopsided the actual scientific consensus—could be used to stymie unwanted reforms and regulation. They observed that for Seitz, "The tobacco road would lead through Star Wars nuclear winter, acid rain, and the ozone hole, all the way to global warming."[30]

Now both Lowell Wood and Frederick Seitz stood before Republican allies in Congress, eager to discredit their SDI opponents. Wood presented a review of the DEW study he had prepared with Gregory Canavan of Los Alamos. Both scientists attacked the APS report as riddled with errors, unreasonably pessimistic, and most damningly, heavily politicized. In an adaptation of his presentation reprinted in the *Wall Street Journal*, Seitz referenced the "extremely negative statement by the APS council" and wrote that "I cannot help but conclude that these actions . . . represent a political as well as scientific declaration." To the Republicans in Congress, however, Seitz reportedly went further. He likened the APS study group to the collaborating scientists under Hitler and Stalin. "Physicists with long memories will recall that when the Nazis came into power in Germany in the 1930s, the German physics journals—which had been, until then, among the finest in the world, as the journals of the American Physical Society are today—began publishing work of questionable quality," he observed. "One also thinks of the unfortunate depths to which some Soviet scientists, especially in genetics, have descended at various times to satisfy the prevailing political thought in the Soviet Union."[31]

Seitz's criticisms outraged the APS leadership. It was not the APS that had been politicized and corrupted, they argued, but Seitz himself, who, as chairman of the Pentagon's SDI Advisory Committee, had proposed the creation of a federally funded SDI think tank for which he would serve as a key manager. His vested interest in the approval of an expanded research program was obvious. The same charge could be leveled at the report prepared by Wood and Canavan, which members of the APS study group characterized as a "polemical statement" rather than a useful review. With the exception of one accurately flagged error, they wrote in a pointed response, the Livermore critique was "incorrect" on "all other technical points of disagreement." Study group cochair Kumar Patel wrote to Val Fitch that the Livermore authors had fallen prey to "extravagant assumptions of the performance of unproven technologies." Such assumptions were surely a self-serving tendency for employees of government weapons labs.[32]

Val Fitch reflected on this problem in a letter to the *Wall Street Journal*. "The Pentagon is, by and large, a consumer of technology; the APS committee represents the originators of technology," he wrote. "Therefore, the committee is in a position to render totally independent judgments." It was an awkward bridging of the polarized arguments from the 1960s: institutional interests must be acknowledged, particularly in the case of the Pentagon, but even so, the APS still represented a neutral, "outsider" voice. "The report is as objective as humanly possible," Fitch explained to skeptical Republican Congressman Curt Weldon. If Weldon sought an example of a "highly charged and emotional" scientist, he should look no further than Seitz himself, whose "bizarre attack" on the APS physicists "has no place in a technical discussion." He reiterated the argument to Rep. John Spratt, emphasizing the impartiality, professionalism, and sacrifice of the report's authors: "These men undertook the heavy burden of this study out of a sense of patriotism and professional responsibility," he wrote. "Their only reward is the knowledge that they have provided an objective basis for informed policy decisions. None stands to gain personally in the slightest from those decisions." Members of the study panel had been chosen based on their expertise and concern for "a balance of conflicting points of view." No one had officially taken a political position on SDI. The same was true for the six-member Review Committee tasked with overseeing the preparation of the report and making a recommendation to the council regarding its acceptance. What men like Lowell Wood had interpreted as opposition was merely a healthy skepticism. "A group of respected American physicists who harbored no reservations about the feasibility of a technological enterprise of this complexity would be difficult to find," Fitch wrote. "It is the first responsibility of a scientist to be skeptical." The APS Council may have extrapolated from the DEW report to issue their recommendation against deployment, which was reasonable and proper, but the DEW study itself was purely technical: "[It] is not an indictment of SDI and its authors are not opponents of SDI."[33]

While high-profile individuals such as Richard Garwin and prestigious organizations such as the APS presented technical and political arguments against SDI, another parallel scientists' movement was also coalescing. In addition to the production of highly critical studies, thousands of scientists demonstrated their opposition to SDI through open petitions and letters and a highly publicized campaign to boycott SDI-related research contracts. By 1984 over fifteen thousand physicists from forty-four countries had signed a nuclear-freeze statement; as *Physics Today* noted proudly, the list of signatories included more than half of the living Nobel Prize winners. The UCS published books and pamphlets arguing the case against Star Wars. And beginning in May 1985, science professors and researchers at Cornell and the University of Illinois began circulating a pledge to boycott SDI research. Within six months, University of Illinois physicist John Kogut was reporting that "at least 54% of the faculty at the nation's top 14 physics departments have pledged to reject [SDI] funds, and the number is rapidly growing." At Princeton, Cornell, Berkeley, the California Institute of Technology (Caltech), Columbia, Carnegie Mellon, Northeastern, and numerous other universities, 50 percent or more of the physics faculty had signed the pledge. As a Committee for a Sane Nuclear Policy press release reported, many of these anti-SDI scientists were researchers who "ordinarily accept other types of military funding." Their opposition to Star Wars was rooted not simply in skepticism about the status of the necessary technology, but in the conviction that SDI "would be destabilizing and dangerous."[34]

A closer examination of both the text of the boycott pledge and the views of its signatories confirms the inclusiveness and intergenerational nature of the movement. Early drafts of the boycott pledge at Cornell emphasized SDI's threat to the institutional neutrality of the university, noting that "participation in SDI by individual Cornell researchers would lend Cornell's name to a program of dubious scientific validity, and give legitimacy to this program at a time when the involvement of prestigious research institutions is being sought to increase Congressional support." A simultaneous effort at the University of Illinois yielded rhetoric containing both moral and technological claims. Pledge language characterized SDI as "technically dubious and politically unwise" and "likely to trigger a nuclear holocaust." When the Cornell and Illinois activists united to produce a joint statement suitable for distribution at universities across the nation, they merged their statements to produce a document that was maximally inclusive, citing the technical problems of feasibility, the moral language of "nuclear holocaust," and the implications for academic freedom and neutrality. It included the Cornell warning that even the participation of individual employees would "lend their institution's name" and boost the "legitimacy" of the SDI program. The actions of individuals in an institutional framework mattered; as the text warned, "Researchers who oppose the SDI program, yet choose to participate in it should therefore recognize that their participation would contribute to political acceptance of SDI."[35]

Over six thousand science and engineering professors, researchers, and graduate students signed the pledge, representing more than one hundred colleges and universities. In a survey of physicist signatories, in which respondents could choose multiple reasons for endorsing the pledge, 40.4 percent identified technical feasibility as a key objection; 37.4 percent said, "SDI would be a destabilizing influence on world peace," and 15.2 percent said their primary objection stemmed from "an absolutist ethical position against conducting research with potential military applications." Older scientists were more likely to choose technical reasons; younger scientists were more likely to choose political or moral reasons. The survey authors noted that "physicists who received their degrees before the dropping of the atomic bomb were most likely to be motivated by technical issues . . . while those with degrees completed after the defeat of the ABM weapons system were least motivated by such concerns . . . Such a finding may reflect the kind of political climate found on many university campuses from 1967 on as a result of anti-Vietnam war activism."[36]

＊ ＊ ＊

John Bardeen observed in the fall of 1985 that "there are few scientists either within or without the administration" who believed Star Wars was "feasible," but "the national laboratories (Los Alamos and Livermore) would have an exciting new project with unlimited funding. The aerospace industry would have a lucrative new program. So, in spite of the doubts, there is no lack of supporters." Major SDI contractors included Livermore, General Motors, General Electric, Lockheed, Los Alamos, MIT's Lincoln Labs, and a host of others. The SDIO formed consortiums of universities and corporate contractors to carry out multiple areas of research, reaching out to elite schools like Caltech and MIT as well as Auburn University in Alabama, SUNY Buffalo, Texas Tech, Georgia Tech, Utah State, and Purdue.[37]

Given the recessionary economics of the 1970s and early 1980s, the issue of defense contracting in the age of Star Wars was a difficult one. Hugh Gusterson has estimated that during the 1980s, 20 to 40 percent of all U.S. scientists and engineers were working on military projects, generally speaking. Los Alamos and Livermore alone employed 6 percent of all U.S. physicists. Though these figures demonstrate that the majority of scientists were not directly working on nuclear weapons or Star Wars, Pentagon funding was still ubiquitous, and the array of SDI-contributing technologies was wide: for example, the computing necessary for remote command-and-control systems or the signal processing required for over-the-horizon radar and other techniques were both popular areas of study along Massachusetts's Route 128. A young Livermore scientist told Gusterson, "You realize how big the military-industrial complex is when you graduate. If you get a degree in physics, there's almost nowhere to get a job where you're not part of the military-industrial complex. Even the universities are getting drawn in. It's too big." Daniel Kevles offered a similar anecdote in his introduction to an updated edition of *The Physicists*, noting that by the 1980s, the pervasiveness of defense funding and influence was such that one recent physics PhD lamented that it seemed "all roads lead to the Pentagon."[38]

The ethical strangeness of the Star Wars debates was such that few scientists believed it would work, yet many simply accepted the available funding and went about their research. It was perhaps an echo of the scenario described by Lincoln Lab employee Paul Easton to the Pounds Panel at MIT: he and his colleagues laughed at the politicians who believed an ABM system was feasible, even as they conducted the relevant research requested of them.

But what is fascinating about the anti-SDI efforts of the 1980s is just the opposite: the large number of scientists and engineers who did not remain silent in their opposition, including not just the councils of academic societies or tenured physics professors but also weapons engineers themselves. Despite the changes in the geography of defense research, despite the rise of new technocratic classes of weapons engineers, and despite the appeal of fabulous funding opportunities, weapons workers were mobilizing in unprecedented ways. This development reflects the culmination of multiple Cold War trends in ethical thinking among scientists: the affinity of many scientists, especially physicists, for arms control; an evolving view of appropriate political behavior by scientists, including among high-level government advisors and laboratory employees; and a heightened sensitivity to the politics of institutions and contracts in the military-industrial-academic complex, even as defense contracting reached new highs during the Reagan boom.

* * *

Almost immediately after Reagan's 1983 speech, skeptical assessments circulated through the nation's weapons laboratories. In 1984 Michael May, an associate director at Livermore, wrote that "the main facts are as follows: a leak-proof or near leak-proof defense against ballistic missiles based in any concept we know about, is not possible." As early as 1984 and 1985, physicists at Los Alamos and Livermore squabbled over errors in calculations connected to the development of an X-ray laser, with an internal report by a Livermore scientist exposing design flaws and other insiders alleging that tests were going ahead due to political concerns rather than technical readiness. In September 1985 a team of visiting Jasons investigated the conflict and sided with the X-ray design critics. Though Livermore officials declined to comment specifically on the outcome of the weapon's underground test, a report in *Science* alleged that it had been a bust. Science reporters began covering the rising factions within Livermore, particularly those who favored ambitious but untested new defensive systems and the more cautious skeptics who demanded more rigorous assessments. The standoff was embodied by the tense relations between Teller, his protégé Lowell Wood, and Roy Woodruff, formerly the lab's associate director.[39]

Roy Woodruff had joined the lab in 1968 and by 1982 had been put in charge of nuclear weapons development. He held a cautiously optimistic attitude toward Teller's space-based directed-energy weapons; though he was skeptical about many technical aspects, he embraced the possibility that such a system could be designed and developed. But as Teller began to promote his vision in grander and less plausible terms, Woodruff grew concerned. He objected when Teller repeatedly offered premature estimates for the X-ray laser's completion to government officials. When early plans for the weapons design fizzled out, Teller quickly proposed a new, super device that could direct multiple intense beams to enhance its destructive capabilities. Woodruff was shocked; to him, Teller was championing a "silver bullet," a magical solution to the problems at hand. It was, he later told the *Los Angeles Times,* a "notional weapon." But when Woodruff expressed his doubts, he was met with the relentless optimism of the young researchers loyal to Teller and Lowell Wood: "You can't prove it's impossible."[40]

When Teller went so far as to cite overly optimistic assessments of the improved X-ray laser to try to halt arms talks with the Soviet Union, Woodruff reached his breaking point. He lodged an internal complaint to the lab's director and, when no changes were forthcoming, he resigned his administrative post in late October 1985. He had been "left with no other ethical choice," he later wrote. As he continued to work in other Livermore divisions, the actions of Teller enraged him. "Seeing someone like Teller use his prestige and reputation to gain access to people and make statements that, to a non-technical person in particular, were misleading is tough," he later explained. In the spring of 1987, Woodruff decided to submit a grievance to the president of the University of California alleging that Teller and Wood "conveyed both orally and in writing overly optimistic, technically incorrect statements regarding this research to the nation's highest policy makers."[41]

Woodruff's skepticism lived on at Livermore despite his demotion. In July 1987 a team of scientists at Livermore issued a deeply skeptical report on the limited value of "kinetic kill weapons," the planned missile-defense components championed by Secretary of Defense Caspar Weinberger as an early means to destroy incoming missiles by impact, rather than chemically or via directed-energy blasts. Due to improvements in Soviet missile technologies, such weapons were unlikely "to provide any real protection against future threats," a top Livermore official told William Broad of the *New York Times.*[42]

That fall, an anonymous source leaked copies of Woodruff's letter to a member of the Southern California Federation of Scientists, which promptly held a press conference and released the letter to the *Los Angeles Times*. A congressional inquiry quickly followed, along with Woodruff's embrace by nuclear-freeze and arms-control activists. In his 1988 *Los Angeles Times* profile, science reporter Robert Scheer portrayed Woodruff as a reluctant whistleblower, a conservative Cold Warrior "not fully comfortable in the company of left-leaning congressmen and academicians who have rallied to [his] cause. . . ."[43]

Uncomfortable or not, Woodruff had reopened new debates about the public roles of scientists. During a visit to Livermore in July 1988, Energy Secretary John Herrington complained about the controversy, explaining: "I think there should be freedom of expression within the laboratory, but I don't favor scientists going public on opposite sides of the issue if it's going to be damaging to the laboratory." He directed his comments pointedly at Woodruff, not Teller or Wood, whose actions he did not find to be so damaging. Instead, he lauded Teller as a trusted "national asset," whereas Woodruff was now being "used" by enemies of the SDI program. But Woodruff responded, in an interview with the *Los Angeles Times*, that it was the responsibility of scientists to do just that: provide the accurate technical information that would be "used" by political decision makers. If anything, he was frustrated at the lack of protections for "academic freedom" provided by the University of California. Though the university had substantiated his claims of unprofessional treatment, administrators had done little to support him publicly. In 1990 he left Livermore for a job advising Los Alamos's arms-control verification program. Three years later the APS Forum on Physics and Society presented him with the Szilard Award "for courageous efforts to provide the government with reliable and objective science advice on critical issues affecting national security and arms control policy."[44]

Woodruff was not the only insider critic suddenly attracting the attention of the national media. In June 1985 computer scientist David Parnas publicly resigned from an SDIO software panel, complaining that the copious funding and engaging "technological puzzles" offered by SDI research encouraged eager scientists to ignore the reality that Star Wars was technically impossible. He recounted pressure from a panel organizer to avoid negative assessments. In a later interview with historian Rebecca Slayton, he described SDIO as a "corrupt" organization. "I found few who believed that SDI would perform as advertised," he recalled. "The vast majority were in it to get the [research] money and most of them said so in private."[45]

* * *

Even in the face of enormously appealing funding opportunities, scientists led an organized, multipronged, successful attack on the deployment of a strategic defense system. By 1988 the APS's harsh report had been published and well publicized, and the American Mathematical Society had voted, in the words of a UCS report, "not to participate in any activities that could be interpreted as supporting the SDI." The professional organizations that in 1969 had fought tooth and nail to avoid public political statements regarding Vietnam now leapt headfirst into the political controversies of Star Wars. This shift in approach was possible because academia and professional societies understood themselves in different ways after the Vietnam years.[46]

The nature of the SDI debate appealed to the physicists who had balked at earlier moments of science activism, including the Vietnam War and the ABM controversy. During the 1970s changes in science advising, education, and defense contracting had constituted a collapse of the old Sputnik order, in which a generation of elite scientists had provided influential expertise and policy guidance to political leaders. By the 1980s many of these former PSAC members looked back on the 1970s with newfound perspective. Hans Bethe reflected that in retrospect, the ABM debate seemed more technical and transparent as opposed to the "secrecy" and "propaganda" of Star Wars. Despite disagreements with presidents Johnson and Nixon, scientists had still held meaningful government advisory posts through which they could review and influence policy. As UCS leader Henry Kendall later observed to a *New York Times* reporter, "[In the 1960s] the controversy was more over pure arms control issues. But now there is much evidence that there is a crusade going on against the Soviet Union. Many scientists feel more estranged from Government, because there is less appreciation for technical criticisms." This estrangement proved catalyzing not just for elite former science advisors, but for weapons engineers who valued technical rigor and who understood SDI to be a threat to long-accepted strategies of mutually assured destruction.[47]

* * *

The inclusiveness and variety of the anti-Star Wars movement signified generational reconciliation after Vietnam and the prevalence of new modes of ethical thinking among scientists, but it did not resolve the problems of academic and institutional neutrality so memorably raised during the 1960s. The outrage of Paul Gray and Marvin Goldberger over attempts to associate MIT and Caltech with SDI reflected their conviction that universities were supposed to travel a separate, pure path—and any attempt by Defense Department officials to tap the prestige of academic institutions constituted abuse. Meanwhile, professional organizations such as the APS and the American Mathematical Society claimed both objectivity and political obligation simultaneously, drawing on their own prestige to boost the case against the SDI program.[48]

The handful of scientists who endorsed SDI picked up on these inconsistencies, revealing a cynical application of the New Left arguments originally intended to force scientists to confront their complicity in the Vietnam War. If neutrality and objectivity were impossible, then what mattered was perception. Conservative supporters of Star Wars, such as Teller and Seitz, played the game both ways. By stoking public perceptions of APS scientists as politicized agents, they could discredit physicists' claims regarding the technical feasibility of Star Wars. When it benefited them, however, they tried to draw on the prestige of academic powerhouses like MIT and Caltech or the respect accorded to Nobelists and government officials. Admittedly, these latter attempts often backfired during the SDI debates. Elite scientists were more attuned to the risks of political exploitation after the chastening lessons of the Vietnam years. When John Bardeen resigned from Reagan's White House Science Council in the aftermath of the SDI announcement, he complained that "he was being used for his name" and resented the implication that he supported the new program. Nevertheless, conservatives had internalized a valuable lesson that could be applied to debates over climate change, environmentalism, and a host of other issues in which regulation was at stake and politicians invited scientific perspective. By rejecting the objectivity of scientists, they could attack the legitimacy of consensus and scientific authority. With political connections and mass media access, any scientific statement could be publicly doubted, and any scientific conclusion with policy implications could be recast as a debate.[49]

Recently, science historians and journalists have demonstrated just how effective the actions of these "merchants of doubt" have been, particularly on the topic of climate change. But the SDI debates of the 1980s offer a reminder of the difference between political rhetoric and scientific content. Funding for SDI-related technologies has continued even after the collapse of the Soviet Union and the end of the Cold War, yielding many promising civilian spin-offs, but the promise of the directed-energy weaponry lauded by Teller and others has not yet been realized. Skeptical physicists within the APS accurately predicted the technical impediments to the space-based components of missile defense, and no amount of posturing by Star Wars champions has been able to overcome this material reality.[50]

Epilogue

Science and Ethics after the Cold War

IN THE SPRING OF 1964, New Mexico senator Clinton Anderson, chair of the Committee on Aeronautical and Space Sciences, gave a revealing interview to a reporter from *International Science and Technology*. He acknowledged that his interest in nuclear-weapons research stemmed mainly from the fact that "it was in my state." When asked how he made legislative decisions about science and technology, he explained that he tended to follow his gut instincts about the personalities of the experts who presented their views to him rather than any deeper knowledge of the relevant topic. He had a good feeling about Los Alamos head Norris Bradbury, whose assessments he would accept "even if I don't know the scientific reason for what he's doing, or couldn't understand the scientific reason if I knew it." Senator Anderson acknowledged the trickiness of his situation. "I do think we pick individuals by use of a queer yardstick," he said. "We pick the ones we trust."[1]

Scientists in the twentieth century may have been continually frustrated at the limits of their influence, but it is also true that government is still dependent on the expertise of scientists and engineers. In his memoir, hydrogen-bomb mathematician Stanislaw Ulam warned of the risks of leaving crucial decisions about weapons science in the hands of "well-meaning but technically uninformed politicians." Ulam's assessment informed his career choices: his professional work stretched from academia to the Pentagon to the research labs of Los Alamos.[2]

This book has provided at least a partial history of how scientists and researchers like Ulam tried to cope with the ethical burdens of expertise. From the postwar embrace of government advisory roles through the contentious years of the Vietnam War to the coalition opposing the Strategic Defense Initiative, scientists have struggled to exercise control over the policy implications of their work. In the era of the Sputnik boom, the implicit trust described by Senator Anderson seemed to run both ways. By the Reagan years, it had dissolved.

The clash between insider and outsider scientists during the Vietnam War was an obvious turning point, as epitomized by the experience of the Jasons. In 1974 the British physicist E. H. S. Burhop wrote in the *Bulletin of the Atomic Scientists* that "the 'Jason group' has often—rightly or wrongly—been taken to symbolize social irresponsibility among scientists and an attitude of callous disregard for the consequences of their work." The Jasons' own correspondence reveals that, far from disregard, many members thought deeply about the consequences of their work but found that they could not control the applications of their own research in the ways that they had expected. Echoing the fears of Norbert Wiener at the end of World War II, they felt their work had been *misused*. George Kistiakowsky had sought—and believed he had received—assurances from Robert Mc-Namara that the electronic barrier would be a tool for de-escalating the war. When the war expanded instead, Kistiakowsky removed himself from his advisory position.[3]

Burhop's essay on the Jasons was written on behalf of the World Federation of Scientific Workers, which asserted that "all scientists have the obligation to examine the likely consequences of their scientific work and, as far as lies in their power, to prevent its use for evil anti-social and destructive ends." Burhop had solicited reflections from the roster of Vietnam-era Jasons and quoted several, anonymously, in the *Bulletin*. The Jasons described feeling misused, ignored, disillusioned, patriotic, regretful, and unapologetic. Some wondered whether it was more effective to work inside or outside "the establishment." But Burhop zeroed in on one response in particular, which he felt identified "the most important ethical problem, namely, the purposes which the scientific advice given to one's government will serve." Burhop quoted the anonymous Jason: "When and where should one draw the line? Designing gas chambers for Auschwitz is clearly over the line, working on the Manhattan project in the same period, clearly (to me at least) not. What about today? Almost all of American science is supported in one way or another by the U.S. government, in many cases directly through the Defense Department. . . . To what extent should one stop cooperating with a government when one disagrees with its policies?"[4]

That final question was the same one asked, in multiple forms, by the students and faculty on the campuses of MIT, Princeton, and countless other universities during the years of the Vietnam War and by Charles Schwartz and his colleagues in SESPA and the American Physical Society. They argued from a deep conviction that not just individuals, but institutions—including those of the U.S. government—must be judged as consequential moral agents. In the 1980s vast numbers of American scientists answered this question in their refusal to accept SDI contracts. They endorsed Hans Bethe's side of the debate over that of Edward Teller. But thousands of other scientists and engineers stood up to receive those same contracts.

This book has responded to Burhop's question by providing a history of scientists' moral reasoning and charting the trajectories of competing ethical arguments, from Sputnik idealism through Vietnam disillusionment, culminating in the broad activism against the Strategic Defense Initiative. But on a deeper level, Burhop's question has no simple answer. The legacy of Cold War science policy and ethics is complex. Our current debates over weapons development, climate change, and network security have engaged scientist activists from multiple ethical vantage points, and across national boundaries, with ambiguous results. Massive government and corporate sponsorship continue to influence the nature and uses of scientific research.[5]

At the same time, the battles of the 1960s and 1970s have contributed to a darker legacy. Many of the major antiscience movements of the early twenty-first century build upon the deep skepticism of the independence and impartiality of science that informed so many Vietnam War-era debates. Climate change deniers accuse scientists of tailoring their conclusions to secure research funding while the antivaccine lobby remains convinced that pharmaceutical companies have suppressed research demonstrating a link between vaccines and autism. In both cases nonexperts have attacked scientific consensus through criticism of the conditions under which scientific knowledge is created and disseminated.

The travails of Cold War scientists reveal the challenge of reconciling a keen awareness of the social forces shaping science with a respect for the value of scientific knowledge and consensus. Responsible policy decisions require both. Scientists cannot expect to control how their work is interpreted and applied by others, but they can and should scrutinize the context of their research and its potential uses, taking action when necessary, working within and without government. Policymakers must also take seriously the warnings and concerns of experts. Most importantly, the maintenance of a just, safe, and humane world depends upon the actions of a well-informed and scientifically literate public. In this quest, our political choices and educational priorities matter greatly. So do the ethics and expertise of scientists.

Notes

PROLOGUE

1. Robert Park, interview by Patrick McCray, 4 March 2001 and 7 March 2001, Niels Bohr Library and Archives, American Institute of Physics, College Park, MD, http://www.aip.org/history/ohilist/4939.html.
2. Victor Weisskopf, *The Joy of Insight: Passions of a Physicist* (New York: Basic, 1991), 121, 128. For a sampling of histories of the Manhattan Project and scientists' experiences, see Jeff Hughes, *The Manhattan Project: Big Science and the Birth of the Atomic Bomb* (New York: Columbia University Press, 2002); Cynthia Kelly, *The Manhattan Project: The Birth of the Atomic Bomb in the Words of Its Creators, Eyewitnesses, and Historians* (New York: Black Dog & Leventhal, 2007); Mary Palevsky, *Atomic Fragments: A Daughter's Questions* (Oakland: University of California Press, 2000); and Sylvan Schweber, *In the Shadow of the Bomb: Oppenheimer, Bethe, and the Moral Responsibility of the Scientist* (Princeton, NJ: Princeton University Press, 2006).
3. Weisskopf, *Joy of Insight,* 147.
4. Lawrence Badash, "American Physicists, Nuclear Weapons in World War II, and Social Responsibility," *Physics in Perspective* 7 (2005): 143–147; Alice Kimball Smith, *A Peril and a Hope: The Scientists' Movement in America, 1945–1947* (Cambridge, MA: MIT Press, 1970). See also Paul Boyer, *By the Bomb's Early Light: American Thought and Culture at the Dawn of the Atomic Age* (New York: Pantheon, 1985).
5. Norbert Wiener, quoted in Peter Galison, "The Ontology of the Enemy: Norbert Wiener and the Cybernetic Vision," *Critical Inquiry* 21, no. 1 (1994): 253–254.

6. See Schweber, *Shadow;* Kelly Moore, *Disrupting Science: Social Movements, American Scientists, and the Politics of the Military, 1945–1975* (Princeton, NJ: Princeton University Press, 2008); and Matthew Wisnioski, *Engineers for Change: Competing Visions of Technology in 1960s America* (Cambridge, MA: MIT Press, 2012).

7. Moore, *Disrupting Science,* 55–95.

8. Eugene P. Wigner and Andrew Szanton, *The Recollections of Eugene P. Wigner, as Told to Andrew Szanton* (New York: Plenum, 1992), 262; Stanislaw Ulam, *Adventures of a Mathematician* (New York: Scribner, 1976); and Palevsky, *Atomic Fragments,* 53, 72. .

9. See, for example, responses by Norman Austern, Louis A. Beach, William Higinbotham, Lewis Slack, R. F. Taschek, Stephen R. White, and Lawrence Wilets in responses to a 1981 History of Nuclear Physics Survey, Niels Bohr Library and Archives American Institute of Physics, College Park, MD, 1981.

10. "Napalm Inventor Honored at Harvard," n.d., ca. 1967, "Retirement dinner, 1967," Papers of Louis F. Fieser and Mary P. Fieser, HUG (FP) 20, box 1, Harvard University Archives.

11. Influential works include biographical and political histories such as Kai Bird and Martin Sherwin, *American Prometheus: The Triumph and Tragedy of J. Robert Oppenheimer* (New York: Random House, 2005); Richard Rhodes, *Dark Sun: The Making of the Hydrogen Bomb* (New York: Simon & Schuster, 1996); and Schweber, *Shadow;* institutional approaches such as those collected in Peter Galison and Bruce Hevly, eds., *Big Science: The Growth of Large-Scale Research* (Stanford: Stanford University Press, 1992); Stuart Leslie, *The Cold War and American Science: The Military-Industrial-Academic Complex at MIT and Stanford* (New York: Columbia University Press, 1993); Sheila Slaughter, *The Higher Learning and High Technology: Dynamics of Higher Education Policy Formation* (Albany: SUNY Press, 1990); and Jessica Wang, *American Science in an Age of Anxiety: Scientists, Anticommunism, and the Cold War* (Chapel Hill: University of North Carolina Press, 1999); analyses of physicists, popular culture, and nuclear weapons development such as Boyer, *Bomb's Early Light* and Spencer Weart, *The Rise of Nuclear Fear* (Cambridge, MA: Harvard University Press, 2012); and anthropological and sociological studies, including Hugh Gusterson, *Nuclear Rites: A Weapons Laboratory at the End of the Cold War* (Oakland: University of California Press, 1996); and Moore, *Disrupting Science.* For works dealing explicitly with the 1960s and 1970s, see, for example, David Kaiser, *How the Hippies Saved Physics* (New York: W. W. Norton, 2011); Moore, *Disrupting Science;* Joy Rohde, *Armed with Expertise: The Militarization of American Social Research*

during the Cold War (Ithaca, NY: Cornell University Press, 2013); Fred Turner, *From Counterculture to Cyberculture: Stewart Brand, the Whole Earth Network, and the Rise of Digital Utopianism* (Chicago: University of Chicago Press, 2006); Wisnioski, *Engineers for Change;* and David Zierler, *Inventing Ecocide: Agent Orange, Vietnam, and the Scientists Who Changed the Way We Think about the Environment* (Athens: University of Georgia Press, 2011). Extending the time frame to include computer scientists and the debate over Star Wars is Rebecca Slayton's *Arguments That Count: Physics, Computing, and Missile Defense, 1949–2012* (Cambridge, MA: MIT Press, 2013).

12. For example, Paul Rubinson, "'Crucified on a Cross of Atoms': Scientists, Politics, and the Test Ban Treaty," *Diplomatic History* 35 (April 2011): 283–319.

13. Gregg Herken, *Cardinal Choices: Presidential Science Advising from the Atomic Bomb to SDI* (New York: Oxford University Press, 1992); Zuoyue Wang, *In Sputnik's Shadow: The President's Science Advisory Committee and Cold War America* (New Brunswick, NJ: Rutgers University Press, 2008).

14. This discussion builds on the work of Leslie, *Cold War.* See also Noam Chomsky et al., *The Cold War and the University: Toward an Intellectual History of the Postwar Years* (New York: New Press, 1998) and Rebecca Lowen, *Creating the Cold War University: The Transformation of Stanford* (Oakland: University of California Press, 1997). On "Big Science" more generally, see Everett Mendelsohn, Merritt Roe Smith, and Peter Weingart, eds., *Science, Technology and the Military* (Boston: Kluwer Academic, 1988) and Galison and Hevly, *Big Science.*

15. Paul Forman, "Behind Quantum Electronics: National Security as Basis for Physical Research in the United States, 1949–1960," *Historical Studies in the Physical and Biological Sciences* 18.1 (1987): 149–229.

16. Daniel Kevles, "Cold War and Hot Physics: Science, Security, and the American State, 1945–56," *Historical Studies in the Physical and Biological Sciences* 20.2 (1990).

17. Thomas Kuhn, *The Structure of Scientific Revolutions: 50th Anniversary Edition* (Chicago: University of Chicago Press, 2012); Bruno Latour, *Laboratory Life: The Social Construction of Scientific Fact* (Princeton, NJ: Princeton University Press, 1977); Donald Mackenzie, *Inventing Accuracy;* and Peter Galison, "Bubble Chambers and the Experimental Workplace," in P. Achinstein and O. Hannaway, eds., *Observation, Experiment and Hypothesis in Modern Physical Science* (Cambridge, MA: Bradford Books, 1985). For insightful reviews of this literature, see John Zammito, *A Nice Derangement of Epistemes: Post-Positivism in the Study of*

Science from Quine to Latour (Chicago: University of Chicago Press, 2004) and Ian Hacking, "Weapons Research and the Form of Scientific Knowledge," supplement, *Canadian Journal of Philosophy* 12 (1986).

18. Hacking, "Weapons Research."

19. Kevles, "Cold War and Hot Physics."

20. Human memory is a frail thing. Rather than interview elderly scientists about events a half century in the past, I have tried, whenever possible, to draw on scientists' stated views as these debates were taking place. When contemporary primary sources were unavailable, I have resorted to reflections made during later writings and interviews and have flagged these instances explicitly in the text.

1. THE SPUTNIK OPPORTUNITY

1. For more detailed discussions of the nuclear strategies of the Truman and Eisenhower administrations, see, for example, David Alan Rosenberg, "The Origins of Overkill: Nuclear Weapons and American Strategy, 1945–1960," *International Security* 7 (Spring 1983): 3–71 and John Lewis Gaddis, *Strategies of Containment: A Critical Appraisal of American National Security Policy during the Cold War* (New York: Oxford University Press, 2005).

2. Rosenberg, "Origins of Overkill." For a broader political perspective on these years, see Campbell Craig and Fredrik Logevall, *America's Cold War: The Politics of Insecurity* (Cambridge, MA: Belknap Press, 2009), 132–138.

3. Rosenberg, "Origins of Overkill," 28; Gaddis, *Strategies of Containment,* 145.

4. Gaddis, *Strategies of Containment,* 162–196.

5. Hugh Gusterson, *Nuclear Rites: A Weapons Laboratory at the End of the Cold War* (Oakland: University of California Press, 1996), 26; Herbert York, "The Origins of the Lawrence Livermore Laboratory," *Bulletin of the Atomic Scientists,* 1 September 1975; Rosenberg, "Origins of Overkill," 23.

6. Rosenberg, "Origins of Overkill," 40.

7. James Killian, *Sputnik, Scientists, and Eisenhower: A Memoir of the First Special Assistant to the President for Science and Technology* (Cambridge, MA: MIT Press, 1977), 68, 70.

8. Daniel Kevles, "K_1S_2: Korea, Science, and the State," in *Big Science: The Growth of Large-Scale Research,* ed. Peter Galison and Bruce Hevly (Stanford: Stanford University Press, 1992); Rosenberg, "Origins of Overkill."

9. Rosenberg writes, "Eisenhower was very much impressed by the TCP report, and from this time on he relied increasingly on the advice of scientists, whom he viewed as honest brokers with regard to the complex and often politicized issues of a nuclear strategy." Rosenberg, "Origins of Overkill," 39; Killian, *Sputnik, Scientists, and Eisenhower,* 79.

10. Killian, *Sputnik, Scientists, and Eisenhower,* 77–79.

11. Killian, *Sputnik, Scientists, and Eisenhower,* xv; "The Big Step" and "Round the World," *New York Times,* 6 October 1957.

12. "Administrative History," n.d., Volume I Administrative History folder, box 1, LBJ Administrative History, Office of Science and Technology, LBJ Presidential Library (hereafter LBJL); Killian, *Sputnik, Scientists, and Eisenhower,* xv.

13. Killian, *Sputnik, Scientists, and Eisenhower,* 35–36; Herbert York, *Making Weapons, Talking Peace: A Physicist's Odyssey from Hiroshima to Geneva* (New York: Basic, 1987), 105.

14. Killian, *Sputnik, Scientists, and Eisenhower,* xix.

15. "Administrative History," n.d., Volume I Administrative History folder; "Major Actions of the President's Science Advisory Committee, November 1957–January 1961," 13 January 1961, "PSAC 1/61–3/61" folder, box 86, President's Office file (hereafter POF), John F. Kennedy Presidential Library and Museum (hereafter JFKL).

16. "Major Actions of the President's Science Advisory Committee," November 1957–January 1961," 13 January 1961.

17. Ibid.

18. Killian, *Sputnik, Scientists, and Eisenhower,* 115, 241. Also quoted in Robert A. Divine, *The Sputnik Challenge* (New York: Oxford University Press, 1993).

19. In 1962, the Kennedy administration would restrict the Air Force SAB's hiring practices to avoid conflicts of interest and would cap its membership at seventy. Not surprisingly, the number of SAB reports, memos, and meetings dropped off significantly between 1962 and 1963. See Thomas A. Sturm, *The USAF Scientific Advisory Board: Its First Twenty Years, 1944–1964* (Washington, DC: Office of Air Force History, 1986), 89, 117; ARPA pamphlet, "The Advanced Research Projects Agency," August 1960, folder 1, box 175, Papers of Lee A. DuBridge, Archives, California Institute of Technology (hereafter DuBridge).

20. IDA Annual Reports, 1956, 1966, folder 3, box 175, DuBridge; Townes to MGM, 20 February 1961, folder 2, box 35, Murray Gell-Mann Papers, 10219-MS, Caltech Archives, California Institute of Technology (hereafter MGM); Gell-Mann's Jason contract, folder 3, box 35, MGM; Wheeler to Gell-Mann, 1958, folder 1, box 57, MGM.

21. Divine, *Sputnik Challenge*, 171; Megan Prelinger, *Another Science Fiction: Advertising the Space Race 1957–1962* (New York: Blast Books, 2010), 139; David Kaiser, "The Physics of Spin: Sputnik Politics and the American Physicists in the 1950s," *Social Research* 74 (December 2006): 1225–1252.

22. Stuart Leslie, *The Cold War and American Science: The Military-Industrial-Academic Complex at MIT and Stanford* (New York: Columbia University Press, 1993); Roger L. Geiger, *Research and Relevant Knowledge: American Research Universities since World War II* (New York: Oxford University Press, 1993), 161–191. Cold War funding was not the sole factor shaping the expansion of the "military industrial academic complex" during these years. See, for example, Rebecca Lowen, *Creating the Cold War University: The Transformation of Stanford* (Oakland: University of California Press, 1997).

23. David Kaiser, "Physics of Spin," 1235; Teller to Daddario, 18 September 1963, folder 4, box 171, DuBridge; Eugene Rabinowitch, "After Missiles and Satellites, What?" *Bulletin of the Atomic Scientists*, December 1957, 346–350.

24. "Major Actions of the President's Science Advisory Committee, November 1957–January 1961," 13 January 1961.

25. Ibid.

26. Kaiser, "Physics of Spin," 1247; Brian Balogh, *Chain Reaction: Expert Debate and Public Participation in American Commercial Nuclear Power, 1945–1975* (Cambridge: Cambridge University Press, 1991), 171–174.

27. Alfred K. Mann, *For Better or Worse: The Marriage of Science and Government in the United States* (New York: Columbia University Press, 2000), 75; "Hydrogen Bomb Blast Amazes Atom Experts," *Los Angeles Times*, 14 March 1954, 1; "AEC Statement on Bomb's Effects," *Los Angeles Times*, 16 February 1955, 14; Martha Smith-Norris, "The Eisenhower Administration and the Nuclear Test Ban Talks, 1958–1960," *Diplomatic History* 27 (September 2003): 506; Gregg Herken, *Cardinal Choices: Presidential Science Advising from the Atomic Bomb to SDI* (New York: Oxford University Press, 1992), 84.

28. L. W. Nordheim, "Tests of Nuclear Weapons," *Bulletin of the Atomic Scientists*, 1 September 1955. Nordheim did not actually oppose testing. Noting that testing was necessary to ensure that the United States maintained a deterrent nuclear force, he wrote that "the dangers of war are so immeasurably greater that the hazards of tests pale in comparison."

29. See, for example, Ralph Lapp, "Radioactive Fall-out," *Bulletin of the Atomic Scientists*, 1 February 1955; "Fall-out and Candor," *Bulletin of the Atomic Scientists*, 1 May 1955; and "Global Fall-out," *Bulletin of the Atomic Scientists*, 1 November 1955. Lapp's worldview was deeply

influenced by his nuclear experiences; in 1965 he published an icono-clastic study of science in American society, dedicated to Szilard and warning of the widening gap between powerful science experts and nonscientist citizens. See Ralph E. Lapp, *The New Priesthood: The Scientific Elite and the Uses of Power* (New York: Harper & Row, 1965).

30. FAS, "Proposal for a United Nations Commission to Study the Problem of H-Bomb Tests," *Bulletin of the Atomic Scientists*, 1 May 1955.

31. David R. Inglis, "Ban H-Bomb Tests and Favor the Defense," *Bulletin of the Atomic Scientists*, 1 November 1954; W. A. Higinbotham, "A Brief History of FAS," *FAS Newsletter*, November 1962, 2. This was not the first call for a test ban; for example, Vannevar Bush had called for an end to thermonuclear testing as early as 1952. See S. S. Schweber, *In the Shadow of the Bomb: Oppenheimer, Bethe, and the Moral Responsibility of the Scientists* (Princeton, NJ: Princeton University Press, 2000), 168.

32. Inglis, "Ban H-Bomb Tests"; Smith-Norris, 507–508; Benjamin P. Greene, *Eisenhower, Science Advice, and the Nuclear Test-Ban Debate, 1945–1963* (Stanford: Stanford University Press, 2007).

33. Greene, *Eisenhower*; "Major Actions of the President's Science Advisory Committee, November 1957–January 1961," 13 January 1961; Los Alamos Lab Policy Committee notes, 7 June 1960 and 10 January 1961, "Los Alamos Scientific Laboratory Policy Committee [minutes, outlines] 1960–61" folder, box 2, series IV, Subject files I–N, Stanislaw Ulam Papers, American Philosophical Society; Pamphlet, "The Advanced Research Projects Agency," August 1960, folder 1, box 175, DuBridge; Grayson Kirk to DuBridge, 13 April 1961, folder 1, box 175, DuBridge.

34. Kelly Moore, *Disrupting Science: Social Movements, American Scientists, and the Politics of the Military, 1945–1975* (Princeton, NJ: Princeton University Press, 2008), 96–129; *Newsletter, Federation of American Scientists*, 17 June 1957 and 11 December 1959; "Background, Origin, and Evolution of the Committee of Principals," August 1963, "ACDA Comm of Principals 8/63–10/63" folder, box 267, NSF, Historical Studies Division, Bureau of Public Affairs, Department of State, JFKL.

35. "Background, Origin, and Evolution of the Committee of Principals," August 1963.

36. Quoted in *Federation of American Scientists Newsletter*, 17 June 1957; Paul Rubinson, "'Crucified on a Cross of Atoms': Scientists, Politics, and the Test Ban Treaty," *Diplomatic History* 35 (April 2011): 313.

37. *Los Angeles Times*, 3 March 1962; Rosenberg, "Origins of Overkill."

38. Dwight Eisenhower, "Military-Industrial Complex Speech," 1961, http://avalon.law.yale.edu/20th_century/eisenhower001.asp; York, *Making Weapons, Talking Peace*, 126.

2. THE MORAL CASE FOR A TEST BAN

1. State Department press release, 27 June 1960, "Disarmament—Nuclear Test Ban Negotiations 6/2/60–12/60 Part I" folder, box 100, President's Office file (hereafter POF), John F. Kennedy Presidential Library and Museum (hereafter JFKL).
2. When quoting the statements of key players according to meeting minutes, I have used single quotation marks. Direct quotations of people and documents retain regular double quotation marks.
3. *FAS Newsletter,* November 1961.
4. *FAS Newsletter,* February 1961.
5. Jerome Wiesner to John F. Kennedy, memorandum, 23 February 1961 (reprint of memo from September 1960), folder 16, box 67, POF, JFKL.
6. Ibid.
7. Ibid.
8. Ibid.
9. Ibid.
10. McNamara statement is cited in "Science and the Citizen," *Scientific American* 249, no. 6, December 1983, 76.
11. Budget review, prepared by Robert McNamara, 21 February 1961, "Department of Defense Review of FY61 and FY62 Military Programs and Budgets 2/21/61" folder, box 273, National Security file (hereafter NSF), JFKL.
12. Ibid.
13. Ibid.
14. Memorandum of conversation, Meeting of Principals, 2 March 1961, "ACDA Disarmament Committee of Principals Memos of Conversation 3/61–11/63" folder, box 267, NSF, JFKL.
15. Ibid.
16. Report of the Ad Hoc Panel on the Technological Capabilities and Implications of the Geneva System, 2 March 1961, "Disarmament—Fisk Panel on Technical Capabilities of the Geneva System 3/61" folder, box 100, POF, JFKL.
17. Ibid.
18. Ibid.
19. Ibid.
20. McCloy to JFK, 8 March 1961, "Disarmament—Fisk Panel on Technical Capabilities of the Geneva System 3/61" folder, box 100, POF, JFKL.
21. Loper to McCloy, 2 March 1961, "Disarmament—Fisk Panel on Technical Capabilities of the Geneva System 3/61" folder, box 100, POF, JFKL.

22. A. L. Latter, R. Latter, E. A. Martinelli, W. G. McMillan, "Some New Considerations Concerning the Nuclear Test Ban," 10 March 1961, "Disarmament—Nuclear Test Ban Negotiations 3/1/61–4/7/61" folder, box 100, POF, JFKL.

23. Ibid.

24. Robert Komer, "The Case for Resumption of Nuclear Tests," "Nuclear Weapons Testing 2/61–4/61" folder, box 299, NSF, JFKL.

25. Joint Chiefs of Staff, "Arms Control Measures Affecting Risk of Surprise Attack," Department of Defense, "Arms Control Measures Affecting Risk of Surprise Attack 5/61" folder, box 273, NSF, JFKL.

26. "Record of Meeting on Nuclear Test Ban Issue," 4 May 1961, "Disarmament—Nuclear Test Ban Negotiations 4/28/61–3/62" folder, box 100, POF, JFKL.

27. McNamara to National Security Council, memorandum, 15 May 1961, "Nuclear Weapons Testing 5/61" folder, box 299, NSF, JFKL.

28. Memorandum of conversation, Committee of Principals, 23 May 1961, "ACDA Disarmament Committee of Principals Memos of Conversation 3/61–11/63" folder, box 267, NSF, JFKL; Komer to Bundy/Rostow, 28 June 1961, "ACDA Disarmament Comm of Principals 4/1/61–7/15/61" folder, box 267, NSF, JFKL; Memorandum of conversation, Committee of Principals, 23 May 1961; Marc Raskin to McGeorge Bundy, 16 June 1961, and Galbraith to JFK, 12 June 1961, "Nuclear Weapons Testing, 6/61" folder, box 299, NSF, JFKL.

29. Norris Bradbury to Wiesner, 17 July 1961, "Nuclear Weapons Panofsky Panel 8/4/61–9/5/61" folder, box 302A, NSF, JFKL.

30. Attachment by Carson Mark, 17 July 1961, "Nuclear Weapons Panofsky Panel 8/4/61–9/5/61" folder, box 302A, NSF, JFKL.

31. "Report of the Ad Hoc Panel on Nuclear Testing," 21 July 1961, "Nuclear Weapons Testing Subjects Rostow File 7/21/61–9/20/61" folder, box 301, NSF, JFKL.

32. Raskin to McGeorge Bundy, 25 July 1961, "Nuclear Weapons Testing, 7/16/61–8/9/61" folder, box 299, NSF, JFKL; George Ball to JFK, 4 August 1961, "Nuclear Weapons Panofsky Panel 8/4/61–9/5/61" folder, box 302A, NSF, JFKL; Seaborg to Wiesner, 4 August 1961, "Nuclear Weapons Testing Subjects Rostow File 7/21/61–8/20/61" folder, box 301, NSF, JFKL.

33. Foster to Wiesner, 2 August 1961, "Nuclear Weapons Panofsky Panel 8/4/61–9/5/61" folder, box 302A, NSF, JFKL; Harold Brown to Wiesner, 3 August 1961, "Nuclear Weapons Testing Subjects Rostow File 7/21/61–9/20/61" folder, box 301, NSF, JFKL; McNamara to McCloy, 28 July 1961, "Nuclear Weapons Testing 7/16/61–8/9/61" folder, box 299, NSF, JFKL,

34. Maxwell Taylor to JFK, 7 August 1961, "Nuclear Weapons Panofsky Panel 8/4/61–9/5/6" folder, box 302A, NSF, JFKL.

35. "Comments of JCS on Report of the Ad Hoc Panel on Nuclear Testing," 2 August 1961, "Nuclear Weapons Testing Subjects Rostow File 7/21/61–9/20/61" folder, box 301, NSF, JFKL.

36. JFK to Maxwell Taylor, 7 August 1961, "Nuclear Weapons Panofsky Panel 8/4/61–9/5/6" folder, box 302A, NSF, JFKL.

37. Minutes of the National Security Council meeting, 8 August 1961 (minutes prepared by McGeorge Bundy, 5 September 1961), "Nuclear Weapons Panofsky Panel 8/4/61–9/5/61" folder, box 302A, NSF, JFKL.

38. Arthur Dean reported that the decision to resume testing was made in JFK's office on 17 August 1961. Arthur Dean to John McCloy, 18 August 1961, "Nuclear Weapons Testing 8/10/61–8/30/61" folder, box 299, NSF, JFKL; Paul Doty, an American physicist attending a joint U.S.-Soviet meeting in England, reported to Walt Rostow that Soviet scientists were "depressed and defensive" about the upcoming Soviet test series. Walt Rostow to General Clifton, 9 September 1961, "Nuclear Weapons Testing Subjects Rostow File 7/21/61–9/26/61" folder, box 301, NSF, JFKL; White House press release, 30 August 1961, "Nuclear Weapons Testing 8/10/61–8/30/61" folder, box 299, NSF, JFKL; Komer to McGeorge Bundy, 31 August 1961, "Nuclear Weapons Testing Subjects Rostow File 7/21/61–9/26/61" folder, box 301, NSF, JFKL; Edward R. Murrow to JFK, 31 August 1961, "Nuclear Weapons Testing 8/10/61–8/30/61" folder, box 299, NSF, JFKL.

39. Wiesner to McGeorge Bundy, 29 September 1961, "Nuclear Weapons Testing 9/27/61–10/10/61" folder, box 299, NSF, JFKL; Memorandum of conversation, Meeting of Principals, 10 October 1961, "ACDA Disarmament Committee of Principals Memos of Conversation 3/61–11/63" folder, box 267, NSF, JFKL.

40. Memorandum of conversation, Meeting of Principals, 10 October 1961; Transcript, "Meet the Press," 20 October 1961, "Nuclear Weapons Testing 10/30/61–10/31/61" folder, box 299, NSF, JFKL.

41. Memorandum of conversation, Meeting of Committee of Principals, 11 November 1961, "ACDA Disarmament Committee of Principals Memos of Conversation 3/61–11/63, sanitized June 2006" folder, box 267, NSF, JFKL.

42. Teller to JFK, 7 December 1961, "Nuclear Weapons Testing 12/7/61–12/18/61" folder, box 299A, NSF, JFKL.

43. Ibid. Teller's relationship with Kennedy was notably poor; Teller later recalled in his memoir that "my interactions with President Kennedy were the low point in my career as a diplomat." See Edward Teller with Judith

Shooler, *Memoirs: A Twentieth-Century Journey in Science and Politics* (Cambridge, MA: Perseus Publishing, 2001), 466.

44. Wiesner to JFK, 19 December 1961, "Nuclear Weapons Testing 12/19/61–12/20/61" folder, box 299A, NSF, JFKL.

45. Raskin to McGeorge Bundy, 20 December 1961, "Nuclear Weapons Testing 12/19/61–12/20/61" folder, box 299A, NSF, JFKL; Arthur Schlesinger Jr. to JFK, 27 December 1961, "Nuclear Weapons Testing 12/21/61–1/8/62" folder, box 299A, NSF, JFKL.

46. McGeorge Bundy to JFK, 30 December 1961, "Nuclear Weapons Testing 12/21/61–1/8/62" folder, box 299A, NSF, JFKL.

47. Bundy informed Kennedy that the report was "probably of high political importance, because men like Bethe and Baker have also signed it" and because it was being circulated among members of Congress. Bethe eventually submitted a strong dissenting addendum, however, objecting to the the alarmist tone of the report and arguing that additional testing was "desirable but not a 'must.'" McGeorge Bundy to JFK, 17 January 1962, "Nuclear Weapons Testing 1/16/62–1/22/62" folder, box 299A, NSF, JFKL; Bethe to Twining, 26 January 1962, "Nuclear Weapons Testing Twining Committee on Military Implications of 1961 Soviet Nuclear Testing Report 1/5/62" folder, box 302A, NSF, JFKL; Wiesner to JFK, 25 January 1962, "Nuclear Testing 1962–1963" folder, box 104, POF, JFKL.

48. Wiesner to JFK, 21 February, "Nuclear Weapons Testing General 2/17/62–4/4/62 and undated" folder, box 300, NSF, JFKL.

49. Memorandum for the record, 28 February 1962 (meeting held 27 February 1962), "Disarmament Eighteen-Nation Disarmament Committee 3/6/62–3/14/62, 11/20/62 and undated" folder, box 100, POF, JFKL.

50. "72 at Cornell Back Kennedy over Tests," *New York Times,* 25 February 1962; *FAS Newsletter,* February 1962; FAS, "Scientists Appraise Atmospheric Tests," *Bulletin of the Atomic Scientists,* 1 April 1962, 33.

51. John F. Kennedy, transcript, 2 March 1962, *New York Times,* 3 March 1962.

52. Ibid.

53. *New York Times,* 3 March 1962; "Sky Tests Win Support of Scientists," *Los Angeles Times,* 3 March 1962, 20.

54. *FAS Bulletin,* April 1962.

55. *FAS Newsletter,* May 1962; Leo Szilard, "Are We on the Road to War?," *Bulletin of the Atomic Scientists,* 1 April 1962, 23–30.

56. In the concurrent discussions of disarmament measures, Wiesner endorsed a "30% across-the-board cut of all armaments applied by type . . . and a complete production cutoff." Memorandum for the record, 6 March 1961, Meeting on Disarmament, 7 March 1962, "Disarmament-Nuclear Test Ban Negotiations 4/28/61–3/63" folder, box 100, POF, JFKL.

57. The 4.75 number had been determined by a 1959 PSAC Panel on Inspection Problems, headed by Robert Bacher. "Major Actions of the President's Science Advisory Committee, November 1957–January 1961," 13 January 1961, "PSAC 1/61–3/61" folder, box 86, POF, JFKL; Wiesner to JFK, 9 March 1962, "Nuclear Testing 1962–1963" folder, box 104, POF, JFKL.

58. Kaysen to JFK, 20 July 1962, "Nuclear Weapons Testing General 4/15/62–7/30/62 and undated" folder, box 300, POF, JFKL; Bundy to JFK, 26 July 1962, "Nuclear Testing 1962–1963" folder, box 104, POF, JFKL.

59. Memorandum of conversation, Meeting of Committee of Principals, 26 July 1962, "ACDA Disarmament Committee of Principles Memos of Conversation 3/61–11/63" folder, box 267, NSF, JFKL.

60. Ibid.; Bundy to JFK, 26 July 1962, "Nuclear Testing 1962–1963" folder, box 104, POF, JFKL.

61. William C. Foster to JFK, 30 July 1962, "Disarmament-Nuclear Test Ban Negotiations, 7/30/62" folder, box 100 (overflow), POF, JFKL; Holifield to JFK, 25 July 1962, "Joint Committee on Atomic Energy Testing Hearings 7/19/62–7/23/62, 7/19/62–8/2/62" folder, box 282, NSF, JFKL; Hosmer press release, 31 July 1962, "Nuclear Weapons Testing General 4/5/62–7/30/62 and undated" folder, box 300, NSF, JFKL; Bundy to JFK, memo attachment, 30 July 1962, "Disarmament-Nuclear Test Ban Negotiations, 7/30/62" folder, box 100 (overflow), POF, JFKL.

62. Sheldon M. Stern, *The Cuban Missile Crisis in American Memory: Myths versus Reality* (Stanford: Stanford University Press, 2012); Spencer Weart, *The Rise of Nuclear Fear* (Cambridge, MA: Harvard University Press, 2012), 153–154.

63. See, for example, Michael Dobbs, *One Minute to Midnight: Kennedy, Khrushchev, and Castro on the Brink of Nuclear War* (New York: Knopf, 2008) and Stern, *Cuban Missile Crisis*, 3, 155; Memorandum of conversation, Meeting of Committee of Principals, 10 November 1962, "ACDA Disarmament Committee of Principals Memos of Conversation 3/61–11/63" folder, box 267, NSF, JFKL.

64. Meeting minutes, 18 February 1963, "Meetings and the President Test Ban Treaty 2/18/63" folder, box 317a, NSF, JFKL; "Report of the Twining Committee: Military Implications of U.S. and Soviet Nuclear Testing," 4 March 1963, "Nuclear Weapons Twining Committee Report to Chief of Staff of U.S. Air Force, Military implications of U.S. and Soviet Nuclear Testing" folder, box [302A], NSF, JFKL.

65. "Report of the Twining Committee: Military Implications of U.S. and Soviet Nuclear Testing," 4 March 1963.

66. Ibid.

67. John Pastore to JFK, 8 March 1963, "Joint Committee on Atomic Energy General 1963 and undated" folder, box 281a, NSF, JFKL; Wiesner to JFK, 13 May 1963, "NSAM 210: Underground Nuclear Tests, 12/12/61, 12/62–12/63 and undated" folder, box 339, NSF, JFKL; Meeting of the Committee of Principals, actions taken on agenda items, and memorandum of conversation, 17 April 1963, "ACDA Disarmament Committee of Principals Memos of Conversation 3/61–11/63" folder, box 267, NSF, JFKL.

68. Statement, 13 May 1963, "Disarmament-Nuclear Test Ban Negotiations 4/62–8/63" folder, box 100, POF, JFKL; Statement, 13 May 1963, "Text of Scientists Statement Supporting Test Ban Treaty" folder, box 100 (overflow), POF, JFKL; "27 U.S. Scientists Urge Test-Ban Pact," New York Times, 13 May 1963; "Education Urged on A-Tests Ban," Washington Post, 27 May 1963.

69. The Baltimore Sun noted that it had "produced no progress" and that the Soviets had evinced "a declining interest in a test ban." See "Scientific Study Asked on Test Ban," Baltimore Sun, 1 June 1963; "New Hope for a Test Ban," New York Times, 11 June 1963; Memorandum of conversation, Nuclear Test Ban Treaty Meeting of Committee of Principals, 14 June 1963, "ACDA Disarmament Committee of Principals Memos of Conversation 3/61–11/63" folder, box 267, NSF, JFKL.

70. Memorandum of conversation, Nuclear Test Ban Treaty Meeting of Committee of Principals, 14 June 1963.

71. Michael Egan, Barry Commoner and the Science of Survival: The Remaking of American Environmentalism (Cambridge, MA: MIT Press, 2007), 11; FAS Newsletter, September 1963; Public statement by PSAC on Nuclear Test Ban Treaty, 24 August 1963, "Disarmament-Nuclear Test Ban Part I Negotiations 7/63 Meeting in Moscow" folder, box 100 (overflow), POF, JFKL; Clipping, "Bradbury Tells Newsmen Treaty to Have Little Effect on Lab," Los Alamos Scientific Laboratory News, 1 August 1963, "Disarmament-Nuclear Test Ban Part I Negotiations 7/63 Meeting in Moscow" folder, box 100 (overflow), POF, JFKL.

72. Fred Dutton to Sorenson, 16 August 1963, "Disarmament—Nuclear Test Ban Part II Negotiations—7/63 Meeting in Moscow" folder, box 100 (overflow), POF, JFKL; Zuoyue Wang, In Sputnik's Shadow: The President's Science Advisory Committee and Cold War America (New Brunswick, NJ: Rutgers University Press, 2008), 9.

73. See Martha Smith Norris, "The Eisenhower Administration and the Nuclear Test Ban Talks: 1958–1960," Diplomatic History 27 (September 2003): 503–541; Harold Jacobson and Eric Stein, Diplomats, Scientists, and Politicians (Ann Arbor: University of Michigan Press, 1966); Robert

Divine, *Blowing on the Wind: The Nuclear Test Ban Debate, 1954–1960* (New York: Oxford University Press, 1978) and *Eisenhower and the Cold War* (New York: Oxford University Press, 1981); Stephen Ambrose, *Eisenhower* (New York: Simon & Schuster, 1984).

74. Benjamin P. Greene, *Eisenhower, Science Advice, and the Nuclear Test-Ban Debate, 1945–1963* (Stanford: Stanford University Press, 2007); Herbert York, *Making Weapons, Talking Peace: A Physicist's Odyssey from Hiroshima to Geneva* (New York: Basic, 1987), 117.

75. Paul Rubinson, "'Crucified on a Cross of Atoms': Scientists, Politics, and the Test Ban Treaty," *Diplomatic History* 35 (April 2011): 283–319.

76. York, *Making Weapons,* 199.

77. Vojtech Mastny, "The 1963 Nuclear Test Ban Treaty: A Missed Opportunity for Détente?" *Journal of Cold War Studies* 10 (Winter 2008): 3–25; Memorandum of conversation, Nuclear Test Ban Treaty Meeting of Committee of Principals, 14 June 1963.

78. Gaddis, *Strategies of Containment,* 217.

79. *Washington Post,* 25 April 1965.

80. James Killian, *Sputnik, Scientists, and Eisenhower: A Memoir of the First Special Assistant to the President for Science and Technology* (Cambridge, MA: MIT Press, 1977), 104.

81. Emma Rothschild, "Continuing Communication," in *Jerry Wiesner: Scientist, Statesman, Humanist,* ed. Walter A. Rosenblith (Cambridge, MA: MIT Press, 2003), 161.

3. THE SCIENCE OF NONNUCLEAR WAR

1. Jerome Wiesner to John F. Kennedy, memorandum, 23 February 1961 (reprint of memo from September 1960), folder 16, box 67, President's Office File (hereafter POF), John F. Kennedy Presidential Library and Museum (hereafter JFKL); George Kistiakowsky to Jerome Wiesner, memorandum, 12 January 1960, "PSAC 1/61–3/61" folder, box 86, POF, FKL; Jerome Wiesner to John F. Kennedy, report, "Major Actions of the President's Science Advisory Committee November 1957–January 1961," 31 January 1960, "PSAC 1/61–3/61" folder, box 86, JFKL.

2. See, for example, John Lewis Gaddis, *Strategies of Containment: A Critical Appraisal of American National Security Policy during the Cold War* (New York: Oxford University Press, 2005); Department of Defense budget review, January 1961, "DoD General, 1/63" folder, box 273, National Security File (hereafter NSF), JFKL; Staff memo, 14 February 1961, "Staff Memoranda Walt W. Rostow Guerrilla and Unconventional Warfare 2/1/61–2/16/61" folder, box 325, NSF, JFKL;

Budget review, 21 February 1961, "Department of Defense Review of FY61 and FY62 Military Programs and Budgets 2/21/61" folder, box 273, NSF, JFKL.

3. Other accounts consider limited war in terms of tactical nuclear weaponry. See Alex Roland, "Technology, Ground Warfare, and Strategy: The Paradox of American Experience," *Journal of Military History*, Vol. 55 (October 1991): 447–468, and Herman Kahn, *On Thermonuclear War*, 2nd ed. (Princeton, NJ: Princeton, 1961), 540–543. For an alternate view, see Seymour Deitchman, *Limited War and American Defense Policy: Building and Using Military Power in a World at War* (Cambridge, MA: MIT Press, 1964); Robert E. Osgood, *Limited War: The Challenge to American Strategy* (Chicago: University of Chicago Press, 1957), 248–249; and Robert E. Osgood, *Limited War Revisited* (Boulder: Westview Press, 1979), 3–5.

4. Deitchman, *Limited War and American Defense*, 3–4.

5. Undated document, "Questions Concerning Counter-Guerrilla Programs," "DoD (B) Subjects Special Warfare 2/61–5/61" folder, box 279, NSF, JFKL; Joint Chiefs of Staff (JCS) report: "Development Status of Military Counterinsurgency Programs, Including Counterguerrilla Forces," "DoD (B) Status of Military Counterinsurgency Programs 9/18/63" folder, box 280, NSF, JFKL; Luther J. Carter, "Vietnam: Jungle Conflict Poses New R&D Problems," *Science* 152 (8 April 1966): 189.

6. Rostow to JFK, memorandum, 23 February 1961, "Staff Memoranda, Walt W. Rostow, Guerrilla and Unconventional Warfare, 2/17/61–2/28/61" folder, box 325, NSF, JFKL; JCS Report, "Development Status of Military Counterinsurgency Programs, Including Counterguerrilla Forces," "DoD (B) Status of Military Counterinsurgency Programs 9/18/63" folder, box 280, NSF, JFKL.

7. Wiesner to JFK, memorandum, 13 April 1961, "PSAC 1/61–3/61" folder, box 86, POF, JFKL.

8. Robert Johnson to Walt Rostow, memorandum, 5 August 1961, "Vietnam, General, 8/61" folder, box 194, NSF, JFKL.

9. For example, see Carter, "Vietnam," 188–189.

10. For histories of the Vietnam War, see Marilyn Young, *Vietnam Wars, 1945–1990* (New York: Harper Perennial, 1991) and John Prados, *Vietnam: The History of an Unwinnable War, 1945–1975* (Lawrence: University Press of Kansas, 2009); Rostow, quoted in William Buckingham Jr., *Operation Ranch Hand: The Air Force and Herbicides in Southeast Asia, 1961–1971* (Washington, DC: Office of Air Force History, 1982); "Vietnam: Concept of Action," 1961, "Department of Defense General 4/61–5/61" folder, box 273, NSF, JFKL.

11. Harold Brown to Ed Lansdale, memorandum, 8 June 1961, and Lansdale to Walt Rostow, memorandum, 21 June 1961, "DoD General 6/61–7/61" folder, box 273, NSF, JFKL; JCS Report, "Development Status of Military Counterinsurgency Programs, Including Counterguerrilla Forces," "DoD (B) Status of Military Counterinsurgency Programs 9/18/63" folder, box 280, NSF, JFKL.

12. William Yarborough to W. B. Rosson, memorandum, 2 May 1962, "DoD(B) Subjects Special Warfare 1962–63" folder, box 279, NSF, JFKL; Lemnitzer to JFK, memorandum, 28 December 1961, "DoD Joint Chiefs of Staff General, 1961" folder, box 276, NSF, JFKL; Report, "RDT&E Annex, Report on General Taylor's Mission to South Vietnam," 3 November 1961, "Vietnam Report on Taylor Mission—November 1961" folder, box 210, Country file, Vietnam, NSF, LBJL; Lt. Gen. Lionel McGarr to Walt Rostow, report, "First Twelve Month Report of Chief MAAG, Vietnam," 1 September 1961, "Vietnam General, McGarr Information Folder for Rostow, 10/25/61" folder, box 194A, NSF, JFKL; "Simulmatics: A Socio-Psychological Study of Regional/Popular Forces in Vietnam," folder 1, box 239, Country file, Vietnam, NSF, LBJL.

13. Report, "RDT&E Annex, Report on General Taylor's Mission to South Vietnam," 3 November 1961; JCS Report, "Development Status of Military Counterinsurgency Programs, Including Counterguerrilla Forces," 18 September 1963, "DoD(B) Status of Military Counterinsurgency Programs 9/18/63" folder, box 280, JFKL; Michael Adas, *Dominance by Design: Technological Imperatives and America's Civilizing Mission* (Cambridge, MA: Belknap Press, 2006), 281–336; For detailed histories of Agent Orange, see Edwin A. Martini, *Agent Orange: History, Science, and the Politics of Uncertainty* (Amherst: University of Massachusetts Press, 2012) and David Zierler, *The Invention of Ecocide: Agent Orange, Vietnam, and the Scientists Who Changed the Way We Think about the Environment* (Athens: University of Georgia Press, 2011).

14. David A. Butler, "Connections: The Early History of Scientific and Medical Research on 'Agent Orange,'" *Journal of Law and Policy* 13, no. 527 (2005); Arthur Galston, "Falling Leaves and Ethical Dilemmas: Agent Orange in Vietnam," in *New Dimensions in Bioethics*, eds. A. W. Galston and E. G. Shurr (Boston: Kluwer Academic Publishing, 2001), 108.

15. Alvin Young, "The History of the U.S. Department of Defense Programs for the Testing, Evaluation, and Storage of Tactical Herbicides," prepared for the Office of the Under Secretary of Defense, U.S. Army Research Office, December 2006. Available at: http://www.dod.mil/pubs/foi/reading_room/TacticalHerbicides.pdf. Funding statistics and Detrick description in Milton Leitenberg, "Biological Weapons," *Scientist and*

Citizen, August–September 1967, 163. Researchers at the U.S. Army Biological Laboratories at Fort Detrick also received information and assistance from other military and industrial research centers; for example, a 1965 Fort Detrick report on Agent Orange acknowledged assistance from the Dow Chemical Company and Edgewood Arsenal (Army Chemical Center). (See Richard Hensen, "Technical Memorandum 74: Physical Properties of Normal Butyl Esters of 2,4-D, 2,4,5-T, and 'Orange,'" August 1965, United States Army Biological Laboratories (Fort Detrick), Alvin L. Young Collection on Agent Orange, National Agricultural Library. Available at: http://www.nal.usda.gov/speccoll /findaids/agentorange/text/00016.pdf; Buckingham, *Operation Ranch Hand.*

16. Donald Barlett and James Steele, "Monsanto's Harvest of Fear," *Vanity Fair,* May 2008; Young, "History."

17. Harold Brown to Lansdale, memorandum, 22 September 1961,"Staff Memoranda Walt W. Rostow Guerrilla and . . . Warfare 9/61" folder, box 326A, NSF, JFKL; "Status Report on the Presidential Program for Viet-Nam," 28 July 1961, "Vietnam, General, Presidential Program Status Reports" folder, box 195A, Countries, NSF, JFKL; Buckingham, *Operation Ranch Hand,* 11, 26; Report, "RDT&E Annex, Report on General Taylor's Mission to South Vietnam," 3 November 1961, "Vietnam Report on Taylor Mission—November 1961" folder, box 210, Country file, Vietnam, NSF, LBJL; Young, "History."

18. Buckingham, *Operation Ranch Hand,* 9–22; "National Security Action Memo 115," November 1961, "NSAM 115 Defoliant Operations in Vietnam" folder, box 332, NSF, JFKL. On the regulation of chemicals: this was distinct from the military use of pesticides in Vietnam, which were subject to approval and regulation by the USDA and other agencies. It was also distinct from the herbicides used on the military bases in Vietnam, which were subject to separate regulatory processes (including USDA regulations) and were not considered "tactical herbicides." See Young, "History."

19. Buckingham, *Operation Ranch Hand,* 43; Report, "RDT&E Annex, Report on General Taylor's Mission to South Vietnam," 3 November 1961, "Vietnam Report on Taylor Mission—November 1961" folder, box 210, Country file, Vietnam, NSF, LBJL. In April 1962 General Harkins reported to Admiral Felt that "there is need to conduct R&D sprayings with changed spray rates and dosages as recommended by Gen. Delmore and technical group from OSD and *I selected the site based on operational considerations.*" [emphasis added] Harkins to Felt, April 1962, "Vietnam, General 4/11/62–4/16/62" folder, box 196, NSF, JFKL.

20. Young, "History"; Buckingham, *Operation Ranch Hand,* 23–44.

21. Buckingham, *Operation Ranch Hand,* 51, 54; Forrestal to JFK, memo-randum, 2 August 1962, "NSAM 178 Destruction of Mangrove Swamps in South Vietnam" folder, box 338, NSF, JFKL.

22. Young, "History" and "Vegetation Analysis of the Pran Buri Defoliation Test Area 1," January 1966, Joint Thai-U.S. Military Research and Development Center, Alvin L. Young Collection on Agent Orange, National Agricultural Library, http://www.nal.usda.gov/speccoll/findaids/agentor ange/text/00019.pdf.

23. Charles Mohr, "U.S. Spray Planes Destroy Rice in Vietcong Territory," *New York Times,* 21 December 1965, 1; Arthur Westing, *Herbicides in War: The Long-Term Ecological and Human Consequences* [Stockholm International Peace Research Institute] (Philadelphia: Taylor & Francis, 1984), 5–6; Buckingham, *Operation Ranch Hand,* 126; Richard Kohn, foreword to Buckingham, *Operation Ranch Hand,* iii–iv. Agent Orange superseded Agent Purple beginning in "late 1964"—see Buckingham, *Operation Ranch Hand,* 122. Buckingham estimates that over 1.6 million acres were sprayed in 1967 alone (Buckingham, *Operation Ranch Hand,* 129); the Stockholm International Peace Research Institute estimates that "the three peak years of herbicide spraying—1967–1969—were about equal in magnitude and together accounted for over three-quarters of the volume of total wartime expenditures." See Westing, *Herbicides in War,* 5; Martini, *Agent Orange,* 147.

24. Buckingham, *Operation Ranch Hand,* 26–28.

25. Gilpatric to JFK, memorandum, 23 November 1961, "NSAM 115 Defoliant Operations in Vietnam" folder, box 332, NSF, JFKL; Rostow to JFK, report, "Viet-Nam Status Report," 21 November 1961, "Vietnam, General 11/18/61–11/20/61" folder, box 195, NSF, JFKL; Rusk to JFK, memo-randum, 24 November 1961, "Vietnam, General, Memos and Reports 11/17/61–11/30/61" folder, box 195, NSF, JFKL. Beginning in the mid-1950s, British military forces in Malaya engaged in aerial spraying of a 2,4,5-T and 2,4-D-containing compound, for the purposes of defoliating communications lines and crop destruction (See Westing, *Herbicides in War,* 4).

26. Matthew Meselson attributed the failure of the U.S. Senate to ratify the Geneva Protocol to organized lobbying by "the American Chemical Society, the Army Chemical Corps, the American Legion, and parts of the chemical industry." See Matthew Meselson, "Controlling Biological and Chemical Weapons," in *March 4: Scientists, Students, and Society,* ed. Jonathan Allen (Cambridge, MA: MIT Press, 1970), 151–160. For more on the ACS, see William J. Bailey, "Introductory remarks," in *Chemical Weapons and U.S. Policy: A Report of the Committee on*

Chemistry and Public Affairs (Washington, DC: American Chemical Society, 1977), Othmer Library, Chemical Heritage Foundation.

27. The United States military refrained from the use of defoliants and gas (to clear underground tunnels) in the Pacific theater, though there is some speculation that chemical crop destruction might have occurred had the Japanese not surrendered after Hiroshima and Nagasaki. Incendiary gels were use in firebombing. For more detailed accounts of CBW use during World War II see Robert Neer, *Napalm: An American Biography* (Cambridge, MA: Belknap Press, 2013); Robert Harris and Jeremy Paxman, *A Higher Form of Killing: The Secret History of Chemical and Biological Warfare* (New York: Random House, 2002); Jonathan Tucker, *War of Nerves: Chemical Warfare from World War I to Al-Qaeda* (New York: Anchor Books, 2006).

28. Quoted in Victor Sidel and Robert Goldwyn, "Chemical and Biologic Weapons—A Primer," *New England Journal of Medicine* 274 (6 January 1966): 21–27.

29. State Department memo, 30 April 1962, "Vietnam General, 4/17/62–4/30/62" folder, box 196, NSF, JFKL; Reports cited in Seymour Hersh, "Our Chemical War, *New York Review of Books,* 25 April 1968.

30. Various claims that unnamed scientists and Red Cross officials have "condemned the criminal acts of the US imperialists . . ." are cited in U.S. Army press reprints, April 1963, "Vietnam General 4/1/63–4/18/63" folder, box 197A, NSF, JFKL; McNamara to JFK, undated memorandum, "NSAM 178 Destruction of Mangrove Swamps in South Vietnam" folder, box 338, NSF, JFKL; JFK, memorandum, "Vietnam, General 7/20/62–7/30/62" folder, box 196A, NSF, JFKL; Embassy telegram, 25 September 1962, "Vietnam, General 9/22/62–9/29/62" folder, box 196A, NSF, JFKL; Telegram, Nolting to State Department, 1 September 1961, "Vietnam General, 9/1/62–9/14/62" folder, box 196A, NSF, JFKL.

31. State Department telegram, 26 September 1962, "Vietnam, General 9/22/62–9/29/62" folder, box 196A, NSF, JFKL; Memorandum for the record, 2 October 1962, "Vietnam General 10/1/62–10/6/62" folder, box 197, NSF, JFKL; State Department airgram, 12 August 1965, "Vietnam Memos (A) vol. XXXVIII 8/1–12/65 [1 of 2]" folder, box 21, Country file, Vietnam, NSF, LBJL; State Department telegram, 15 March 1963, "Vietnam, General 3/1/63–3/19/63" folder, box 197, NSF, JFKL; Nolting memorandum #547, 26 November 1962, "Vietnam General 11/26/62–11/30/62" folder, box 197, NSF, JFKL; see also Buckingham, *Operation Ranch Hand,* 79–80.

32. Buckingham, *Operation Ranch Hand,* 81–82; Hersh, "Our Chemical War"; Memorandum of conversation, 4 April 1963, "Vietnam General 4/1/63–4/18/63" folder, box 197A, NSF, JFKL.

33. Memorandums, April 1963, "Vietnam General 4/1/63–4/18/63" folder, box 197A, NSF, JFKL.

34. Buckingham, *Operation Ranch Hand,* 82; *New York Times,* April 9, 1963; "One Man's Meat," in "The Week," *New Republic* 148 (23 March 1963): 3–7.

35. Adam Rome, "'Give Earth a Chance': The Environmental Movement and the Sixties," *Journal of American History* (1 September 2003); Ralph Lutts, "Chemical Fallout: Rachel Carson's *Silent Spring,* Radioactive Fallout, and the Environmental Movement," *Environmental Review* 9 (Autumn 1985): 210–225.

36. State Department airgram, 12 August 1965; Buckingham, *Operation Ranch Hand,* 99.

37. Robert McNamara, statement before the Democratic Platform Committee, 17 August 1964, "DoD 11/63 vol I, 1 of 2" folder, box 11, Agency file, NSF, LBJL.

38. Donald Hornig, speech to the engineers and physicians for Johnson and Humphrey, 23 October 1964, "Addresses and Remarks by Donald Hornig, 1964" folder, box 8, Papers of Donald Hornig, LBJL.

39. Cost-plus contracts guarantee payment for all project costs plus profit minimums. Robert McNamara, statement before the Democratic Platform Committee, 17 August 1964; "Cost Reduction Program" pamphlet, Department of Defense, "DoD 11/63 vol I, 1 of 2" folder, box 11, Agency file, NSF, JFKL; Carter, "Vietnam," 187.

40. Buckingham, *Operation Ranch Hand,* 117–119; State Department telegram, 5 August 1965, "Vietnam Cables vol. XXXVIII 8/1–12/65 [1 of 2]" folder, box 21, Country file, Vietnam, NSF, LBJL; State Department airgram, 12 August 1965, "Vietnam Cables vol. XXXVIII 8/1–12/65 [1 of 2]" folder, box 21, Country file, Vietnam, NSF, LBJL; Buckingham, *Operation Ranch Hand,* 101; State Department telegram, 13 August 1965, "Vietnam Cables vol. XXXIX 8/13–31/65" folder, box 21, Country file, Vietnam, NSF, LBJL. Seymour Hersh argued in 1968, based on internal documents and conversations with sources, that at the key moments of decision making, the State Department tended to take a more cautious stance than the Pentagon in regard to both the use of defoliants and tear gases. Hersh describes the "bitter" opposition by Averill Harriman to Pentagon defoliation testing in Thailand, for example. (See Hersh, "Our Chemical War.")

41. Quoted in Buckingham, *Operation Ranch Hand*, 94; Mohr, "U.S. Spray Planes Destroy Rice" and "Defoliation Unit Lives Perilously," *New York Times*, December 20, 1965.

42. This chapter addresses the ethical debates concerning the particular tear gases used in Vietnam. For a Cold War history of nerve gas and other lethal chemicals, see Tucker, *War of Nerves*.

43. "RDT&E Annex, Report on General Taylor's Mission to South Vietnam," 3 November 1961.

44. Westmoreland, undated report from Saigon, "Gas, Vol I" folder, box 194, Country file, Vietnam, NSF, LBJL; McGeorge Bundy, memorandum, 26 March 1965, "Gas, Vol I" folder, box 194, Country file, Vietnam, NSF, LBJL; Seymour Hersh, "Poison Gas in Vietnam," *New York Review of Books*, 9 May 1968.

45. Hersh, "Poison Gas in Vietnam"; Report, 23 March 1965, "Gas Vol I" folder, box 194, Country file, Vietnam, NSF, LBJL.

46. These incidents were described in some detail in Westmoreland, undated report from Saigon, "Gas, Vol I" folder; Department of Defense cable, January 1965, "Gas, Vol I" folder, box 194, Country file, Vietnam, NSF, LBJL.

47. Hersh, "Our Chemical War"; William M. Hammond, *Public Affairs: The Military and the Media, 1962–1968* (Washington, DC: Center of Military History, U.S. Army, 1988); Howard Margolis, "Police-Type Gas Used in Saigon," *Washington Post*, March 23, 1965, A15; Quoted in Margolis, "Police-Type Gas." The *Los Angeles Times* published a similar account on 23 March 1965; Dean Rusk, statement, 24 March 1965, "Gas, Vol II" folder, box 194, Country file, Vietnam, NSF, LBJL; Bundy to LBJ, memorandum, 23 February 1965, "Gas, Vol I" folder, box 194, Country file, Vietnam, NSF, LBJL; Hersh, "Poison Gas in Vietnam."

48. Hornig's office estimated the duration of incapacitation at up to four hours. *Los Angeles Times*, 23 March 1965; "The Perils of Even a Humane Gas," *Guardian*, 23 March 1965; Max Frankel, "U.S. Reveals Use of Nonlethal Gas against Vietcong," *New York Times*, 23 March 1965.

49. Quoted in Sidel and Goldwyn, "Chemical and Biological Weapons," 21–27; Letter, Kastenmeier et al. to LBJ, 25 March 1965, "Gas, Vol I" folder, box 194, Country file, Vietnam, NSF, LBJL.

50. Office of the Deputy Secretary of Defense to Cong. Robert Kastenmeier, letter, 30 March 1965, "Gas, Vol I" folder, box 194, Country file, Vietnam, NSF, LBJL.

51. Ibid.; Report, 23 March 1965, "Gas, Vol I" folder, box 194, Country file, Vietnam, NSF, LBJL. Strikeout in original.

52. Telecom, 18 September 1965, "Gas, Vol II" folder, box 194, Country file, Vietnam, NSF, LBJL; D. Hank Ellison, *Chemical Warfare during the Vietnam War: Riot Control Agents in Combat* (New York: Routledge, 2011), 33–34; Cable, 9 September 1965, "Gas, Vol II" folder, box 194, Country file, Vietnam, NSF, LBJL; Wheeler to McNamara, memorandum, 11 September 1965, "Gas, Vol II" folder, box 194, Country file, Vietnam, NSF, LBJL; McGeorge Bundy to LBJ, memorandum, 23 September 1965, McNamara to LBJ, memorandum, 22 September 1965, and circular, 8 October 1965, all in "Gas, Vol II" folder, box 194, Country file, Vietnam, NSF, LBJL; Editorial, *New York Times,* 11 September 1965; Hersh, "Poison Gas in Vietnam."

53. Goldberg, State Department cable, 15 September 1965, "Gas, Vol II" folder, box 194, Country file, Vietnam, NSF, LBJL. Seymour Hersh quoted a State Department source in 1968 who recalled that the State Department, historically skeptical of gas use since the 1925 agreement, had only with great reluctance confirmed the legality of gas use in 1964 for a Pentagon review (see Hersh, "Poison Gas in Vietnam").

54. Hornig to McGeorge Bundy, memorandum, 17 September 1965, "Gas, Vol II" folder, box 194, Country file, Vietnam, box 194, LBJL.

55. McNamara to McGeorge Bundy, memorandum, 22 September 1965, "Gas, Vol II" folder, box 194, Country file, Vietnam, NSF, LBJL.

4. INTO THE ETHICAL HOT POT

1. Arthur Galston, "An Accidental Plant Biologist," *Plant Physiology* 128 (March 2002): 786–787; Arthur W. Galston, interview by Shirley K. Cohen, 8 October 2002, Pasadena, CA, October 8, 2002, Oral History Project, California Institute of Technology Archives, http://resolver.caltech.edu/CaltechOH:OH_Galston_A; Arthur Galston, "Falling Leaves and Ethical Dilemmas: Agent Orange in Vietnam," in *New Dimensions in Bioethics*, ed. A. W. Galston and E. G. Shurr (Boston: Kluwer Academic, 2001).

2. Galston, "Accidental Plant Biologist," 786–787; Galston, "Falling Leaves."

3. Galston, "Falling Leaves," 116–117. But Buckingham reports that in July 1965, Ambassador Lodge had requested authority "to change the May 1963 guidelines to allow crop destruction operations in more populated and less remote areas of South Vietnam," which resulted in a liberalization of policy. See William Buckingham Jr., *Operation Ranch Hand: The Air Force and Herbicides in Southeast Asia, 1961–1971* (Washington, DC: Office of Air Force History, 1982), 113–114.

4. BAFGOPI letter, 1962, "Boston Area Faculty Group on Public Issues #1" folder, box Series IIa, Subject files An–Ce, Salvador E. Luria Papers, American Philosophical Society.

5. Seymour Hersh, "Our Chemical War, *New York Review of Books,* 25 April 1968; Galston, "Falling Leaves," 118.

6. Briefing transcripts, 8 March 1966, "Vietnam Fasting by Robert Nichols against Vietnam policies" folder, box 197, Country file, Vietnam, NSF, LBJL.

7. *New York Times,* 20 September 1966; Seymour Hersh, "Poison Gas in Vietnam," *New York Review of Books,* 9 May 1968.

8. Quoted in "22 Scientists Bid Johnson Bar Chemical Weapons in Vietnam," *New York Times,* 20 September 1966.

9. "Brom" to Rostow, memorandum, 20 September 1966, and accompanying draft dated 9 March 1966, "10: Chemical and Biological Weapons" folder, box 212, Country file, Vietnam, NSF, LBJL; Benjamin Welles, "Pentagon Backs Use of Chemicals," *New York Times,* 20 September 1966; Buckingham, *Operation Ranch Hand,* 112.

10. Matthew Meselson, "Controlling Biological and Chemical Weapons," in *March 4: Scientists, Students, and Society,* ed. Jonathan Allen (Cambridge, MA: MIT Press, 1970), 151–160; Victor Sidel and Robert Goldwyn, "Chemical and Biological Weapons—A Primer."

11. J. L. Collins et al., "Medical Problems of South Viet Nam, January 1967, Prepared for the Physicians' Committee of Social Responsibility," "The Committee of Responsibility, Inc." folder, Series IIa, box Ch–F, Papers of Salvador Luria, American Philosophical Society.

12. *Scientist and Citizen,* January 1967; *Scientist and Citizen,* February 1967; For a discussion of Commoner's "liberal" science activism, see Kelly Moore, *Disrupting Science: Social Movements, American Scientists, and the Politics of the Military, 1945–1975* (Princeton, NJ: Princeton University Press, 2008), 96–129.

13. John T. Edsall, "Introduction," *Scientist and Citizen,* August–September 1967: 114.

14. Jean Mayer, "Starvation as a Weapon: Herbicides in Vietnam I," *Scientist and Citizen,* August–September 1967: 114. Mayer's analyses, beginning in 1966, spawned a RAND study that concluded that crop destruction was ultimately a propaganda boon to the Viet Cong. See Cecil, quoted in Wilbur Scott, *Vietnam Veterans since the War* (Norman: University of Oklahoma Press: 2004), 81.

15. Arthur W. Galston, "Changing the Environment: Herbicides in Vietnam II," *Scientist and Citizen,* August–September 1967: 122–129. (Many print runs of this issue contained printing errors that omitted the last half of this article.)

16. Ibid.

17. Philip M. Boffey, "Defense Issues Summary of Defoliation Study," *Science* 159, 9 February 1968, 613; J. B. Neilands, "Vietnam: Progress of the Chemical War," *Asian Survey* 10 (March 1970); *Science* 161, No. 3838, 19 July 1968.

18. *Science* 161, 19 July 1968; Sheldon Novick, "The Vietnam Herbicide Experiment," *Scientist and Citizen*, January–February 1968, 21. For a more detailed account of this trip, see David Zierler, *The Invention of Ecocide: Agent Orange, Vietnam, and the Scientists Who Changed the Way We Think about the Environment* (Athens: University of Georgia Press, 2011), 119–121.

19. E. W. Pfeiffer and G. H. Orians, "Mission to Vietnam Part 2," *Scientific Research* 23 (June 1969); Neilands, "Vietnam: Progress," 229.

20. Ernest C. Pollard, "Call for Scientific Help," *Science* 157, no. 3790, 18 August 1967, 755–756.

21. Salvador Luria and Albert Szent-Gyorgyi, "Vietnam: A National Catastrophe," *Science* 158, no. 3797, 6 October 1967, 47; Clipping, *MIT Tech*, 17 October 1967, "Vietnam #2" folder, box Series IIa: Subject files, V–W & Series IIb: Personal Material A–L, Papers of Salvador Luria, American Philosophical Society. Other similar exchanges occurred in the letters pages of *Science:* see G. H. Orians, E. W. Pfeiffer, and Clarence Leuba, "Defoliants: Orange, White, and Blue," *Science* 165, 1 August 1969, 442–443.

22. Alvin L. Young to Branch Chief BCW, memorandum, 4 March 1969, "Trip Report: Las Vegas Nevada, 10–14 Feb 69," Alvin L. Young Collection on Agent Orange, National Agricultural Library, accessed September 2010, http://www.nal.usda.gov/speccoll/findaids/agentorange/text/03656.pdf 12; Neilands, "Vietnam: Progress"; also described in *Federation of American Scientists Bulletin*, May 1967; Hornig to LBJ, 2 June 1967, "Donald Hornig Chronological File: April–June 1967, box 5, Papers of Donald Hornig, LBJL.

23. George R. Harvey and Jay D. Mann, "Picloram in Vietnam," *Scientist and Citizen*, September 1968, 165–171.

24. See Hersh, "Our Chemical War"; Buckingham, *Operation Ranch Hand*, 101, 145–146. Buckingham attributes at least some of this moderation to diplomatic concern. Other reports included Minarik's 1969 assessment of damage to Cambodian rubber plantations, which he concluded had not, as had been initially suspected, been caused by defoliation drift from Tay Ninh operations but were most likely due to direct overhead spraying in Cambodia, which he attributed to "an unknown party." See Charles Minarik, "Report of Cambodian Rubber Damage," 11 December 1969,

Alvin L. Young Collection on Agent Orange, National Agricultural Library, http://www.nal.usda.gov/speccoll/findaids/agentorange/text/03124.pdf. For an alternate account of Tschirley's trip and conclusions, see Zierler, *Invention of Ecocide*, 117–118.

25. Fred H. Tschirley, "Defoliation in Vietnam," *Science* 163 (21 February 1969): 779–786. The *Science* article was based on Tschirley's September 1968 report to the State Department.

26. Ibid.

27. Ibid.

28. Arthur W. Galston and Edwin O. Willis, "Lesser of Two Evils," *Science* 164 (25 April 1969): 373–375; Buckingham, *Operation Ranch Hand*, 134–135, 151. The JCS review was a response to a 1967 RAND study concluding that the crop destruction program had been a failure thus far, alienating allies and failing to diminish VC food supplies.

29. Peter H. Schuck, *Agent Orange on Trial: Mass Toxic Disasters in the Courts* (Cambridge, MA: Belknap Press, 1986). Information about the 1964 Dow standard and the 1965 meeting was revealed during the 1970s Agent Orange class action lawsuit, in the form of a 1965 memo from Dow officials describing the meeting.

30. Scott, *Vietnam Veterans*, 168; Butler, "Connections."

31. Robert Smith, "U.S. Curbs Use of Weed Killer That Produces Rat Deformities," *New York Times*, 30 October 1969; Morton Mintz, "Wide Used Herbicide Tied to Birth Defects," *Washington Post*, 8 April 1970.

32. Joel Kramer, "Yesterday Cyclamates, Today 2,4,5-T, Tomorrow DDT?" *Science* 166 (7 November 1969): 724; Buckingham, *Operation Ranch Hand*, 164; "Contaminant in Pesticide Linked to Defects in Mice," *New York Times*, 7 February 1970.

33. Galston, "Falling Leaves," 114, 119. Galston explained: "[Agent Orange] is synthesized by combing 2,4,5-trichlorophenol with a modified acetic acid under alkaline conditions. But during synthesis at elevated temperatures designed to make the reaction proceed more quickly, an unwanted side reaction occurs. Two molecules of the chlorinated phenols react with each other to form a tricyclic planar compound with four chlorine atoms at the periphery of the plane. Such compounds are able to insert themselves into the groove between the two complementary chains of the duplex DNA molecule, thereby interfering with basic replicative processes essential to the cell. These inadvertently produced dioxins, such as 2,3,7,8-tetrachloro-para-dibenzodioxin (TCDD), turned out to be extremely toxic to both humans and animals."

34. Buckingham, *Operation Ranch Hand*, 166.

35. Matthew Meselson, testimony before Senate Committee on Foreign Relations, 30 April 1969, U.S. Congress, Senate, Committee on Foreign Relations, *Chemical and Biological Warfare*, 91st Cong, 1st Sess., 30 April 1969, LexisNexis, accessed 7 May 2011.

36. Buckingham, *Operation Ranch Hand*, 161; Quoted in William J. Bailey, "Introductory Remarks," in *Chemical Weapons and U.S. Policy: A Report of the Committee on Chemistry and Public Affairs*, 1977, American Chemical Society, Washington, DC, Othmer Library, Chemical Heritage Foundation.

37. Buckingham, *Operation Ranch Hand*, 170–171.

38. Buckingham, *Operation Ranch Hand*, 171–174; Young, "The History of the U.S. Department of Defense Programs for the Testing, Evaluation, and Storage of Tactical Herbicides."

39. For a brief memoir by Lang omitting the details of his Vietnam work, see Anton Lang, "Some Recollections and Reflections, *Annual Review of Plant Physiology* 31 (June 1980): 1–27; Buckingham, *Operation Ranch Hand*, 189–191. As the NAS researchers had not conducted dioxin testing during their initial visits to Vietnam, they later sent a research team to the Thai spraying sites. See "Vegetation Analysis of the Pran Buri Defoliation Test Area 1," Joint Thai-U.S. Military Research and Development Center, January 1966, and "The Effects of Herbicides in South Vietnam, National Academy of Sciences, 1974, both in the Alvin L. Young Collection on Agent Orange at the National Agricultural Library, http://www.nal.usda.gov/speccoll/findaids/agentorange/text/00019.pdf; Deborah Shapley, "Herbicides: DoD Study of Viet Use Damns with Faint Praise," *Science* 177, no. 4051, 1 September 1972, 776–779.

40. Shapley, "Herbicides."

41. Robert Baughman and Matthew Meselson, "An Analytical Method for Detecting TCDD (Dioxin): Levels of TCDD in Samples from Vietnam," *Environmental Health Perspectives* 5 (September 1973): 27–35; Arthur Westing, *Herbicides in War: The Long-Term Ecological and Human Consequences* [Stockholm International Peace Research Institute] (Philadelphia: Taylor & Francis, 1984).

42. Scott, *Vietnam Veterans*, 87–89.

43. Galston, "Falling Leaves," 121; For a skeptical view of the connection between Agent Orange and many of the ailments now recognized by the federal government as related, see Michael Gough, ed., "The Political Science of Agent Orange and Dioxin," *Politicizing Science: The Alchemy of Policymaking* (Stanford: Hoover Institute Press, 2003).

44. *Science* 161, no. 3838, 19 July 1968.

45. Sidel and Goldwyn, "Chemical and Biological Weapons."

46. Testimony before Senate Committee on Foreign Relations, Matthew Meselson, 30 April 1969; Hanson Baldwin, "After Fifty Years the Cry of Ypres Still Echoes," *New York Times,* 18 April 1965; Clipping, *Sunday Herald Traveler,* 13 November 1967, "Retirement dinner, 1967" folder, Papers of Louis F. Fieser and Mary P. Fieser, box 1, HUG (FP) 20, Harvard University Archives.

47. Seymour Hersh, "Your Friendly Neighborhood MACE," *New York Review of Books,* 27 March 1969. For a third assessment, see Eugene Rabinowitch, "Living With H-Bombs," *Bulletin of the Atomic Scientists,* 1 January 1955; Alvin Young, the air force colonel who later wrote a detailed account of the herbicide research program, observed this trend, from a critical perspective, in a 1989 interview. He recalled: "The agenda [of the 1969 visiting AAAS members] was not an agenda that talked about the health of the Vietnamese people. Their agenda was . . . they were wanting the military of the United States out of Vietnam." Quoted in Scott, *Vietnam Veterans,* 82.

48. Galston, "Falling Leaves," 108; Milton Leitenberg, "Biological Weapons," *Scientist and Citizen,* August–September 1967, 166; Sidel and Goldwyn, "Chemical and Biological Weapons."

49. Michael Newton and L. T. Burcham, "Defoliation Effects on Forest Ecology," *Science* 161, 12 July 1968, 109; E. W. Pfeiffer and Gordon H. Orians, "Ecological Effects of the War in Vietnam," *Science* 168, no. 3931, 1 May 1970, 552; For a full account of the development of the concept of "ecocide," see Zierler, *Invention of Ecocide;* Arthur Westing, *Ecological Consequences of the Second Indochina War* (Stockholm: Stockholm International Peace Research Institute, 1976), 80–89; Barry Weisberg, *Ecocide in Indochina: The Ecology of War* (San Francisco: Canfield Press, 1970). This kind of "deep ecology" can itself be considered a new kind of ethics. See David B. Resnick, *The Ethics of Science: An Introduction* (New York: Routledge, 1998).

50. Westing, *Herbicides in War,* 9–15.

51. Galston, "Falling Leaves," 122.

52. Paul Cecil, *Herbicidal Warfare: The Ranch Hand Project in Vietnam* (New York: Prager, 1986), 1; Neilands cites this group specifically in his assertion that opposition to chemical use in Vietnam came largely from two sources: "the science community and the Congress." Neilands, "Vietnam: Progress"; Buckingham, *Operation Ranch Hand,* 160–169.

53. Quoted in a report from Saigon Embassy, 1968, *Herbicide Policy Review,* Alvin L. Young Collection on Agent Orange, National Agricultural Library, http://www.nal.usda.gov/speccoll/findaids/agentorange/text/03124.pdf.

5. DISASTER AND DISILLUSIONMENT IN VIETNAM

1. Spurgeon Keeny to McGeorge Bundy, memorandum, 24 March 1965, "Gas, Vol I" folder, box 194, Country file, Vietnam, National Security File (hereafter NSF), LBJ Presidential Library (hereafter LBJL).

2. For example, see "A Free Zones Policy for Vietnam," March 1966; "Draft Paper by GBK," 23 February 1966; "My Involvement in DCPG," June 1968; and Bundy to Kistiakowsky, 25 February 1966, "Vietnam, 1963–1968" folder, HUG (FP) 94.18 (Vietnam War, ca. 1963–1973), Papers of George B. Kistiakowsky, Harvard University Archives (hereafter Kistiakowsky).

3. Paul Dickson, *The Electronic Battlefield* (Bloomington: Indiana University Press, 1976), 20–21; Gordon MacDonald, speech delivered at the Jasons' twenty-fifth anniversary celebration, "JASON and the DCPG—Ten Lessons, 30 November 1984, folder 12, box 37, MGM; Galbraith to LBJ, 19 April 1966, "Cambridge Discussion group" folder, HUG (FP) 94.18, Kistiakowsky. Kistiakowsky's 13 January 1966 letter in the same file recommended that the United States "pursue an essentially defensive strategy that would rest on the establishment and securing of suitable enclaves along the cast and around Saigon."

4. MacDonald, "JASON and the DCPG," MGM.

5. Goldberger correspondence, September 1964, folder 5, box 35, Murray Gell-Mann Papers, 10219-MS, Caltech Archives, California Institute of Technology (hereafter MGM); MacDonald, "JASON and the DCPG," MGM. Other members of the Nierenberg panel included Murray Gell-Mann, Seymour Deitchman, Ithiel de Sola Pool of MIT, Bernard Fall, and Theodore Vallance of the Special Operations Research Office, among others. For discussion of the parallel travails facing social scientists such as Pool, see Joy Rohde, *Armed with Expertise: The Militarization of American Social Science during the Cold War* (Ithaca: Cornell University Press, 2013).

6. Kenneth Watson, interview by Finn Aaserud, 10 February 1986, Niels Bohr Library and Archives, American Institute of Physics, College Park, MD, http://www.aip.org/history/ohilist/4939.html, accessed 2 May 2011.

7. Goldberger to Jason members, memorandum, 30 September 1964, folder 5, box 35, MGM.

8. Katcher to Nierenberg, memorandum, 31 March 1965, folder 6, box 35, MGM.

9. Katcher to Jason steering committee, 22 January 1966, folder 1, box 36, MGM; Sharp and Wheeler to Goldberger and Ruina, 14 July 1965, folder 1, box 36, MGM; Kenneth Watson, interview by Finn Aaserud.

10. "Jason Publications" 1965–1966, folder 6, box 35, MGM. In MGM documentation, S-255 is referred to only as ". . . Barrier," but the complete report, declassified in 1990, is available at the LBJ Library at http://www .dtic.mil/cgi-bin/GetTRDoc?AD=ADB954899&Location=U2&doc=GetTR Doc.pdf.

11. Zacharias to Gell-Mann, 15 April 1966, folder 1, box 36, MGM.

12. Ibid.

13. Rathjens to Kistiakowsky, 6 April 1966, "Cambridge Discussion group" folder, HUG (FP) 94.18, Kistiakowsky.

14. Kistiakowsky to Rathjens, 21 April 1966, "Cambridge Discussion group" folder, HUG (FP) 94.18, Kistiakowsky.

15. Ruina to MGM, 3 May 1966, folder 1, box 36, MGM; Dickson, *Electronic Battlefield*, 22; Lederman to Zacharias et al., memorandum, 24 June 1966, "Vietnam, 1963–68" folder, HUG (FP) 94.18, Kistiakowsky.

16. John Lewis Gaddis, *Strategies of Containment, Revised and Expanded Edition* (New York: Oxford University Press, 2005), 245. For broad histories of the Vietnam War, see Marilyn Young, *Vietnam Wars, 1945–1990* (New York: Harper Perennial, 1991) and John Prados, *Vietnam: The History of an Unwinnable War, 1945–1975* (Lawrence: University Press of Kansas, 2009).

17. Jason report, "U.S. Bombing in North Vietnam," 29 August 1966, "Vietnam—the Effects of U.S. Bombing on NVN's Ability to Support Military Operations in SVN and Laos" folder, box 192, Country file, Vietnam, NSF, LBJL.

18. Ibid.; Section also quoted in U.S. Senate Committee on Foreign Relations, *Bombing as a Policy Tool in Vietnam: Effectiveness* (Washington, DC: U.S. Government Printing Office, 1972).

19. Ibid.

20. Ibid.

21. Ginsburgh to Rostow, memorandum, 13 September 1966, "Vietnam—the Effects of U.S. Bombing on NVN's Ability to Support Military Operations in SVN and Laos" folder, box 192, Country file, Vietnam, NSF, LBJL.

22. Oleg Hoeffding, draft report, "Bombing North Vietnam: An Appraisal of Economic and Political Effects," October 1966, with accompanying note, "To RG from W," "Vietnam Rand Report: Bombing North Vietnam: An Appraisal of Economic and Political Effects (U)" folder, box 192, Country file, Vietnam, NSF, LBJL; Westmoreland to Rostow, memorandum, 24 October 1966, "10: History File 18–29 October 1966" folder, box 9, Papers of William C. Westmoreland, LBJ Library; U.S. Senate Committee on Foreign Relations, *Bombing as a Policy Tool in Vietnam*.

23. Ibid.

24. Jason report, Institute for Defense Analysis Jason Division, Study S-255, "Air Supported Anti-Infiltration Barrier (C)," August 1966, "Vietnam Barrier, 2D, 9/66–9/68 [1 of 2]" folder, box 74, Country file, Vietnam, NSF, LBJL.

25. Ibid.

26. Dickson, *Electronic Battlefield,* 27; Jason report, Study S-255, "Air Supported Anti-Infiltration Barrier (C)."

27. Jason report, Study S-255, "Air Supported Anti-Infiltration Barrier (C)."

28. Ibid.

29. Ibid; Kistiakowsky to McNamara, 23 June 1966, "Cambridge Discussion group" folder, HUG (FP) 94.18, Kistiakowsky.

30. For insight into McNamara's "technowar" and quantification, see Nick Turse, *Kill Anything That Moves* (New York: Picador, 2013), 41–75 and Michael Adas, *Dominance by Design: Technological Imperatives and America's Civilizing Mission* (Cambridge, MA: Belknap Press, 2006), 281–336; Jason report, Study S-255, "Air Supported Anti-Infiltration Barrier (C)."

31. Jason report, Study S-255, "Air Supported Anti-Infiltration Barrier (C)."

32. Ginsburgh to Rostow, 26 September 1967," Vietnam Barrier, 2D, 9/66–9/68 [1 of 2]" folder, box 74, Country file, Vietnam, NSF, LBJL.

33. MacDonald, "JASON and the DCPG," MGM.

34. Kistiakowsky, "My Involvement in DCPG," June 1968, "Vietnam, 1963–68" folder, HUG (FP) 94.18, Kistiakowsky; McNamara to Starbird, memorandum, 15 September 1966, "8 (History File, 17 July–17 Sept 66)" folder, box 9, Papers of William C. Westmoreland, LBJL; MacDonald, "JASON and the DCPG," MGM.

35. General Westmoreland's Historical Briefing, 6 October 1966, "9 (History File 19 Sep–17 Oct 66)" folder, box 9, Papers of William C. Westmoreland, LBJL.

36. General Westmoreland's Historical Briefing, 6 November 1966, #11 History file, 30 Oct–12 Dec 66 folder, box 10, Papers of William C. Westmoreland, LBJL.

37. Westmoreland to Starbird, 17 December 1966, "12 History File 13 Dec 66–26 Jan 67" folder, box 10, Papers of William C. Westmoreland, LBJL.

38. Kistiakowsky, "My Involvement in DCPG," June 1968. Emphasis added.

39. McMillan to Westmoreland, memorandum, 18 March 1967, "17 History File 1–31 May 67" folder, box 12, Papers of William C. Westmoreland, LBJL.

40. Rostow to LBJ, 15 July 1967, "Vietnam Barrier, 2D, 9/66–9/68 [1 of 2]" folder, box 74, Country file, Vietnam, NSF, LBJL; Kistiakowsky memo-

randum, 12 July 1967, "Cambridge Discussion group" folder, HUG (FP) 94.18, Kistiakowsky.

41. Richard Garwin, interview by Finn Aaserud, 24 June 1991, IBM Research Lab, Croton-Harmon, NY, American Institute of Physics, College Park, MD, http://www.aip.org/history/ohilist/5075.html; Confirmation of the February 1968 trip by Garwin, Kendall, and others in Ann Finkbeiner, *The Jasons: The Secret History of Science's Postwar Elite* (New York: Penguin, 2006), 98.

42. Neustadt memo, 27 June 1967, "Vietnam Barrier, 2D, 9/66–9/68 [1 of 2]" folder, box 74, Country file, Vietnam, NSF, LBJL.

43. Steven Weinberg, "What is JASON?" Nautilus Institute for Security and Sustainability, http://www.nautilus.org/projects/foia/essentially-annihilated/what-is-jason-author-steven-weinberg/, accessed 25 December 2010.

44. Freeman Dyson, *Disturbing the Universe* (New York: Basic, 1979), 150.

45. Dyson, *Disturbing the Universe*, 148–149.

46. Peter Hayes and Nina Tannenwald, "Nixing Nukes in Vietnam," *Bulletin of the Atomic Scientists*, May/June 2002.

47. F. Dyson, R. Gomer, S. Weinberg, and S. C. Wright, "Tactical Nuclear Weapons in Southeast Asia, Study S-266," Jason Division, Institute for Defense Analysis, March 1967, http://www.nautilus.org/projects/foia/essentially-annihilated/dyson67.pdf/view; Dyson, *Disturbing the Universe*, 149.

48. Weinberg, "What is JASON?," Nautilus Institute for Security and Sustainability; Dyson, Gomer, Weinberg, and Wright, "Tactical Nuclear Weapons, Study S-266."

49. Ibid.

50. Ibid.

51. Ibid.

52. Dyson, *Disturbing the Universe*, 149; Peter Hayes and Nina Tannenwald, "Nixing Nukes in Vietnam," *Bulletin of the Atomic Scientists*, May/June 2002.

53. Hayes and Tannenwald, "Nixing Nukes in Vietnam"; State Department memo, June 1968, "Vietnam 7F (3) 4/68–10/68 Congressional Attitudes and Statements [1 of 2]" folder, box 102 [2 of 2], Country file, Vietnam, NSF, LBJL; Dyson, *Disturbing the Universe*, 149.

54. Andrew Hamilton, "Vietnam-Fencing in the North," *New Republic*, 8 July 1967; Joseph Kraft, "Policy of Creeping Gavinism Affects U.S. Vietnam Posture," *Washington Post*, 7 September 1967.

55. McNamara to LBJ, memorandum, 11 September 1967, "Vietnam Barrier, 2D 9/66–9/68 [2 of 2]" folder, box 74, Country file, Vietnam, NSF, LBJL. The Kraft article was the first of several major leaks concerning the barrier.

In late October, Jack Robertson's detailed account of the barrier in *Electronic News* prompted Donald Hornig to write to Walt Rostow, "This is a shocking commentary on our security." Dean Rusk reiterated that technical information about Muscle Shoals should only be shared with those "with true need to know," because "effectiveness will in large measure depend upon enemy ignorance of how it works."

56. McNamara to LBJ, memorandum, 11 September 1967.

57. V. L. Fitch and L. Lederman, abstract, "Air-Sown Mines for the Massive Barrier" (Alexandria, VA: Institute for Defense Analysis, Jason Division, May 1967), http://www.nps.edu/Library/Research/Bibliographies/Land-Mines/LandMinesBibTechRptsDH.html; Kistiakowsky to Polly Yates, memorandum, 6 November 1967, "Cambridge Discussion group" folder, HUG (FP) 94.18, Kistiakowsky.

58. Jack Robertson, "Viet 'Wall' Will Sense Enemy, Flash Warning to Main HQ," *Electronic News,* 30 October 1967.

59. Dickson, *Electronic Battlefield,* 32–35.

60. Kistiakowsky, "My Involvement in DCPG," June 1968; Jason study, January 1968, "Vietnam Four Volume Study of the Bombing of North Vietnam 1/8/68 [1 of 2]" folder, box 247, Country file, Vietnam, NSF, LBJL.

61. Kistiakowsky to Pimentel, 15 November 1967, "Vietnam, 1963–68" folder, HUG (FP) 94.18, Kistiakowsky.

62. Zachariasen to Kistiakowsky, first letter, 7 June 1968, "Vietnam, 1963–68" folder, HUG (FP) 94.18, Kistiakowsky. DCPG attendees at the 9 November 1967 meeting with McNamara included: David Caldwell, Richard Garwin, Murray Gell-Mann, Henry Kendall, George Kistiakowsky, Harold Lewis, Jack Ruina, Leonard Sheingold, F. Zachariasen. Kistiakowsky to Yates, memorandum, 6 November 1967, "Vietnam, 1963–68" folder, HUG (FP) 94.18, Kistiakowsky.

63. Kendall to Kistiakowsky, 10 June 1968, "Vietnam, 1963–68" folder, HUG (FP) 94.18, Kistiakowsky; Kistiakowsky, "My Involvement in DCPG," June 1968; Kistiakowsky to McNamara, 11 December 1967, "Vietnam, 1963–68" folder, HUG (FP) 94.18, Kistiakowsky.

64. Kistiakowsky, "My Involvement in DCPG," June 1968.

65. Zachariasen to Kistiakowsky, first and second letters, 7 June 1968, "Vietnam, 1963–68" folder, HUG (FP) 94.18, Kistiakowsky.

66. See letters to Kistiakowsky from David Caldwell, 1 February 1968; Leonard Sheingold, 2 February 1968; Jack Ruina, 17 January 1968; William Nierenberg, 13 March 1968; all in "Vietnam, 1963–68" folder, HUG (FP) 94.18, Kistiakowsky.

67. Letters from Max Delbrück, 2 February 1968; Norman Davidson, 8 February 1968; Jesse Greenstein, 1 March 1968; George Hammond,

21 February 1968; John Roberts, 31 January 1968; Mark Kac, 9 February 1968; George Fraenkel, 26 January 1968; Kenneth Pitzer, 27 January 1968; Emanuel Piore, 25 January 1968; Wolfgang Panofsky, 7 February 1968; all in "Vietnam, 1963–68" folder, HUG (FP) 94.18, Kistiakowsky.

68. Letters from Emanuel Piore, 25 January 1968; Wolfgang Panofsky, 7 February 1968; all in "Vietnam, 1963–68" folder, HUG (FP) 94.18, Kistiakowsky.

69. George Pimentel to Kistiakowsky, 23 January 1968, "Vietnam, 1963–68" folder, HUG (FP) 94.18, Kistiakowsky.

70. Kistiakowsky to Fisher, 14 March 1968, and Kistiakowsky to Mather, 14 March 1968, Vietnam, 1963–68" folder, HUG (FP) 94.18, Kistiakowsky; Kistiakowsky to MacDonald, 7 February 1968, "Vietnam, 1963–68" folder, HUG (FP) 94.18, Kistiakowsky.

71. D. S. Greenberg, "Kistiakowsky Cuts Defense Department Ties over Vietnam," *Science* 158, March 1968, 958; Evert Clark, "Top Scientist Cuts All Links to War," *New York Times,* 1 March 1968.

72. Case, Lederman, and Ruderman to Goldberger, 29 August 1966, folder 1, box 36, MGM.

73. Jason Steering Committee meeting minutes, 1967, folder 1, box 36, MGM.

74. Martin to Jason members, 13 November 1967, folder 1, box 36, MGM; Martin to MGM, 3 January 1968, folder 1, box 37, MGM.

75. Department of Defense messages, 4 April 1968 and February 1968, "Vietnam Barrier, 2D 9/66–9/68 [2 of 2]" folder, box 74, Country file, Vietnam, NSF, LBJL; Earl Young to Bromley Smith, memorandum, 9 September 1968, and attached report, "Southeast Asia Analysis report for April 1968," "Vietnam Barrier, 2D 9/66–9/68 [2 of 2]" folder, box 74, Country file, Vietnam, NSF, LBJL.

76. Wheeler to LBJ, 17 June 1968, and PSAC report, "Vietnam Material including a report by the President's Science Advisory Committee on the effects of the bombing of NVN 4/68–5/68" folder, box 250, Country file, Vietnam, NSF, LBJL.

77. Dickson, *Electronic Battlefield,* 44–45.

78. MacDonald, "JASON and the DCPG," MGM.

79. Deborah Shapley, "Jason Division: Defense Consultants Who Are Also Professors Attacked," *Science* 179, 2 February 1973; Kenneth Watson, interview by Finn Aaserud; Minutes of 14 January 1968 Jason Steering Committee Meeting, 18 January 1968, folder 1, box 37, MGM; Minutes of 24 April 1968 Jason Steering Committee meeting, 2 May 1968, folder 1, box 37, MGM.

80. These Vietnam-related topics are listed on the February summary of the May agenda, but not on the April summary; Meeting is mentioned in a

later memo from Hal Lewis to the Jasons, 22 May 1970, folder 3, box 37, MGM; Hal Lewis's report on the meeting is quoted in Ann Finkbeiner, *The Jasons: The Secret History of Science's Postwar Elite* (New York: Penguin, 2006), 91; Minutes of August and September Jason Steering Committee meetings, folder 2, box 37, MGM. The missing attendees were Blankenbecler, Drell, Dyson, Foley, Goldberger, Kendall, Ruderman, Salpeter, Sands, and Weinberg.

81. Salpeter to Lewis, 12 May 1970, folder 3, box 37, MGM; Finkbeiner, *The Jasons*, 91.

82. Lewis to Jason members, 22 May 1970, folder 3, box 27, MGM.

83. Lewis to MGM, 12 April 1973, folder 7, box 37, MGM.

84. Shapley, "Jason Division."

85. Shapley, "Jason Division"; SESPA, "Science against the People," December 1972, folder 3, box 5, Papers of Brian Schwartz, 1966–1977, Niels Bohr Library & Archives, American Institute of Physics, College Park, MD.

86. Shapley, "Jason Division."

87. SESPA, "Science against the People," December 1972.

88. Ibid.

89. Ibid.

90. Ibid.

91. Ibid; For additional discussion of SESPA and Science for the People in the context of the Jasons, see Kelly Moore, *Disrupting Science: Social Movements, American Scientists, and the Politics of the Military, 1945–1975* (Princeton, NJ: Princeton University Press, 2008), 170–175.

92. Partial listing of recent Jason reports available through the Federation of American Scientists, http://www.fas.org/irp/agency/dod/jason/; Finkbeiner, 115.

6. INSTITUTIONAL RECKONINGS AT MIT

1. Jonathan Allen, ed., *March 4: Scientists, Students, and Society* (Cambridge, MA: MIT Press, 1970); SDS, "MIT and the Warfare State," 1967, folder 21, box 1, Records of the MIT Review Panel on Special Laboratories 1969–1971, Massachusetts Institute of Technology (hereafter MIT).

2. For overview histories of MIT's military and industrial ties, see relevant sections of Noam Chomsky et al., *The Cold War and the University: Toward an Intellectual History of the Postwar Years* (New York: New Press, 1998); David Kaiser, ed., *Becoming MIT: Moments of Decision* (Cambridge, MA: MIT Press, 2010); Stuart Leslie, *The Cold War and American Science: The Military-Industrial-Academic Complex at MIT*

and Stanford (New York: Columbia University Press, 1993); Richard Freeland, *Academia's Golden Age: Universities in Massachusetts, 1945–1970* (New York: Oxford University Press, 1992); and David Noble, *America by Design: Science, Technology, and the Rise of Corporate Capitalism* (New York: Knopf, 1977).

3. Dorothy Nelkin, *The University and Military Research: Military Research at MIT* (Ithaca: Cornell Press, 1972), 18; Leslie, *Cold War,* 14.

4. Stuart Leslie writes that Bush "kept his relationships with universities and industry contractual, restricted in time and in scope . . . He always considered OSRD [Office of Scientific Research and Development] a strictly emergency operation and dismantled it promptly at the end of the war, despite considerable opposition from the Pentagon and the White House." Leslie, *Cold War,* 7, 29; Victor Weisskopf, *The Joy of Insight: Passions of a Physicist* (New York: Basic, 1991), 166.

5. Leslie, *Cold War,* 8; Janet Martin-Nielson, " 'It Was All Connected': Computers and Linguistics in Early Cold War America," in *Cold War Social Science: Knowledge Production, Liberal Democracy, and Human Nature,* ed. Mark Solovey and Hamilton Cravens (New York: Palgrave MacMillan, 2012), 63–78; Walter Sullivan, "Fighting the Misuse of Knowledge," *New York Times,* 9 February 1969; "Data on Universities," folder 8, box 1, MIT; Nils Gilman, *Mandarins of the Future: Modernization Theory in Cold War America* (Baltimore: Johns Hopkins University Press, 2004).

6. Leslie, *Cold War,* 32–33; Roger L. Geiger, *Research and Relevant Knowledge: American Research Universities since World War II* (New York: Oxford University Press, 1993), 63–67.

7. Everett to Pounds, 6 May 1969, folder 3, box 1, MIT.

8. "Background on Research at Massachusetts Institute of Technology," 4 March 1969, folder 8, box 1, MIT; Herlin testimony transcript, 1 May 1969, folder 21, box 2, MIT; "MIT Directory of Current Research," 1969, folder 10, box 1, MIT.

9. "MIT Directory of Current Research," 1969; Lincoln/ARPA Project PRESS pamphlet, "Text for MIT's President's Report, July 1968," folder 18, box 1, MIT; *Boston Globe,* 3 June 1969.

10. Leslie, *Cold War,* 81. For a deeper history of Draper's role in the development of guidance systems, see Donald Mackenzie, *Inventing Accuracy: An Historical Sociology of Nuclear Missile Guidance* (Cambridge, MA: MIT Press, 1990); Woodbury testimony transcript, 30 April 1969, folder 19, box 2, MIT.

11. Leslie reports, "In the post-Sputnik missile buildup (1957–63) the Air Force pumped $9 million a year into the Instrumentation Laboratory for

ballistic missile guidance research and development." Leslie, *Cold War,*
92–98.

12. "Background on Research at the Massachusetts Institute of Technology," 4
March 1969; Woodbury testimony transcript, 30 April 1969, folder 19,
box 2, MIT.

13. For short, useful summaries of MIRV and MIT, see Nelkin, *University,*
48–53, and for MIRV generally, see George Rathjens and George
Kistiakowsky, "The Limitation of Strategic Arms," *Scientific American*
222, no. 1, January 1970; "Poseidon Program Organization" chart collection, April 1969, folder 12, box 1, MIT.

14. "Instrumentation Laboratory" booklet, April 1968, folder 13, box 1, MIT.
Nelkin estimates total lab employment at 2248 in 1969; John Walsh, "MIT:
Panel on Special Labs Asks More Nondefense Research," *Science* 164, no.
3885, 13 June 1969, 1264; Transcript, 29 April 1969, folder 18, box 2, MIT.

15. Quoted in Nelkin, *University,* 20; Karl Compton, "Annual Report of the
President: The War Record of the Institute," 1945, folder 3, box 2, MIT.

16. Leslie, *Cold War,* 43; Quoted in "The Physicist with a Cause—An
Interview with Jay Orear," *Physics Today,* May 1968, 85–91, folder 7,
box 4, Papers of Brian Schwartz, 1966–1977 (hereafter BSP), Niels Bohr
Library and Archives, American Institute of Physics, College Park, MD.

17. Jerome Wiesner, clipping, "The Federal Presence at MIT," *Technology
Review,* April 1967, folder 25, box 1, MIT.

18. Transcript, 15 May 1969, folder 5, box 3, MIT; SDS, "MIT and the
Warfare State," 1967, folder 21, box 1, MIT.

19. See Walsh, "MIT: Panel on Special Labs," 1264–1265. Walsh observes
that "there is little sign of the radicalization of the undergraduates at
MIT" and "by tradition and still dominantly in atmosphere, MIT is an
engineering school, and engineering students have been a conspicuously
inert group in most universities during the current upheavals." He credits
much of the activism on campus to graduate students, liberal faculty
members, and outside groups.

20. Murray Eden, "Historical Introduction," in Allen, *March 4.*

21. Murray Eden, "Historical Introduction," in Allen, *March 4;* Kelly Moore,
*Disrupting Science: Social Movements, American Scientists, and the
Politics of the Military, 1945–1975* (Princeton, NJ: Princeton University
Press, 2008), 138.

22. Ibid.; Moore, *Disrupting Science,* 142; *Boston Globe,* 21 January 1969.
The letter to DuBridge complained about the militarization of science,
charging that funding emphases on military applications had led to the
creation of "unnecessary" and dangerous antiballistic missile and chemical
and biological weapons technology. Despite the SACC-UCS split, the

original DuBridge letter was signed by numerous members of MIT's faculty, including Feshbach, Luria, Jerome Lettvin, and Noam Chomsky. Of all the MIT departments represented, physics provided the largest number of signers, followed by biology and mathematics. See also *New York Times*, 21 January 1969 and *Boston Globe*, 24 January 1969.

23. Murray Eden, "Historical Introduction," in Allen, *March 4*, xi; Clippings, *New York Times*, *Science News*, 22 February 1969, and *Catalyst*, March 1969, folder 8, box 4, BSP; SACC flyer, n.d., folder 8, box 4, BSP.

24. Union of Concerned Scientists, "Faculty Statement," in Allen, *March 4*, xii–xxiii.

25. Clippings, *Electronic News*, 17 February 1969 and 24 February 1969, folder 7, box 4, BSP; Clipping, *Industrial Research*, March 1969, folder 8, box 4, BSP.

26. Clippings, *New York Times*, 5 March 1969, *The Tech*, 7 March 1969, *Science News* 95, 15 March 1969, 257, folder 8, box 4, BSP.

27. Jonathan Kabat, "Proposals for Further Action," in Allen, *March 4*, 125.

28. Weisskopf, "Intellectuals in Government," in Allen, *March 4*, 29.

29. Noam Chomsky, *American Power and the New Mandarins* (New York: Pantheon, 1969), 324, 339.

30. Noam Chomsky, "Responsibility," in Allen, *March 4*, 8–14.

31. W. G. McMillan, "The Scientist in Military Affairs," in Allen, *March 4*, 17–18.

32. Ibid.

33. D. S. Dayton, "Problems and Possibilities in Reconversion," in Allen, *March 4*, 42–48; L. W. Gruenberg, "Reconversion within Government Laboratories," in Allen, *March 4*, 49–50.

34. R. F. Probstein, "Reconversion and Academic Research," in Allen, *March 4*, 34. For an overview history of the Fluid Mechanics Lab in this era, see also Matthew Wisnioski, *Engineers for Change: Competing Visions of Technology in 1960s America* (Cambridge, MA: MIT Press, 2012), 103–106.

35. "Discussion," in Allen, *March 4*, 51–56.

36. SACC pamphlet, n.d., folder 2, box 5, BSP.

37. *Thursday* no. 2, 24 April 1969, folder 8, box 4, BSP; Nelkin, *University*, 67.

38. Transcript and minutes, 27 April 1969, folder 16, box 2, MIT.

39. Transcript, 29 April 1969, folder 18, box 2, MIT.

40. Ibid.

41. Ibid.

42. Ibid.

43. Transcript, 2 May 1969, folder 22, box 2, MIT; Review Panel abstracts, folder 13, box 3, MIT.

44. Woodbury testimony transcript, 30 April 1969, folder 19, box 2, MIT.
45. Ibid.
46. Ibid.
47. Forter and Houston testimony transcripts, 30 April 1969, folder 19, box 2, MIT.
48. Porter to Panel, 8 May 1969, folder 19, box 1, MIT.
49. Review Panel abstract, 3 May 1969, folder 13, box 3, MIT.
50. Transcript, 1 May 1969, Lincoln Laboratory, folder 21, box 2, MIT.
51. Ibid.
52. Transcript, 2 May 1969, folder 23, box 2, MIT. For example, researchers assured Kabat that Lincoln's satellite technology, purchased by the Defense Department, was not being used in connection with the war in Vietnam.
53. "Fact Sheet on Lincoln Laboratory," folder 18, box 1, MIT.
54. "Lincoln Laboratory Staff submission to the Panel on Special Laboratories," 15 May 1969, folder 18, box 1, MIT.
55. Review Panel abstract, 4 May 1969, folder 13, box 3, MIT.
56. Transcript, 15 May 1969, folder 5, box 3, MIT.
57. Ibid.
58. UCS statement, 8 May 1969, folder 2.3, box 2, MIT; *Electronic News*, 5 May 1969, folder 25, box 1, MIT.
59. Poll results, folder 32, box 3, MIT. Committee staff sorted response letters into separate position categories, from strong defense of labs to strong opposition to weapons-related research. By my count, reactions supporting the labs or status quo numbered 173 individual letters and 254 petition signatories; opponents 25 letters plus over a hundred signatories of the UCS statement. About 28 letters complained of other, indirectly related matters, such as the tactics of student protesters. Allen to Pounds, 7 May 1969; Broxmeyer statement, 8 May 1969; Casey statement, 6 May 1969; Position Papers, folder 33, box 3, MIT.
60. Instrumentation Laboratory documentation, April 1968, folder 13, box 1, MIT; Broxmeyer position statement, folder 3, box 1, MIT, emphasis added.
61. Review Panel abstract, 8 May 1969, folder 13, box 3, MIT.
62. Lee DuBridge, *Bulletin of the Atomic Scientists,* May 1969, reprinted in the *Boston Globe,* 11 May 1969.
63. Transcript, 13 May 1969, folder 2, box 3, MIT.
64. Ibid.
65. Chomsky to Pounds, folder 26, box 1, MIT.
66. Ibid.
67. Ibid.
68. Ibid.

69. MIT Office of Public Relations press release, 2 June 1969, folder 19, box 1, MIT; "First Report of the Review Panel on Special Laboratories," 31 May 1969, folder 14, box 4, MIT.

70. The history of militarized social science research runs parallel to the narrative of this book. See Joy Rohde, *Armed with Expertise: The Militarization of American Social Research during the Cold War* (Ithaca: Cornell University Press, 2013); Chomsky statement, Appendix II, "First Report on the Review Panel on Special Laboratories," 31 May 1969, folder 14, box 4, MIT.

71. *New York Times*, 4 June 1969.

72. Nelkin, *University*, 86–87, 90–92, 120.

73. Alsop clippings, folder 2, box 2, MIT; William Leavitt, "A Triumph of Reverse McCarthyism," *Air Force/Space Digest*, December 1969, 46; Signs mentioned in Nelkin, *University*, 96 and *The Tech* (MIT), 24 October 1969.

74. *The Tech* (MIT), 26 September and 24 October 1969.

75. Leslie, *Cold War*, 239; "Statement by President Howard Johnson on the Special Laboratories," 20 May 1970, Appendix II; Nelkin, *University*, 181; Victor McElheny, "MIT Administration Makes Public Its Intentions on Disposition of Draper and Lincoln Laboratories," *Science* 168, 29 May 1970, 1074–1075.

76. *The Tech* (MIT), 22 May 1970; Nelkin, *University*, 145.

77. Clauser to Ruina, 17 September 1969, folder 18, box 1, MIT.

78. *The Tech* (MIT), 20 March 1970; Leslie, *Cold War*, 249–250

79. Leslie reports that "about a quarter of Lincoln's budget [came] from the Strategic Defense Initiative." Leslie, *Cold War*, 250; *Boston Globe*, 24 April 1986.

80. Margaret Pugh O'Mara, *Cities of Knowledge: Cold War Science and the Search for the Next Silicon Valley* (Princeton, NJ: Princeton University Press, 2005; Leslie, *Cold War*, 254.

81. The trend of suburbanized "research parks" predated Vietnam War-era protests; See O'Mara, *Cities of Knowledge*, 28–45; Joy Rohde, "From Expert Democracy to Beltway Banditry: How the Antiwar Movement Expanded the Military-Academic-Industrial Complex" in *Cold War Social Science: Knowledge Production, Liberal Democracy, and Human Nature*, ed. Mark Solovey and Hamilton Cravens (New York: Palgrave MacMillan, 2012), 137–153.

82. Nelkin, *University*, 155–56.

83. Jerome Wiesner, "Commencement Address," 2 June 1980, in *Jerry Wiesner: Scientist, Statesman, Humanist*, ed. Walter A. Rosenblith (Cambridge, MA: MIT Press, 2003), 365–366.

7. THE NEW LEFT ASSAULT ON NEUTRALITY

1. Harland G. Bloland and Sue M. Bloland, *American Learned Societies in Transition* (New York: McGraw Hill, 1974), 87; Krane to Schwartz; 12 June 1973; folder 2, box 2, Papers of Brian Schwartz, 1966–1977 (hereafter BSP), Niels Bohr Library and Archives, American Institute of Physics (hereafter AIP), College Park, MD.
2. Charles Schwartz, interview by Finn Aaserud, 5 May 1987, Niels Bohr Library and Archives, AIP, http://www.aip.org/history/ohilist/5053.html (hereafter Schwartz interview transcript).
3. Ibid.
4. Ibid.
5. *Physics Today* clipping, December 1967, folder 1, box 3, BSP; "Physicists and Public Policy," *Physics Today*, December 1967, 128, folder 1, box 3, BSP.
6. "Physicists and Public Policy," *Physics Today*, December 1967; C. H. Blanchard, "Letters to the Editor," *Physics Today*, April 1968. For historical perspectives on German scientists during the Nazi era, see Robert Proctor, "Nazi Science and Nazi Medical Ethics: Some Myths and Misconceptions," *Perspectives in Biology and Medicine* 43 (2000): 335–346 and Alan Beyerchen, *Scientists under Hitler: Politics and the Physics Community in the Third Reich* (New Haven: Yale University, 1977).
7. Editors, "Physicists and Public Policy," *Physics Today*, December 1967.
8. Ibid.
9. Eugene Wigner, "Letters to the Editor," *Physics Today*, December 1967, 69; Frederick Seitz, "Letters to the Editor," *Physics Today*, January 1968, 17; Edward Teller, "Letters to the Editor," *Physics Today*, January 1968, 17–18.
10. Goetz Oertel and others, "Letters to the Editor," *Physics Today*, February 1968.
11. Schwartz interview transcript.
12. Eugene Saletan, "Letters to the Editor," *Physics Today*, February 1968.
13. Clipping, *Physics Today*, May 1968, folder 7, box 4, BSP; Schwartz interview transcript.
14. Martin Perl, "Letters to the Editor," *Physics Today*, March 1968, 9.
15. Bloland and Bloland, *Learned Societies*, 90; F. Jona, "Letters to the Editor," *Physics Today*, March 1968, 9. For additional discussion of the Schwartz Amendment in the context of ethics, see Kelly Moore, *Disrupting Science: Social Movements, American Scientists, and the Politics of the Military, 1945–1975* (Princeton, NJ: Princeton University Press, 2008), 149–151.

16. Barry Casper, "Physicists and Social Responsibility: A New Role for the APS," 1973–1974, folder 3, box 5, BSP.

17. Vonnegut, quoted in *Electronic News* clipping, 2 October 1969, folder 7, box 4, BSP; Brian Schwartz and Emanuel Maxwell, "Letter to the Editor," *Scientific Research*, 12 March 1969, folder 1, box 1, BSP.

18. Schwartz and Maxwell, "Letter to the Editor," *Scientific Research*. For a retrospective reflection on the history of the forum, see David Hafemeister, "Forum History at Forty (2012)," Forum on Physics & Society, American Physical Society, http://www.aps.org/units/fps/history.cfm.

19. SSPA flyer, folder 7, box 4, BSP.

20. Casper, "Physicists"; AP clipping, folder 7, box 4, BSP; Schwartz interview transcript; Moore, *Disrupting Science*, 151–152.

21. Casper, "Physicists." Late in his life, Hans Bethe would publicly endorse the idea of a Hippocratic oath as well. See S. S. Schweber, *In the Shadow of the Bomb: Oppenheimer, Bethe, and the Moral Responsibility of the Scientists* (Princeton, NJ: Princeton University Press, 2000), 171.

22. Schwartz interview transcript. Scanned issues of *Science for the People* available at: http://science-for-the-people.org/sftp-resources/magazine/.

23. Schwartz interview transcript; Earl Callen, "Report of the Chairman," *Newsletter of the Forum on Physics and Society* 1, July 1972, folder 8, box 1, BSP; "July 1971–June 1972 budget for APS's Forum on Physics and Society," folder 5, box 1, BSP.

24. Hafemeister, "Forum History."

25. "Forum Symposium on Physicists and Public Affairs," 25 April 1972, folder 6, box 1, BSP.

26. *New York Times*, 1 May 1971; Robert B. Semple Jr., "Kissinger in Talk with 3 Berrigan Case Figures, *New York Times*, 13 March 1971.

27. William Davidon, "Significance of the Harrisburg Trial for Scientific Workers" conference abstract, folder 7, box 1, BSP.

28. Callen to Havens, 22 February 1972, and Orear to Callen, 28 March 1972, folder 6, box 1, BSP.

29. APS paper, Pierre Noyes, folder 7, box 1, BSP.

30. Pierre Noyes, press conference text, 25 April 1972, folder 7, box 1, BSP.

31. Victor Cohn, "Scientists Face Their Frankensteins," *Washington Post*, 27 June 1971.

32. Advertisements, *Science for the People*, September 1971 and May 1971.

33. "CSRE Statement of Purpose," *Spark* 1, no. 2, Fall 1971, folder 3, box 5, BSP. For more on CSRE see Matthew Wisnioski, *Engineers for Change: Competing Visions of Technology in 1960s America* (Cambridge, MA: MIT Press, 2012), 111–121; *Spark* 1, no. 2, Fall 1971, folder 3, box 5, BSP.

34. Robert Gillette, "ACS: Disgruntled Chemists Seek New Activist from Old Society," *Science* 173, no. 4003, 24 September 1971, 1218–1220.

35. "Dissent Blooms at AAAS Circus," *Nature* 229, 8 January 1971, 81–82, folder 1, box 5, BSP.

36. Ibid.

37. "Dissent Blooms"; for a history of the Wisconsin bombing, see Tom Bates, *Rads: The 1970 Bombing of the Army Math Research Center and Its Aftermath* (New York: HarperCollins, 1992).

38. "Dissent Blooms."

39. Ibid; Philip H. Abelson, "The Chicago Meeting," *Science*, 22 January 1971.

40. "1970 Chicago AAAS Actions: Review and Critique," *Science for the People*, February 1971, 8–11.

41. Victor Cohn, "Scientists Face Their Frankensteins"; for further discussion of the strategies of Science for the People, see Kelly Moore, *Disrupting Science*, 153–155, 161–169.

42. *New York Times*, 1 May 1970.

43. Clippings, *Chemical &Engineering News*, 1 June 1970 and 18 May 1970, folder 3, box 3, BSP.

44. Leaflet, "Why Strike? What Princeton Can Do as an Institution," Special Committee on Sponsored Research Records (hereafter Kuhn Papers), folder 4, box 2, University Archives, Department of Rare Books and Special Collections, Princeton University Library.

45. Council Resolution, 12 May 1970, folder 3, box 2, Kuhn Papers.

46. *New York Times*, 19 May 1970. For additional perspectives on engineers and antiwar activism at Princeton, see Wisnioski, *Engineers for Change*, 106–111.

47. Stanislaw Ulam, *Adventures of a Mathematician* (New York: Scribner, 1976).

48. Kuhn to Hitch, 12 June 1970, folder, box 5, Kuhn Papers; Kuhn to Christy, 2 July 1970, folder 4, box 5, Kuhn Papers.

49. Summerfield to Kuhn, 13 November 1970, folder 2, box 1, Kuhn Papers.

50. Summerfield to Fleming, 27 October 1970, folder 2, box 1, Kuhn Papers; Minutes, 22 September 1970, folder 7, box 2, Kuhn Papers. For more on the particularities of the problems for engineers posed by the political crises of the late 1960s, see Matthew Wisnioski, "Inside 'The System': Engineers, Scientists, and the Boundaries of Social Protest in the Long 1960s," *History and Technology* 19, no. 4 (2003): 313–333 and Wisnioski, *Engineers for Change.*

51. "Researching Research," folder 2, box 4, Kuhn Papers; *Daily Princetonian* letter; 15 April 1970, folder 4, box 2, Kuhn Papers.

52. Summerfield to Jacobs, 13 May 1970, folder 4, box 2, Kuhn Papers; Jacobs to Kuhn; 21 May 1970, folder 4, box 2, Kuhn Papers.
53. Minutes, 15 September 1970, folder 7, box 2, Kuhn Papers; Frosch speech, 29 September 1960, folder 1, box 5, Kuhn Papers.
54. *Daily Princetonian,* 23 September 1970, folder 12, box 2, Kuhn Papers.
55. Report draft, 1970, folder 18, box 1, Kuhn Papers.
56. Minutes, 16 October 1970, folder 7, box 2, Kuhn Papers.
57. Kuhn to Hitch, 12 June 1970, folder 3, box 5, Kuhn Papers, emphasis added.
58. Cooke to Kuhn, 30 July 1970, folder 9, box 5, Kuhn Papers; Hitch to Kuhn, 23 August 1970, folder 3, box 5, Kuhn Papers.
59. Kuhn to committee, 17 May 1971, folder 6, box 3, Kuhn Papers.
60. For institutional studies of this phenomenon, see Peter Galison and Bruce Hevly, eds., *Big Science: The Growth of Large-Scale Research* (Stanford: Stanford University Press, 1992) and Stuart Leslie, *The Cold War and American Science: The Military-Industrial-Academic Complex at MIT and Stanford* (New York: Columbia University Press, 1993).
61. For an account on the parallel trajectory of militarized social science, see Joy Rohde, *Armed with Expertise: The Militarization of American Social Science during the Cold War* (Ithaca: Cornell University Press, 2013); "Sponsored Research: Do You Control It or Does It Control You?" *College Management,* April 1969, MIT 2.3; and Tom Bates, *Rads: The 1970 Bombing of the Army Math Research Center at the University of Wisconsin and Its Aftermath* (New York: HarperCollins, 1992).
62. "Sponsored Research: Do You Control It or Does It Control You?" *College Management,* April 1969, MIT 2.3; Panofsky speech to students, 18 April 1969, MIT 2.9; Panofksy, "Statement concerning SRI to Trustees Panel on April 30, 1969," MIT 1.19.
63. Chomsky, also quoted in John Walsh, "MIT: Panel on Special Labs Asks More Nondefense Research," *Science* 164, 13 June 1969, 1264–1265.
64. Union of Concerned Scientists, "Faculty Statement," in *March 4: Scientists, Students, and Society,* ed. Jonathan Allen (Cambridge, MA: MIT Press, 1970), xxii–xxiii.
65. For example, see Jeremy Suri, "The Rise and Fall of an International Counterculture, 1960–1975," *The American Historical Review* 14 (February 2009): 45–68.

8. COLLAPSE OF THE SPUTNIK ORDER

1. Narrative statement, "Volume II Documentary Supplement [1 of 3]" folder, Papers of Donald Hornig, LBJ Presidential Library (hereafter LBJL).

2. "Activities Related to Vietnam," "Volume II Documentary Supplement [1 of 3]" folder, Papers of Donald Hornig, LBJL.
3. Ibid.
4. Ibid.
5. Hornig to LBJ, 28 December 1965, "Donald Hornig Chronological File October–December 1965" folder, box 3, Papers of Donald Hornig, LBJL; "Intellectuals to Johnson: War's the Rub," *New York Times*, 22 May 1967; Hornig to LBJ, 29 May 1967, "Donald Hornig Chronological File: April–June 1967" folder, box 5, Papers of Donald Hornig, LBJL. Harold Brown, physicist and secretary of the air force, was in attendance.
6. Hornig to Mansel Davies, 1967, "Donald Hornig Chronological File: July–September, 1967" folder, box 5, Papers of Donald Hornig, LBJL.
7. Hornig to Rostow, memorandum with attached meeting schedule, 13 December 1967, box 69, Country file, Vietnam, National Security file (hereafter NSF), LBJL; "Activities Related to Vietnam" narrative statement, "Volume II Documentary Supplement [1 of 3]" folder, box 1, Office of Science and Technology, Administrative History, LBJL.
8. Memo to members of the PSAC Ad Hoc Vietnam Group, 25 August 1967, "Donald Hornig Chronological File: July–September, 1967" folder, box 5, Papers of Donald Hornig, LBJL.
9. Draft, "The Effect of Air Strikes in North Vietnam and Laos: A Report by a Special Subpanel of PSAC," 26 April 1968, "Vietnam Material including a report by the President's Science Advisory Committee on the effects of the bombing of NVN 4/68–5/68" folder, box 250, Country file, Vietnam, NSF, LBJL.
10. Ibid.
11. Clifford to LBJ, 21 June 1968, and Wheeler to LBJ, 17 June 1968, "Vietnam Material including a report by the President's Science Advisory Committee on the effects of the bombing of NVN 4/68–5/68" folder, box 250, Country file, Vietnam, NSF, LBJL.
12. Wheeler to LBJ, 17 June 1968, Rostow to LBJ, 22 June 1968, and LBJ to Rostow, 4 June 1968, "Vietnam Material including a report by the President's Science Advisory Committee on the effects of the bombing of NVN 4/68–5/68" folder, box 250, Country file, Vietnam, NSF, LBJL.
13. Drell to Hornig, 30 December 1967, "Vietnam 6A 1/66–3/68 Bombing Pauses in Viet Nam" folder, box 93, Country file, Vietnam, NSF, LBJL.
14. Ibid.; Hornig to Rostow, 18 January 1968, "Vietnam 6A 1/66–3/68 Bombing Pauses in Viet Nam" folder, box 93, Country file, Vietnam, NSF, LBJL.
15. A 1969 survey of over sixty thousand academic faculty members, for example, found that physicists were among "the most liberal in the natural

sciences," including in their antiwar stance: 67 percent opposed the war. When the pool of physicists was limited to just those deemed "achievers"—i.e., productive scientists employed at elite universities, the number opposing the war rose to 80 percent. "Survey Finds Physicists on the Left," *Physics Today*, October 1972, 61–62; Mrs. Joseph Mather to Rep. E. S. Walker, 1 April 1968, "Vietnam, 1963–68" folder, HUG (FP) 94.18, Kistiakowsky; See, for example, Hugh Gusterson, *Nuclear Rites: A Weapons Laboratory at the End of the Cold War* (Oakland: University of California Press, 1996); Sharon Traweek, *Beamtimes and Lifetimes: The World of High Energy Physicists* (Cambridge, MA: Harvard University Press, 1988); Bruno Latour and Steve Woolgar, *Laboratory Life: The Construction of Scientific Facts* (Princeton, NJ: Princeton University Press, 1986).

16. Peter D. Hart Research Associates, *A Survey of Physicists' Attitudes toward the Strategic Defense Initiative,* folder 4, box 2, Records of Directed Energy Weapons (DEW) Study, 1983–1988, American Physical Society, American Institute of Physics, Niels Bohr Library and Archives, College Park, MD.

17. Kistiakowsky quoted in James Everett Katz, *Presidential Politics and Science Policy* (New York: Praeger, 1978), 179. For a breakdown of SAB membership from 1946 to 1964, see Appendix C in Thomas A. Sturm, *The USAF Scientific Advisory Board: Its First Twenty Years, 1944–1964* (Washington, DC: Office of Air Force History, 1986). John W. Finney, "Pentagon Scored on Scientist Loss," *New York Times,* 21 May 1968.

18. William G. McMillan, interview by James J. Bohning, 25 March 1992, Los Angeles, CA, Oral History Transcript #0104, Chemical Heritage Foundation, Philadelphia.

19. Ibid.

20. Ibid.

21. Ibid.; William Westmoreland, *A Soldier Reports* (Garden City, NY: Doubleday, 1976), 267–268.

22. W. G. McMillan, "The Scientist in Military Affairs," in *March 4: Scientists, Students, and Society,* ed. Jonathan Allen (Cambridge, MA: MIT Press, 1970).

23. William G. McMillan, interview by James J. Bohning, 25 March 1992.

24. Ibid.

25. John D. Baldeschwieler, interviewed by David C. Brock and Arthur Daemmrich, 13 June 2003, Oral History Transcript #0280, Chemical Heritage Foundation, Philadelphia; John D. Baldeschwieler, interview by Shirley K. Cohen, January–February, 2001, Pasadena, CA, Oral History Project, California Institute of Technology Archives, http://oralhistories.library.caltech.edu/154/01/Baldeschwieler_OHO.pdf.

26. John D. Baldeschwieler, interview by Shirley K. Cohen, January–February, 2001.

27. Ibid.

28. Ibid.

29. Ibid.

30. Joel Primack and Frank von Hippel, *Advice and Dissent: Scientists in the Political Arena* (New York: Basic Books, 1974), 25.

31. For a brief discussion of anti-ABM activism, see Kelly Moore, *Disrupting Science: Social Movements, American Scientists, and the Politics of the Military, 1945–1975* (Princeton, NJ: Princeton University Press, 2008), 135–137; Richard Garwin and Hans Bethe, "Anti-Ballistic Missile Systems," *Scientific American* 218 (March 1968).

32. Hornig to LBJ, memorandum, 16 January 1969, "Donald Hornig Chronological File: January 1969" folder, box 6, Papers of Donald Hornig, LBJL.

33. Graham Spinardi, "The Rise and Fall of Safeguard: Anti-Ballistic Missile Technology and the Nixon Administration," *History and Technology* 26 (December 2010): 316; Rick Perlstein, *Nixonland: The Rise of a President and the Fracturing of America* (New York: Scribner, 2008), 390. The *New York Times* reported that Nixon "consulted few Senators, even fewer scientists." *New York Times,* 19 March 1969. See also Wang, *In Sputnik's Shadow,* 290–308; Thomas Halstead, "Lobbying against the ABM, 1967–1970," *Bulletin of the Atomic Scientists,* 1 April 1971.

34. James Killian, *Sputnik, Scientists, and Eisenhower: A Memoir of the First Special Assistant to the President for Science and Technology* (Cambridge, MA: MIT Press, 1977), 23; DuBridge, quoted in Primack and von Hippel, *Advice and Dissent,* 21.

35. James M. Naughton, "DuBridge, a Quiet Man at White House, Stirs Worry among Scientists That His Views Are Not Heard," *New York Times,* 1 March 1970.

36. John D. Baldeschwieler, interview by David C. Brock and Arthur Daemmrich, 13 June 2003.

37. Mary McGrory, undated clipping (1969?), folder 8, box 4, Papers of Brian Schwartz, 1966–1977 (hereafter BSP), Niels Bohr Library and Archives, American Institute of Physics, College Park, MD; *FAS Newsletter,* April 1969.

38. Mary McGrory, undated clipping (1969?), BSP; Thomas Halstead, "Lobbying against the ABM."

39. Clipping, *Chemical and Engineering News,* 1 June 1970, folder 3, box 3, BSP.

40. For more in-depth analyses of the dismantling of PSAC, see Gregg Herken, *Cardinal Choices: Presidential Science Advising from the Atomic*

Bomb to SDI (New York: Oxford University Press, 1992); Richard Lawrence Garwin, interview by Finn Aaserud, 23 October 1986, Yorktown Heights, NY, Niels Bohr Library and Archives, American Institute of Physics, College Park, MD; Sheila Slaughter, *The Higher Learning and High Technology: Dynamics of Higher Education Policy Formation* (Albany: SUNY Press, 1990), 42; Primack and von Hippel, *Advice and Dissent,* 25, 36–37; Frank von Hippel, *Citizen Scientist* (New York: American Institute of Physics, 1991), vii; Wiesner quoted in "Nixon v. the Scientists," *Time Magazine,* 26 February 1973, http://www.time.com/time/magazine/article/0,9171,910590,00.html, accessed 5 May 2011.

41. For further assessment of science advising during the 1970s, see Herken, *Cardinal Choices*; Gerhard Sonnert, *Ivory Bridges: Connecting Science and Society* (Cambridge, MA: MIT Press, 2002); and Zuoyue Wang, *In Sputnik's Shadow: The President's Science Advisory Committee and Cold War America* (New Brunswick, NJ: Rutgers University Press, 2008).

42. Katz, *Presidential Politics,* 156; Thomas A. Sturm, *The USAF Scientific Advisory Board: Its First Twenty Years, 1944–1964* (Washington, DC: Office of Air Force History, 1986); Thomas C. Reed, *At the Abyss: An Insider's History of the Cold War* (New York: Ballantine, 2004), 208; Alfred K. Mann, *For Better or Worse: The Marriage of Science and Government in the United States* (New York: Columbia University Press, 2000), 134; Arthur L. Norberg, "Changing Computing: The Computing Community and DARPA," *IEEE Annals of the History of Computing* 18, no. 2 (1996).

43. Sheila Slaughter, *Higher Learning and High Technology,* 42–43; For a useful overview of the rise and fall of Route 128, see AnnaLee Saxenian, *Regional Advantage: Culture and Competition in Silicon Valley and Route 128* (Cambridge, MA: Harvard University Press, 1994) and David Lampe, ed., *The Massachusetts Miracle: High Technology and Economic Revitalization* (Cambridge, MA: MIT Press, 1988); Berkeley Rice, "Down and Out along Route 128," *New York Times,* 1 November 1970; Deborah Shapley, "Route 128: Jobless in a Dilemma about Politics, Their Professions," *Science* 127, no. 3988, June 11, 1971, 1116–1118.

44. Science for the People flyer, 25 February 1970, folder 1, box 5, Papers of Brian Schwartz, 1966–1977, Niels Bohr Library and Archives, American Institute of Physics, College Park, MD.

45. For example, see Wang, *In Sputnik's Shadow,* 290–308; Report, "Presidential Policy on Strengthening Academic Capability for Science throughout the Nation," n.d., "Volume II Documentary Supplement [2 of 3]" folder, box 1, Administrative History, Office of Science and Technology, LBJL.

46. Ibid.; Katz, *Presidential Politics*, 155. For a brief discussion of THEMIS and the 1970 Mansfield Amendment, see Roger L. Geiger, *Research and Relevant Knowledge: American Research Universities since World War II* (New York: Oxford University Press, 1993), 194–195; Margaret Pugh O'Mara, *Cities of Knowledge: Cold War Science and the Search for the Next Silicon Valley* (Princeton, NJ: Princeton University Press, 2005), 217–222; on the "southern and westward tilt" of army and air force contracting, see Ann Markuson, Peter Hall et al., *The Rise of the Gunbelt: The Military Remapping of Industrial America* (New York: Oxford University Press, 1991), 16–17; for discussion of two academic examples, the University of Arizona and Georgia Tech, see Geiger, *Research*, 273–296.

47. Saxenian, *Regional Advantage*, 67–68.

48. Asa Knowles, "Cooperative Education—A Timely Concept" presented to the Council of Executives on Company Contributions, 4 June 1970, folder 1963, box 44/49, Northeastern University Office of the President (Knowles), Records 1939–1983, Northeastern University; Antoinette Frederick, *Northeastern University: An Emerging Giant, 1959–1975* (Boston: Northeastern University, 1982), 220–240. See also Richard Freeland, *Academia's Golden Age: Universities in Massachusetts, 1945–1970* (New York: Oxford University Press, 1992), 98, 260–268, 295–297.

49. Slaughter, *Higher Learning and High Technology*, 42; Joy Rohde, "From Expert Democracy to Beltway Banditry: How the Antiwar Movement Expanded the Military-Academic-Industrial Complex" in *Cold War Social Science: Knowledge Production, Liberal Democracy, and Human Nature*, Mark Solovey and Hamilton Cravens, ed. (New York: Palgrave MacMillan, 2012), 137–153; Matthew Wisnioski, *Engineers for Change: Competing Visions of Technology in 1960s America* (Cambridge, MA: MIT Press, 2012), 3–4. For a prominent example of this sentiment, see Theodore Roszak, *The Making of a Counter Culture: Reflections on the Technocratic Society and Its Youthful Opposition* (Oakland: University of California Press, 1969).

50. Brian Balogh, *Chain Reaction: Expert Debate and Public Participation in American Commercial Nuclear Power, 1945–1975* (Cambridge, MA: Cambridge University Press, 1991), 244, 323.

9. A UNITED FRONT AGAINST STAR WARS

1. Commencement address, Paul Gray, 3 June 1985, folder 2, box 1, Records of Directed Energy Weapons (DEW) Study, 1983–1988, American

Physical Society, American Institute of Physics, Niels Bohr Library and Archives, College Park, MD (hereafter APS DEW).

2. Recording, March 4 Teach-In, 1989, AC212, Institute Archives and Special Collections, MIT Libraries, Cambridge, MA.

3. For histories of the Strategic Defense Initiative (SDI), see Frances Fitzgerald, *Way Out There in the Blue* (New York: Simon & Schuster, 2000) and Rebecca Slayton, *Arguments that Count: Physics, Computing, and Missile Defense, 1949–2012.*

4. For histories of the nuclear-freeze movement, see David S. Meyer, *A Winter of Discontent: The Nuclear Freeze and American Politics* (New York: Praeger, 1990) and Lawrence Wittner, *Toward Nuclear Abolition: A History of the World Nuclear Disarmament Movement, 1971–Present* (Stanford: Stanford University Press, 2003). Some films include: *On the Beach* (1957 novel, 1959 film), *Fail-Safe* (1964), *Dr. Strangelove* (1964), *Planet of the Apes* (1968), *The China Syndrome* (1979), *The Atomic Café* (1982), and *The Day After* (1983). For a thorough bibliography of nuclear-themed literature through 1984, see Paul Brians, *Nuclear Holocausts: Atomic War in Fiction, 1895–1984,* (Kent, OH: Kent University Press, 1987). For books published in the early 1980s, see Edward Kennedy and Mark O. Hatfield, *Freeze! How You Can Help Prevent Nuclear War* (New York: Bantam, 1982); Jonathan Schell, *The Fate of the Earth* (New York: Knopf, 1982); Solly Zuckerman, *Nuclear Illusion and Reality* (New York: Viking, 1982); Robert Scheer, *With Enough Shovels: Reagan, Bush and Nuclear War* (New York: Random House, 1982). Vote results in PCNAC/CANW pamphlet list, "Merger with Citizens against Nuclear War" folder, box 2, Series C: Administrative/Programmatic Efforts, Professionals' Coalition for Nuclear Arms Control Records (DC 164), Swarthmore College Peace Collection.

5. Hugh Gusterson, *Nuclear Rites: A Weapons Laboratory at the End of the Cold War* (Oakland: University of California Press, 1996), 67.

6. Gregg Herken, "The Earthly Origins of Star Wars," *Bulletin of the Atomic Scientists,* 1 October 1987; Edward Teller with Judith Shoolery, *Memoirs: A Twentieth-Century Journey in Science and Politics* (Cambridge, MA: Perseus Publishing, 2001), 509; Teller, quoted in Robert Scheer, "The Man Who Blew the Whistle on 'Star Wars,'" *Los Angeles Times,* 17 July 1988; "Science and the Citizen," *Scientific American* (December 1983), 76. See also Fitzgerald, *Way Out There.*

7. Quoted in SANE press release, 13 September 1985, DEW box 1, folder 2, APS DEW; Lillian Hoddeson and Vicki Daitch, *True Genius: The Life and Science of John Bardeen* (Washington, DC: Joseph Henry Press, 2002), 269; Quoted in Herken, "Earthly Origins."

8. Ronald Reagan, "Address to the Nation on National Security," 23 March 1983, *Federation of American Scientists,* accessed 29 July 2013, http://www.fas.org/spp/starwars/offdocs/rrspch.htm.

9. For an account of Reagan's thinking on Star Wars, see Fitzgerald, *Way Out There;* Gerald Yonas, "The Strategic Defense Initiative," *Daedalus* 114 (Spring 1985): 73–90.

10. Ibid.

11. Clipping, "Academy Members Skeptical on SDI," *Science* 234, 14 November 1986, 816, box 1, folder 8, APS DEW; Richard Garwin, "Star Wars: Shield or Threat?," *Journal of International Affairs* 3 (Summer 1985).

12. Garwin, "Star Wars: Shield or Threat?"

13. Ibid.

14. Ashton B. Carter, "OTA Background Paper," April 1984, folder 3, box 2, APS DEW; Union of Concerned Scientists "Issue Backgrounder," August 1988, folder 3, box 1, APS DEW.

15. Kenneth Ford, interview by Alexei Kojevnikov, 22 November 1997, Niels Bohr Library and Archives, American Institute of Physics, College Park, MD, http://www.aip.org/history/ohilist/24701_2.html; *Newsletter of the Forum on Physics and Society* 2 (March 1973), http://www.aps.org/units/fps/newsletters/upload/march73.pdf.

16. "DEW Study Proposal," 1983, folder 10, box 2, APS DEW; "The American Physical Society Issues Statement on Nuclear Arms Limitation," *Physics and Society* 12 (April 1983): 8; *Physics and Society* 12 (October 1983).

17. "DEW Study Proposal," 1983.

18. Press release, 29 June 1987, folder 6, box 3, APS DEW; Correspondence, 1984, folder 1, box 3, APS DEW; Hebel to APS Executive Committee, 20 June 1985, folder 3, box 3, APS DEW.

19. Havens to Patel et al., 1984, folder 1, box 3, APS DEW; Hebel to APS Executive Committee, 20 June 1985; Schewe to Park, 20 January 1987, folder 4, box 3, APS DEW.

20. Report, *"Science and Technology of Directed Energy Weapons: Report of the American Physical Society Study Group,"* folder 11, box 4, APS DEW.

21. "SDIO Comment," n.d. (1987?), folder 6, box 3, APS DEW; "The Naysayers' Report," *Wall Street Journal,* 1 May 1987.

22. Orear to APS Council, 10 December 1985, folder 3, box 3.

23. Park to Orear, 22 January 1986, folder 4, box 3, APS DEW; Ford to Drell, 10 January 1986, folder 4, box 3, APS DEW; Ford to Orear, 11 January 1986, folder 4, box 3, APS DEW.

24. Orear to Park, 2 January 1986, folder 4, box 3, APS DEW.

25. Moss to APS Council, 14 April 1986, folder 4, box 3, APS DEW.

26. Jaccarino to Fitch, 10 June 1987, folder 6, box 3, APS DEW; Press release, 28 April 1987, folder 6, box 3, APS DEW.

27. Press release, 28 April 1987, folder 6, box 3, APS DEW; Avizonis et al., "Letters," *Physics Today*, October 1987; Johnson to Fitch, 8 June 1987, folder 6, box 3, APS DEW.

28. Fitch, "Letters," *Physics Today*, October 1987.

29. William J. Broad, *Star Warriors: A Penetrating Look into the Lives of the Young Scientists behind Our Space Age Weaponry* (New York: Simon & Schuster, 1985), 181–182; Broad, *Star Warriors*, 205.

30. Naomi Oreskes and Erik Conway, *Merchants of Doubt* (New York: Bloomsbury, 2010), 27–28; Dennis Hevesi, "Frederick Seitz, 96, Dies; Physicist Who Led Skeptics of Global Warming," *New York Times*, 6 March 2008; Oreskes and Conway, *Merchants of Doubt*, 35.

31. Frederick Seitz, "Notable and Quotable," *Wall Street Journal*, 2 June 1987; Press release, 1 July 1987, folder 7, box 3, APS DEW.

32. "The Naysayers Respond," unsigned draft, 10 May 1987, folder 6, box 3, APS DEW; "APS Directed Energy Study Group Responses to Critiques by Wood and Canavan," n.d. (1987?), folder 6, box 3, APS DEW; Patel to Fitch, 18 June 1987, folder 6, box 3, APS DEW.

33. Fitch to *Wall Street Journal*, May 1987, folder 6, box 3, APS DEW; Fitch to Weldon et al., 19 June 1987, folder 6, box 3, APS DEW; Fitch to Spratt, 20 October 1987, folder 7, box 3, APS DEW.

34. Clipping, *Physics Today*, January 1984, folder 6, box 1, APS DEW; SANE press releases, 17 October 1985 and 13 May 1986, folder 2, box 1, APS DEW.

35. Quoted in Rebecca Slayton, "Discursive Choices: Boycotting Star Wars between Science and Politics," *Social Studies of Science* 37 (February 2007): 27–66. Though Slayton argues that the University of Illinois pledge was stripped of its impassioned moral language during the revision process, phrases such as "nuclear holocaust" imply a moral imperative to halt SDI.

36. Michael R. Nusbaumer, Judith A. DiIorio, and Robert D. Baller. "The Boycott of 'Star Wars' by Academic Scientists: The Relative Roles of Political and Technical Judgement," *Social Science Journal* 31, no. 4: 376–88

37. Quoted in SANE press release, 13 September 1985, folder 2, box 1, APS DEW; *FAS Bulletin*, April 1987; SDI press releases, 28 February 1985, 32 February 1985, and 23 May 1985, folder 2, box 1, APS DEW; Peter Westwick, "The International History of the Strategic Defense Initiative: American Influence and Economic Competition in the Late Cold War," *Centaurus* 52 (November 2010): 339.

38. Gusterson, *Nuclear Rites,* 43; Daniel Kevles, *The Physicists: The History of a Scientific Community in Modern America* (New York: Vintage, 1979).

39. Quoted in Garwin, "Star Wars: Shield or Threat?"; Lee Dye, "Research Scientist Resigns Administrative Post at Weapons Lab," *Los Angeles Times,* 3 November 1985; See, for example, Robert Scheer, "Lab Plans Nuclear Trial Despite Charges of Flaw in Data," *Los Angeles Times,* 12 November 1985; William Broad, "Turmoil: Turmoil, Turmoil, Turmoil," *New York Times,* 9 October 1988, http://search.proquest.com.ezproxy.lib .calpoly.edu/docview/110452897/fulltextPDF/13F89305E025CBE8AF9/1 ?accountid=10362; Ta Heppenheimer, "New Director Shifts Balance of Power at Livermore Lab," *The Scientist,* 7 August 1989, http://www.the -scientist.com/?articles.view/articleNo/10529/title/New-Director-Shifts-B alance-Of-Power-At-Livermore-Lab/.

40. Scheer, "Man Who Blew the Whistle."

41. Ibid.

42. William Broad, "Space Weapon is Disparaged by Missile Lab," *New York Times,* 13 August 1987.

43. Scheer, "Man Who Blew the Whistle."

44. Dan Morain, "Energy Secretary Warns Weapons Scientists Not to Disagree in Public," *Los Angeles Times,* 23 July 1988; Scheer, "Man Who Blew the Whistle"; *Physics and Society* 22 (April 1993), http://www.aps .org/units/fps/newsletters/1993/april/napr93.html.

45. Rebecca Slayton, "Speaking as Scientists: Computer Professionals in the Star Wars Debate," *History and Technology* 19, no. 4 (2003): 335–364.

46. Union of Concerned Scientists "Issue Backgrounder," August 1988, folder 3, box 1, APS DEW; Clipping, *Science News,* 2 April 1988, folder 8, box 1, APS DEW.

47. Wayne Biddle, "Scientists Compare 'Star Wars' to ABM Debates," *New York Times,* 30 May 1985.

48. Clipping, 5 June 1985, folder 7, box 1, BAPS DEW.

49. Hoddeson and Daitch, *True Genius,* 269.

50. Oreskes and Conway, *Merchants of Doubt;* Alfred K. Mann, *For Better or Worse: The Marriage of Science and Government in the United States* (New York: Columbia University Press, 2000), 215.

EPILOGUE

1. "The Politics of Science," *International Science and Technology,* April 1964, "Clinton P. Anderson, #1," folder, Stanislaw Ulam Papers, box 1, American Philosophical Society.

2. Stanislaw Ulam, *Adventures of a Mathematician* (New York: Charles Scribner's Sons, 1983), 222.

3. E. H. S. Burhop, "Scientists and Soldiers," *Bulletin of the Atomic Scientists,* 1 November 1974.

4. Burhop, "Scientists and Soldiers."

5. See, for example, Daniel Lee Kleinman, *Impure Cultures: University Biology and the World of Commerce* (Madison: University of Wisconsin Press, 2003); Sheila Slaughter and Larry L. Leslie, *Academic Capitalism: Politics, Policies, and the Entrepreneurial University* (Baltimore: Johns Hopkins University Press, 1997); Sheila Slaughter and Gary Rhoades, *Academic Capitalism and the New Economy: Markets State, and Higher Education* (Baltimore: Johns Hopkins University Press, 2004); Jennifer Washburn, *University, Inc: The Corporate Corruption of Higher Education* (New York: Basic, 2005).

Acknowledgments

This project would have been impossible without the aid of many people and institutions. *Scientists at War* began life at Columbia University, shaped in no small part by the wise advice and encouragement of Eric Foner and Betsy Blackmar. They have been model scholars and mentors. I am also thankful for the support and expertise of David Rosner, Dan Kevles, Alan Brinkley, Marilyn Young, the late Michael Nash, and my fellow graduate students and friends in New York.

Research and writing require travel and time, and so I am profoundly grateful for the financial assistance for both provided by Cal Poly's College of Liberal Arts, the Columbia History Department and Graduate School of Arts and Sciences, the Doris G. Quinn Foundation, New York University's Center for the United States and the Cold War, the Philadelphia Area Center for History of Science, the LBJ Foundation, the Kennedy Library Foundation, the Caltech Archives, the Chemical Heritage Foundation, the Princeton University Library, and the American Institute of Physics' Center for History of Physics.

At a critical moment, the Society of American Historians enabled me to bypass some onerous hurdles and advance straight to Harvard University Press. For this I am indebted to Andie Tucher and the members of the Nevins Prize committee. At HUP, I have benefited from the guidance of Joyce Seltzer, the patience of Brian Distelberg, and the invaluably detailed comments of Andrew Jewett and an anonymous reader. Thank you so much. Others who kindly read drafts and offered advice and encouragement include Betsy Blackmar, Noam Chomsky, George Cotkin, Bob Fenichel, Amanda Katz, Amy Offner, Ellen Schrecker, and Perrin Selcer.

Since 2011, I have had the great fortune to find a welcoming home at Cal Poly, and I am grateful for my wonderful colleagues in the History Department, particularly Andrew Morris, Lewis Call, Matt Hopper, and, at the eleventh hour,

Molly Loberg and James Tejani. My work has been enriched by the thoughtful comments of the students in my Cold War Science and Society seminars and by the able research assistance of Wendy Myren. David Hafemeister offered valuable perspective and support. And of course, nothing would have been accomplished without the tireless efforts of Kim Barton and Sherrie Miller.

Many, many friends, colleagues, archivists, conference commenters, and even chatty strangers have contributed to this work—too many to name individually. I owe particularly large debts to Daniela Gerson for her constant hospitality and generosity, Laurie Weymann for her care, and Alison and Joshua Bridger for their support and humor. I strive to be more like Amy Offner and Rebecca Onion, and I think we all strive to be more like George Cotkin.

Fifteen years ago, the legendary Jack Thomas encouraged me on this path, and I am very grateful for his advice and example. His memory haunts the pages of this book, as do the memories of my grandparents, Homer Bridger and Lillian Simon. There are so many questions I wish I had asked them when I had the chance.

Finally, I owe the largest debt of all to my parents, Mark and Maxine Bridger, without whose support and care I surely would have abandoned this project years ago.

Trying to recover the past is a beautiful burden but a far less agonizing one than the grim ethical calculus that challenged and paralyzed so many Cold War scientists. They struggled over high-stakes dilemmas in research and politics with care and great vulnerability. I hope this book does right by them.

Index

A-1 sight, 159
AAAS. *See* American Association for the Advancement of Science
Aaserud, Finn, 147
Abelson, Philip, 209
Aberdeen Proving Ground, 65, 74
ABM systems. *See* Antiballistic missile (ABM) systems
Abrams, Creighton, 231–23
Academic Council (MIT), 188
Academic neutrality: debates at Princeton, 7, 212–213, 216–219; debates at MIT, 163–164, 176, 181, 187, 192–193, 219; and university policy changes, 243. *See also* Political neutrality of scientists
ACDA. *See* Arms Control and Disarmament Agency
AC Spark Plug, 159
Activism, 4, 7, 162–163, 207, 245–246, 252, 267
Adamsite (diphenylaminechloroarsine, DM), 81–86, 95, 223
Ad Hoc Group on Vietnam, 224
Ad Hoc Panel on Nuclear Testing, 40–42
"Ad Hoc Panel on the Technological Capabilities and Implications of the Geneva System" (Fisk report), 35–38
Ad Hoc Panel on Vietnam, 223
Advanced Research Projects Agency (ARPA): creation of, 20–21; and VELA program, 26, 51, 158; and Project

AGILE, 65, 69–70, 117; limited-war research conducted by, 66; and defoliation efforts, 73, 74, 94, 98; and Joint Research and Test Activity, 80; and Jasons, 144; MIT contracts with, 157; and Project PRESS, 158; William McMillan and, 232; DARPA name change, 240; shifted focus of, in 1970s, 240
AEC. *See* Atomic Energy Commission
Aeronautical Laboratory (Cornell University), 218
"Agenda Days," 170
Agent Blue, 72–75, 99, 100
Agent Green, 73, 75
Agent Orange: Arthur Galston and, 9, 88–89; and development of 2,4,5-T, 71–72; test spraying, 74; initial use in Vietnam, 75; ecological effects of, in Vietnam, 98–100; and dioxin, 100; biological effects of, 100–101; policies regarding use of, 101–102, 104, 110; postwar controversies and lawsuits, 100, 105–106, 113; and Vietnam veterans, 106
Agent Orange: Vietnam's Deadly Fog (television documentary), 106
Agent Orange Policy Group, 106
Agent Pink, 73, 75
Agent Purple, 71, 73–75
Agent White, 74, 97–100

Agnew, Harold, 232
Agricultural Research Service, 98
AIP (American Institute of Physics), 194
Air defense, 33–34, 148, 156, 157, 162
Air Force. *See* U.S. Air Force
Air Force Scientific Advisory Board (SAB), 20, 53, 92, 157, 230, 240
Air-Force/Space Digest, 189
Alamagordo test (Trinity), 2
Algeria, 220
Allen, John, 181–182
Alperovitz, Gar, 166
Alsop, Joseph, 189
Alvarez, Luis, 66, 120, 148, 152, 195
American Association for the Advancement of Science (AAAS), 192; and defoliation, 89, 94–96, 101, 103, 106–107, 113–114; and Agent Orange use, 113; protest at 1970 meeting of, 207–209. *See also* Herbicide Assessment Commission
American Association of Physics Teachers
American Chemical Society, 207
American Federation of Labor-Congress of Industrial Organizations, 57
American Institute of Physics (AIP), 194
American Mathematical Society, 206, 267, 268
American Medical Association, 109
American Physical Society (APS): 1981 survey conducted by, 8; and "neutrality" of science, 193; founding and development of, 194–195; Charles Schwartz and, 195–196, 198–202; and opposition to Vietnam War, 196–206; 1969 poll conducted at annual meeting of, 238; New Right attacks on, 246–247; directed-energy weapons report of, 252–261; and DEW study group, 254–261
American Society for Microbiology, 95, 97
American Society of Plant Physiologists, 89
American University, 204
Amrine, Michael, 49
Anderson, Clinton, 270, 271

Anesthetic dart gun, 144
Antiballistic missile (ABM) systems: Decoy problem, 19, 158, 236, 251–252; and decoy problem, 19, 158, 180; PSAC and, 19, 56; Zeus missile, 19; Fisk panel and, 36, 38; and limited test ban, 53; research on, 60; MIT and, 162, 167, 174, 177, 186, 190, 200, 201; Sentinel system, 177, 236; Richard Nixon and, 235–239; Safeguard ABM system, 236, 238. *See also* Strategic Defense Initiative (SDI)
Anti-Ballistic Missile Treaty (1972), 238
Anticommunism, 16, 66, 259
Anti-infiltration project. *See* Electronic barrier
Antinuclear movement, 8, 247–248
Antinuclear protest (New York, 1982), 247
Antiscience sentiment, 235, 244, 272
Antisubmarine warfare, 20, 223, 225
Antivaccine lobby, 272
Antiwar movement, 7, 81, 121, 163–168
Apollo 13 mission, 191
Apollo Group, 173
Apollo program, 160, 191
Applied Electronics Laboratory (Stanford), 22, 157
Applied research, 11, 22, 174, 191, 250
APS. *See* American Physical Society
Argonne National Laboratory, 25, 54, 165, 203
Armalite AR-15 rifle, 69
Arms control and disarmament, 253, 254, 257, 264, 266; Manhattan Project veterans and, 5–6; during Eisenhower administration, 13, 20, 21, 24–27, 30; during Kennedy administration, 30–31, 34–41, 43, 44, 46–49, 52, 54–61, 64; during Johnson administration, 80, 87, 115, 166, 183; Jasons and, 151, 152; during Nixon administration, 219, 238; during Reagan administration, 245–250, 253–254, 266, 267
Arms Control and Disarmament Agency (ACDA), 31, 47, 50–52, 56, 90, 141
Arms Limitation and Control panel, 26
Army Chemical Corps, 76

Army Corps of Engineers' Strategic Study
    Group (ESSG), 104
Army Limited War Laboratory, 138
Army Math Research Center (University
    of Wisconsin), 9, 208, 219
Arnett, Peter, 83
ARPA. *See* Advanced Research Projects
    Agency
ARPANET, 34
Aspin, Les, 104
Asymmetry, nuclear, 14
Atlas ICBM, 32
Atmospheric studies, 158
Atmospheric testing, 41, 43–49, 51, 58,
    231
Atomic Energy Commission (AEC):
    creation of, 3; and post-Sputnik boom,
    18, 22; and nuclear testing, 24, 26–28,
    35, 41–42, 48, 50, 52, 54; and Princeton
    contracts, 216–217; Richard Garwin
    and, 236
AT&T, 155, 158
Auburn University, 263
Auschwitz, 271
Auxins, 70
Avco, 160

BAFGOPI (Boston Area Faculty Group
    on Public Issues), 89, 163
Baldeschwieler, John, 230, 233–235,
    237–239
Baldwin, Hanson, 108–109
Ball, George, 41
Ballistic Missiles Panel (of PSAC), 19
Ban the Bomb movement, 43
Bardeen, John, 195, 201, 248–249, 263,
    268
Barrier project. *See* Electronic barrier
Basic research, 10–11, 23, 148, 174, 184,
    186, 188, 215, 218, 250
Bell, Beverly, 204
Bell Labs, 35, 104, 164, 198, 237, 254
Bendix, 159–160, 175
Bennett, Ivan, Jr., 104
Berkeley. *See* University of California,
    Berkeley
Berkner, Lloyd, 66
Bernstein, Leonard, 55
Berrigan, Daniel, 203

Berrigan, Philip, 203
Bethe, Hans, 6, 18, 26, 35, 40, 54, 58,
    163, 166, 236, 238, 254, 267, 272
Betts, Austin, 35
Bien Hoa, Vietnam, 73, 80–81
Bikini Atoll, 24
Biological Warfare Laboratories (Fort
    Detrick), 71
Biological weapons. *See* Chemical and
    biological weapons (CBW)
Biology, 3, 95–96
Bionetics herbicide study, 101, 105
Bionetics Research Laboratories
    (Bethesda, Maryland), 100, 101
Birth defects, 101, 105, 106, 113
Bishop, Robert, 173
"Black boxes," 52
"Black Warrior" automatic pilot system,
    159
Blanchard, C. H., 196
Blimps, 232
Bloembergen, Nicolaas, 254, 258
Boeing, 160
Boi Loi woods, South Vietnam, 91
Bombing campaigns, 121–124, 226
Bootstrapping, 14
Boston area: Cambridge Discussion
    Group, 116–117, 119–121, 139, 147;
    Route 128 companies, 156, 191,
    240–242, 263; Union of Concerned
    Scientists, 165; growth of Northeastern
    University, 242
Boston Area Faculty Group on Public
    Issues (BAFGOPI), 89, 163
Botany and botanists, 9, 70–73, 90, 94,
    103, 109, 213
Bourne, Randolph, 22
Boycott, of SDI-related research
    contracts, 261–262
Bradbury, Norris, 40, 42, 43, 57, 270
Branfman, Fred, 150
BRAVO test, 24, 25, 26
British, 78
Broad, William, 258–259, 265
Brookhaven Laboratory, 206
Brown, Harold, 35, 39, 41, 45, 231
Brown, James W., 73
Broxmeyer, Charles, 182–183, 189, 220
Buckingham, William, 113

Building Service International Union, 181

Bulletin (APS), 204

Bulletin of the Atomic Scientists, 3, 13, 22, 25, 26, 49, 54, 90, 133, 136, 184, 197, 271

Bundy, McGeorge, 39, 43, 45–46, 51, 74, 83, 86, 115, 116

Burcham, Levi, 73

Burhop, E. H. S., 271–272

Burlington, Massachusetts, 160

Bush, Vannevar, 156

Butler, David, 100

Button bomblets, 126

Cacodylic acid, 72, 75

Caldwell, David, 138, 142

California Institute of Technology (Caltech), 15, 21, 88, 101, 156, 210, 212, 218, 261, 263, 268

Callen, Earl, 204

Caltech. *See* California Institute of Technology

Cambodia, 75, 110, 146, 149, 210, 228, 238–239

Cambridge Discussion Group, 116–117, 119–121, 139, 147

Camp Detrick, 70, 71, 76, 88. *See also* Fort Detrick

Canavan, Gregory, 259, 260

Carleton College, 200

Carnegie Corporation, 254

Carnegie Institution, 156

Carnegie-Mellon University, 191, 261

Carson, Rachel, 79, 100

Carter, Ashton, 251, 252

Carter, Jimmy, and administration, 171, 239–240

Case, Kenneth, 143–144

Case Institute of Technology, 21

Casey, Thomas, 182

Casper, Barry, 200, 201

CDTC. *See* Combat Development and Test Centers

Cecil, Paul, 112

Center for International Studies, 157

Central Intelligence Agency (CIA), 18, 26, 39, 127, 157

Cha La, Vietnam, 81

Chemical and biological weapons (CBW), 63–65, 76, 81–82, 89–92, 102, 107–108, 110, 144, 179, 186

Chemical and Engineering News, 164, 210

Chemical Corps, 73

Chemical defoliants. *See* Defoliants

Chemical warfare, 67, 75, 84

Chemistry, 21

Chemistry departments, 141, 210

China, 59, 102, 122, 125, 133–135

China Lake, 65

Chloracne, 100, 101, 105, 106

Chlorine gas, 75

Chloromalonitrile (CN). *See* Tear gas

Chodos, Alan, 163

Chomsky, Noam, 89, 156, 163, 165–168, 171, 172, 176, 185–187, 219, 220

Christy, R. F., 212

CIA. *See* Central Intelligence Agency

Citizens for Peace in Vietnam, 229

Civil defense, 25, 33, 79, 89

Clauser, Milton, 177–178

Clifford, Clark, 227

Climate change, 272

Clinton, Bill, and administration, 191

"Cluster bombs," 146

CN. *See* Tear gas

Cold War, 4, 10, 11, 16, 22, 24, 30, 59, 158, 162

College de France, 150

Colorado State University, 210

Columbia University, 21, 150, 151, 166, 194, 211, 261

Combat Development and Test Centers (CDTC), 69, 70, 72, 73

Command and control, 33–34, 148, 250, 252, 263

Committee for a Sane Nuclear Policy (SANE), 27, 43, 57, 261

Committee for Nonviolent Action, 203

Committee for Social Responsibility in Engineering, 206

Committee of Principals, 28, 30, 31, 35, 39, 43, 44, 50–52, 54, 55, 61, 65

Committee on Problems of Physics and Society, 201

Commoner, Barry, 27, 43, 56, 92, 95, 106, 107
Compton, Karl, 161, 162
Computer memory, 158
Computer People for Peace, 166, 206
Computer science, 266
Conant, James, 22, 217
"Concept of Action," 69
Congo, 64
Congress, 31, 51–52, 54, 57, 76, 78, 104, 186, 258. *See also* U.S. Senate
Constable, John, 103
Containment, 13–14
Convair, 160
Conway, Erik, 259
Cook, Robert, 103
Cooke, W. D., 217, 218
Copland, Aaron, 55
Cornell Air War Study, 203
Cornell University, 47, 194, 205, 216, 218, 261, 262
Cosmos Club, 140
Council for a Livable World, 49
Council of the Princeton University Community, "special committee" of, 211–217
Counterforce, 41, 45
Counterinsurgency, 65, 117, 118, 222
Credibility, 78, 107, 109, 132
Crop destruction, 70–72, 77–81, 88–91, 93, 100, 104–105, 109
Cross, James, 145
CS. *See* Super tear gas
Cuba, 59, 65
Cuban Missile Crisis, 52
Cusick, Paul, 179–180
Cybernetics, 3, 156

Daedalus, 250
Daily Princetonian, 214, 215
"Daisy cutters," 146
Dana Hall summer study, 121, 125, 131, 132, 143, 145, 224
DARPA (Defense Advanced Research Projects Agency), 240. *See also* Advanced Research Projects Agency (ARPA)
Dart gun, anesthetic, 144
David, Edward, 104, 205–206, 237

Davidon, William, 9, 203–205, 207
Davidson, Norman, 142
Davidson, Thomas, 204
Davison, Michael, 89
Dayton, David S., 169
DCPG. *See* Defense Communications Planning Group
DCPG Advisory Committee, 138, 140–143
DDT, 79
Dean, Arthur, 39
Decoy problem, 19, 158, 236, 251–252
Deep Submergence Systems Group (I-Lab), 182
De-escalation, in Vietnam, 7, 116, 119–120, 125, 130–132, 139–142, 146, 150, 151, 210
Defense Advanced Research Projects Agency (DARPA), 240. *See also* Advanced Research Projects Agency (ARPA)
Defense Atomic Support Agency, 42
Defense Communications Planning Group (DCPG), 129, 130, 132, 138, 141, 146, 147, 226, 230
Defense contracting, 4, 7, 10, 65, 66, 168, 187, 189, 216, 219, 240, 242, 243, 261–262, 263–264
Defense Electronics, 138
Defense Research and Engineering Ad Hoc Weapons Effects Group, 53
Defense Research and Engineering Department, 117, 127
Defense Science Board, 252
Defense Science Seminar, 231
Defensive Technology Study Team (Fletcher Panel), 250
Defoliants, 70–78, 80–81, 88–91, 93–99, 101–107, 109, 111–113, 207. *See also* Agent Orange
Deitchman, Seymour, 64, 117–120, 133, 135
DeLauer, Richard, 250
Delbrück, Max, 142
DELILAH system, 232
Delmore, Fred, 73
Democratic Convention (1968), 201

Democratic Platform Committee (1964), 79

Deniability, 127–128

Department of Defense (DOD), 18, 20, 22, 26, 48, 76; and limited war, 64, 65; "Concept of Action" of, 69; and gas use, 84–85; and herbicide use, 104–105; and Cambridge Discussion Group, 119; and "Muscle Shoals" system, 145; and MIT, 157, 164, 174, 186, 219–220; Lincoln Lab funding by, 158; I-Lab funding by, 161; subcontracting by, 171; research support from, 184; Stewart Udall on, 208; and Princeton University, 211, 213–216. *See also* Advanced Research Projects Agency (ARPA)

Department of Defense Reorganization Act of 1958, 20

Department of the Army, 70

Department of Transportation, 148

Department of Veterans Affairs, 112

Detection. *See* VELA nuclear test detection program

Détente, 102

Deterrence, 14, 33–34, 37, 131, 252

Deutsch, John, 191

Deutsch, Martin, 161

DEW (Directed Energy Weapons) study group, 254–261

Diamond Shamrock, 75

Dickson, Paul, 138–139, 146

Diem, Ngo Dinh, 68, 69, 71

Dinneen, Gerald, 173

Dioxin, 100, 101, 103, 105, 106, 113

Diphenylaminechloroarsine (DM). *See* Adamsite

Directed-energy weapons, 250, 252, 254–261

Directed Energy Weapons (DEW) study group, 254–261

Disarmament. *See* Arms control and disarmament

Disturbing the Universe (Dyson), 132

DM. *See* Adamsite

DOD. *See* Department of Defense

"Dominic" atmospheric tests, 49

"Doves and hawks," 152

Dow Chemical Corporation, 71, 72, 75, 97, 100, 101, 163, 242

"Dr. Strangelove Award," 208, 209

Draft records, destruction of, 203, 204

Draper, Charles Stark, 157, 159–160, 174, 176, 189

"Draper Fellows," 191

Draper Lab (MIT). *See* Instrumentation Lab

Drell, Sidney, 150, 153, 228, 238, 251, 254

"Drift," 216, 219

DuBridge, Lee, 15, 23, 27, 101, 102, 104, 113, 156, 184, 201, 235, 237

Dugway, Utah, 71

Dulles, Allan, 28, 42

Dulles, John Foster, 14, 27, 32

DUMP TRUCK, 145

DuPont, 72, 106, 155, 242, 247

Dutton, Fred, 57

Dyson, Freeman, 54, 56, 90, 132–136, 148

Earthquakes, 35, 51

Easton, Paul, 180–181, 263

Ecocide, 111

Ecocide in Indochina: The Ecology of War (Weisberg), 111

Ecology, 93–94

Eden, Murray, 163, 164

Edison Electric, 169

Edsall, John, 89, 90, 92–95, 97, 100

Eglin Air Force Base, 65, 71, 74, 97

Eisenhower, Dwight D., and administration: and PSAC, 6, 18–20, 66, 237; Truman's policy vs., 13–14; research and education funding under, 14–15, 21–22; New Look policy of, 14–16, 24, 29, 33, 34, 41; and Sputnik launch, 17; science advising under, 18–20; and arms control efforts, 26–29, 58, 59; farewell speech of, 29, 61; Kennedy's nuclear policy vs., 64; and gas use, 84; and George Kistiakowsky, 116; and I-Lab, 159

Elderfield Committee, 218–219

Electronic barrier, 7, 116–117, 125–132, 136–139, 145–146, 148, 150, 152

Electronic battlefield, 150

Electronic News, 138, 165, 181, 201

Elgin Air Force Base Limited War Group, 138

Engineers and Physicians for Johnson and Humphrey, 80
Engineers' Strategic Study Group (ESSG), 104–105
Environment, 92
Environmentalism, 247
Environmental movement, 75, 79
Environmental Protection Agency (EPA), 106
ESSG (Engineers' Strategic Study Group), 104–105
Ethiopia, 108
Ethylenes, 70

Faas, Horst, 83
Fall, Bernard, 117, 152
Fallout, radioactive, 24–25, 47
Falmouth, Massachusetts, 144
FAS. *See* Federation of American Scientists
Fassnacht, Robert, 208
Federal Council for Science and Technology, 23, 241
Federal Laboratories, 82
Federally Funded Research and Development Center, 158, 191
Federation of American Scientists (FAS), 3, 25–27, 31, 47, 49, 56, 58, 161–162, 197, 199, 238
Federation of American Societies for Experimental Biology, 95
Federation of Responsible Scientists, 165
Feigenbaum, Joel, 163, 166
Feld, Bernard, 90, 115, 161, 163
Fermi, Enrico, 60
Feshbach, Herman, 164
Fieser, Louis, 9, 108–109
Finkbeiner, Ann, 153
Firebombing, 76
First-strike capability, 32, 33
Fisher, Roger, 116
Fisher Laboratory, 82
Fisk, James B., 35, 40
Fisk panel (Fisk report). *See* "Ad Hoc Panel on the Technological Capabilities and Implications of the Geneva System"
Fitch, Val, 138, 258, 260–261
Fletcher, James, 250
Fletcher report, 251

"Flexible response" strategy, 6, 34
Fluid-flow fields, 158
Fluid Mechanics Laboratory (MIT), 169, 189
Fluid mechanics research, 169–170
Food and Drug Administration, 101
Ford, Gerald, and administration, 103, 239
Ford, Kenneth, 253, 256
Ford Foundation, 217
Ford Motor Company, 33
Foreign Relations Committee. *See* Senate Committee on Foreign Relations
Forman, Paul, 10–11
Forrestal, Michael, 74
Fort Detrick, 71, 72, 74, 97, 214
Fort Drum, 71
Forter, Samuel, 175
Fort Ritchie, 71
Forum on Physics and Society, 201–206, 253, 266
Foster, John S., 38, 40–42, 53, 95, 129, 204, 231, 232
Foster, William C., 31, 44, 51, 52, 54
Fraenkel, George, 142
Fragmacord linear mines, 232
Francis Bitter Laboratory (MIT), 200
Frankel, Max, 84
Frosch, Robert, 215
Fukuryu Maru, 24

Gaddis, John Lewis, 121–122
Galbraith, John Kenneth, 40, 116
Galison, Peter, 11
Galston, Arthur, 9, 88–89, 93–95, 99, 101–103, 106, 109, 110–113
Garwin, Richard, 120, 131, 138, 223, 224, 236–237, 239, 251–252, 261
Gases: "nonlethal," 81; public opposition to use of, 90–92, 103, 106–110. *See also* Tear gas
Gell-Mann, Murray, 21, 118–120, 138, 141, 145, 150
General Dynamics, 138, 160
General Electric, 155, 160, 175, 176, 234, 263
General Motors, 160, 263
Geneva Conference on the Discontinuance of Nuclear Weapons Tests, 28

Geneva Protocol, 69, 76, 78, 83, 91,
102–103, 113
George C. Marshall Institute, 259
Georgetown University, 203
Georgia, 74
Georgia Tech, 191, 263
Germany, 59, 75
Gilliland, Edwin, 173
Gilpatric, Roswell, 72, 75
Ginsburgh, Robert, 124–125, 128, 129
Glaser, Donald, 54, 55, 148, 152
Goldberg, Arthur, 86
Goldberger, Marvin, 118, 138, 143–145,
152, 215, 226, 238, 253, 268
Goldhaber, Maurice, 201
Goldhaber, Michael, 201
Goldwater, Barry, 80
Goldwyn, Robert, 91, 93, 108, 109
Gomer, Robert, 133
Gore, David Ormsby, 78
GPS, 191
Graduated (gradual) escalation, 34, 87,
121–123, 125, 195
Gravel, Mike, 203
Gravel mines, 126
Gray, Paul, 245, 268
Gray, Peter, 171, 173
Great Depression, 156, 159
Greater St. Louis Citizens' Committee for
Nuclear Information, 27
Greenberg, Daniel, 143
Greenstein, Jesse, 142
Ground Warfare Panel, 225, 228
Group 35, 180
Group Delta, 163
Gruenberg, Leonard W., 169, 170
Grumman, 160
Guerrilla warfare, 64, 66–67
Guidance systems, inertial, 160, 174–175
Gulf War, first, 233
Gusterson, Hugh, 247–248, 263
Gyroscopes, 159

Hacking, Ian, 11
Haiphong, North Vietnam, 124, 225,
226
Halberstam, David, 152
Hamilton, Andrew, 137
Hamlet pacification program, 234

Hammond, George, 142, 210, 239
Hanscom Air Force Base, 157
Hanscom Field, 157
Harvard Medical School, 91
Harvard University, 89, 115–116, 156,
194, 211, 240–241
Haverford College, 203
Hawaii, 74
Haworth, Leland, 50
H-bomb. See Hydrogen bomb
Herbicide Assessment Commission, 89,
103–104
Herbicide Policy Review Committee
(State Department), 100
Herbicides. See Defoliants
Herbicides in War (ed. Westing), 111
Hercules, 75
Heritage Foundation, 248
Herrington, John, 266
Hersh, Seymour, 90, 109
HEW, 211
Hicks, Nancy, 208
Hildebrand, J. H., 28
Hill, Harold, 118
Hippocratic oath, 110, 202, 209
Hiroshima and Nagasaki bombings, 2–3,
5, 13, 25, 189
Hitch, Charles, 212, 217, 218
Hitler, Adolf, 208, 260
H'mong, 112
Hoag, David, 173
Ho Chi Minh, 107
Ho Chi Minh City, Vietnam, 111
Ho Chi Minh Trail, 117, 121, 234
Hoeffding, Oleg, 125
Holifield, Chet, 51
Honeywell, 159, 160, 175
Hornig, Donald, 61, 80, 85–87, 95, 97,
103, 223–225, 227, 228, 235, 236, 238,
241
Hosmer, Craig, 51
Hughes (defense firm), 160
Hungary, 91
Huston, John, 55
Hydrogen bomb (H-bomb), 16, 24, 26, 53

IBM. See International Business
Machines
IDA. See Institute for Defense Analyses

IEEE (Institute of Electrical and Electronics Engineers), 206
"Igloo White" system, 145–146
I-Lab (MIT). *See* Instrumentation Lab
Incendiaries, 76. *See also* Napalm
Industrial Research, 165
Inertial guidance systems, 160, 174–175
Inglis, David, 25–26, 54
Insider activism, 59–60, 98, 142, 264, 266, 271
Inspections, nuclear-site, 52, 55
Institute for Defense Analyses (IDA), 6, 21, 27, 54, 64, 66, 68, 117, 118, 120, 145, 147, 149, 195, 210–211. *See also* Jasons
Institute of Electrical and Electronics Engineers (IEEE), 206
Instrumentation Lab (I-Lab, Draper Lab) (MIT), 7, 22, 157, 159–161, 171, 173–175, 181, 182, 184, 186, 189–191, 218
Intercontinental ballistic missiles, 15, 19
International Business Machines (IBM), 158, 160, 199, 236
International Chemical Workers Union, 247
International Science and Technology, 270
"International Symposium on Herbicides and Defoliants in War," 111
Interrupt, 206
Ionization, 158
Ionizing radiation, 231
Ionson, James, 250
Italy, 108

Jackson State University, 210
Jacobs, William P., 213–214, 219
Jahn, Robert, 213
Japan, 13, 24. *See also* Hiroshima and Nagasaki bombings
Jason Division, 21
"Jason East," 121
Jasons: and guerrilla combat/counterinsurgency, 6, 117–119; recruitment for, 21; and VELA, 27; creation of, 54, 117; summer studies, 120–121, 125, 130, 131, 132, 143, 145, 224; and Vietnam bombing, 121–125; and electronic

barrier, 125–132, 136–139; and tactical nuclear weapons, 132–136; and development of nonlethal weapons, 144; morale problems among, 144–145; "secret" Jason group, 145; and "Muscle Shoals" system, 146; ambivalence toward Vietnam War, 147–149; resignations from, 148–150; and release of Pentagon Papers, 149–153; New Left opposition to, 150; and SESPA, 151–154; and APS, 201, 203; and PSAC, 226–228; and X-ray laser, 264; E. H. S. Burhop's reflections on, 271
Jason Steering Committee, 144, 147–148
Jastrow, Robert, 259
Jet Propulsion Lab (Caltech), 218
Johnson, Howard, 181, 189–190
Johnson, Lyndon B., and administration, 7, 45, 96, 115, 192, 222, 241, 267; Robert McNamara and, 33; and limited-war weapons technologies, 79–80, 83, 84; and chemical defoliants, 89; and public opposition to use of gases and defoliants, 89, 91; and use of chemical weapons, 97; and Cambridge Discussion Group, 116; and electronic barrier, 130, 131, 137; and tactical nuclear weapons, 133, 136; decision of, not to run for reelection, 141; George Pimentel and, 142; APS invitation to, 199; and PSAC, 224, 226, 227; and ABM system, 236
Johnson, Robert, 66–67
Johnson, Thomas, 258
Johnson, U. Alexis, 72
Joint Chiefs of Staff, 14, 15, 18, 26, 39, 42, 53–56, 64–65, 69, 100, 104, 105, 122–123, 125, 129, 136, 137, 146
Joint Committee on Atomic Energy, 51
Joint Operations Evaluation Group, 69
Joint Research and Test Activity, 80
Jona, F., 199
Journal of International Affairs, 251
Jupiter missile, 19

Kabat, Jonathan, 163, 172, 177–178
Kac, Mark, 142
Kahn, Herman, 167
Kaiser, David, 23

Kastenmeier, Robert, 78, 84, 85
Katcher, David, 118
Katsiaficas, George, 173, 189
Katz, James, 240–242
Kaysen, Carl, 115, 116, 190
Kazan, Elia, 55
Keeny, Spurgeon, 115–116, 223
Kelly Air Force Base, 74
Kendall, Henry, 138, 140, 170, 267
Kendall Square (Cambridge, Massachusetts), 157
Kennedy, John F., and administration, 6, 17, 20, 29, 30, 192; arms control as key goal of, 31; Edward Teller and, 44–45; letter from Jerome Wiesner to, 63; and limited war capability, 63–64
Kennedy, Ted, 247
Kent State, 210
Kevles, Daniel, 10–11, 263
Khe Sanh, Vietnam, 136, 145, 146, 234
Khrushchev, Nikita, 26, 39, 49, 55, 56, 59
Killian, James, 15–21, 28, 29, 57, 61, 66, 115, 120, 136, 159, 162, 237, 238
Kinetic-energy weapons, 250, 265
Kirk, Grayson, 27
Kissinger, Henry, 203, 204
Kistiakowsky, George: and electronic barrier, 7, 125–126, 129–131, 130–132, 137, 139, 146; and PSAC, 18, 19, 40, 57, 63–64, 66; and SIOP, 28–29; and Limited Test-Ban Treaty, 60, 61; on "limited war," 63; and "Scientists and Engineers for Johnson," 80, 115; and Vietnam policy, 115–116, 120–121, 132, 255, 271; background of, 116; and Jasons, 125–127, 147; and DCPG, 129, 130, 138, 140–141; and "flexible response," 137; de-escalation pushed by, 139–141; resignation of, from DCPG, 140–143; and "Muscle Shoals" system, 146; and MIT antiwar activism, 163; on military funding of research, 184–185; on Nixon's ousting of academic science advisors, 230; John Baldeschwieler and, 233; resignation of, from Jasons, 236; and Safeguard ABM system, 238
Knowles, Asa, 242–243
Koch, Ed, 112, 206
Kogut, John, 261

Komer, Robert, 39, 43
Korean War, 13, 64, 71, 85, 109, 157, 159
Kosygin, Alexei, 102
Kraft, Joseph, 137
Krakatau, 99
Kuhn, Thomas, 11, 193, 211–218, 220
Kuhn Committee, 211–217, 239
Kumar, Chandra, 254
Kwajelein, 180

Labeling agents, 70
LaCouture, Jean, 152
Laird, Melvin, 104
Lake Erie Company, 82
Lamont, Nicholas, 211
Land, Edwin, 15
Land-based lasers, 251
Land mines, 126, 137, 138, 146, 227, 232
Lang, Anton, 104
Laos, 64, 65, 75, 110, 128, 145, 146, 228
Lapp, Ralph E., 25
Laser weapons, 251, 264, 265
Latour, Bruno, 11
Latter, Richard, 35
Lawrence Livermore Laboratory, 1, 8, 14, 22, 38, 54, 157, 190, 192, 201, 205, 247–248, 252, 254, 260, 263–266
Lebanon, 64
Lebow, Irwin, 173
Lederman, Leon, 138, 143–144
Leitenberg, Milton, 92, 109
LeMay, Curtis, 14, 15, 28, 32, 33, 46, 53
Lemnitzer, Lyman, 42
Lemnos, William, 177
Lerman, Jerry, 173
Leslie, Stuart, 159, 191
Levine, A. J., 214
Lewis, E. B., 49
Lewis, Harold, 138, 148
Lexington, Massachusetts, 157, 190
Libby, Willard F., 27
Licklider, J. C. R., 170
Lightweight weapons, 69
Limited war (limwar), 63, 117, 118
Limited War Panel, 64
Lincoln Laboratory (MIT), 22, 157–159, 161, 170, 172, 176–181, 186, 187, 190, 191, 263

Lincoln Laboratory Ballistic Range, 190
Littauer, Raphael, 203
Lockheed, 26, 160, 263
Lomax, Alan, 31
Long, Franklin, 116, 238, 239
Loper, Herbert, 36, 37
Los Alamos, 14, 26, 31, 40, 54, 57, 60,
    201, 229, 233, 254, 263
Los Angeles Times, 84, 266
Low, Francis, 54
Low-yield tests, 50
Lunar surface research, 158
Luria, Salvador, 54, 89, 96, 163

MacArthur Foundation, 254
McCain, John, Jr., 112
McCall, David, 198
McCarthy, Eugene, 163
McCarthy, Richard, 112
McCarthyism, 16, 58. *See also*
    Anticommunism
McCloskey, Robert, 90
McCloy, John, 35, 37, 39
McCone, John, 28, 45
McConnell, Alan, 206
MacDonald, Gordon, 147, 226
Mace, 109
McGovern, George, 166
McGrory, Mary, 238, 239
Mackenzie, Donald, 11
McLelland Air Force Base, 74
McMillan, William, 53, 130, 136,
    166–168, 224, 230–234
McNamara, Robert: and PSAC, 6;
    security objectives of, 33–34; and arms
    control, 39, 41–43, 52; on capability for
    limited war, 64; and science advisors,
    67; and language of science, 68; and
    defoliant use, 72–73, 77, 78, 89, 95; on
    need for new weapons technologies,
    79–80; and gas use in Vietnam, 85–87;
    and Cambridge Discussion Group,
    116–117; and Jasons, 121, 125, 127, 147,
    151; graduated escalation strategy of,
    122; and barrier project, 129, 130,
    137–139; and tactical nuclear weapons,
    133, 135–136; resignation of, 140, 227;
    Max Delbruck on, 142; George
    Kistiakowsky's letter to, 143; and MIT

research, 180–181; scientific ignorance
    of, 233; and nuclear weapon use, 248
McNaughton, John, 116, 129, 135
McRae, Vincent, 223
MAG, 206
Malaya, 75
Mangroves, 73–75, 77, 98, 104, 111–112
Manhattan Project, 2, 3, 5–7, 13, 16,
    18, 23, 61, 66–67, 109, 110, 116, 161,
    176, 182, 192, 230, 243, 245
Mansfield, Mike, 78
March 4 protest, 155, 163–171, 192, 245
Mark, J. Carson, 35, 40
Mark-14 gunsight, 159
Marshak, Robert, 254
Marshall Islands, 24
Marshall Plan, 13
Martin, John, 145
Masers, 158
Massachusetts Institute of Technology
    (MIT), 218–220; Union of Concerned
    Scientists at, 7; and Institute of Defense
    Analyses, 21; ties to government and
    industry, 22, 161–163; federal funding
    for academic science at, 61–62; and
    Cambridge Discussion Group, 116, 119;
    March 4 protest, 155, 163–171, 192,
    245; Lincoln Lab at, 157–159, 176–180,
    190, 191; I-Lab (Draper Lab) at,
    159–161, 174–175, 188–191; antiwar
    movement at, 163–168; "reconversion"
    of research at, 169–170; SACC at,
    170–174; MIRV research at, 175–176,
    181, 188; middle tiers of research
    staffers at, 180–181; individual guilt vs.
    institutional responsibility of re-
    searchers at, 181–183; Chomsky's call
    for shift to civilian-oriented research,
    185–187; decision of, not to sever ties
    with research labs, 187–188; and
    Defense Department funding,
    219–220; "spin-offs" from, 240–241;
    support for nuclear freeze at, 247; and
    SDI, 263, 268
"Massachusetts Miracle," 192
Massive retaliation, 32–33
Mastny, Vojtech, 59
Maxwell, Emanuel, 200
May, Michael, 264

Mayer, Jean, 93
Mead, Margaret, 208
Media, 17, 51, 77, 91, 107, 168, 247, 266, 268
"Medical Problems of South Viet Nam" (report), 92
Mekong Delta, 228
Melcher, James, 245
Melman, Seymour, 206
Meredith, James, 109
Meselson, Matthew, 54, 90–92, 94, 97, 100–103, 105, 106, 108, 207
Middle East, 135
Midwest Research Institute (MRI), 94–95, 107
"Mighty Mite," 86
Military-industrial-academic complex, 10, 24, 61, 150, 264
Military-industrial complex, 4, 29, 61, 68, 152–154, 157, 206, 263
Miller, Bruce, 255
Miller, Charles, 189
Minarik, Charles E., 73, 98, 113
Mines, 126
Minneapolis Tribune, 78
Minuteman Intercontinental Ballistic Missile Program, 19
Minuteman missile, 32, 33, 60, 191
MIRV. *See* Multiple independently-targetable reentry vehicle
Missile defense. *See* Antiballistic missile (ABM) systems
"Missile gap," 29
MIT. *See* Massachusetts Institute of Technology
MIT Alumni Advisory Council, 171
MITRE Corporation, 138–139, 158, 160, 171, 179–180, 230
MIT Review Panel on Special Laboratories, 171
MIT Tech, 96
Modernization theory, 157
Mohawk aircraft, 69
Mohr, Charles, 81
Monsanto, 71, 72, 75, 97, 100, 106, 242, 247
Moral arguments, 9, 58–59, 113, 167
Moratorium, testing, 27–28, 31, 58
Morison, Elting, 171

Morrison, Philip, 161
Morse, Philip, 195
Morse, Wayne, 84
Mosher, Harry, 210
Moss, Thomas, 257
MRI (Midwest Research Institute), 94–95, 107
MUD RIVER, 145
Mu Gia Pass, 133
Multiple independently-targetable reentry vehicle (MIRV), 160, 175–176, 181, 186, 188
Murrow, Edward R., 43, 51
"Muscle Shoals" system, 145–146
Mustard gas, 75
Mutually assured destruction, 236, 246, 248, 267

Nader, Ralph, 207
Napalm, 9, 65, 76, 91, 92, 108
NASA. *See* National Aeronautics and Space Administration
National Academy of Sciences (NAS), 91, 103, 104, 197, 207–208, 251, 253
National Aeronautics and Space Administration (NASA), 22, 215, 217
National Cancer Institute, 100–101
National Defense Education Act of 1958, 22
National Defense Research Committee, 17, 156
National Institutes of Health, 169
National Liberation Front, 68
National Research Council, 155
National Science Foundation (NSF), 22, 210, 211, 214, 215, 217, 238, 239
National Security Council (NSC), 18, 28, 42
NATO (North Atlantic Treaty Organization), 13, 102
Nature, 207–209
Naval Air Development Center, 182
Naval Warfare Panel, 225
Nazi Germany, 2, 109–110, 196, 205, 208, 225, 260
Neilands, J. B., 95, 96, 102, 112
Nelkin, Dorothy, 188, 190, 192
Neustadt, Richard, 116, 131–132, 145

Neutrality. *See* Academic neutrality; Political neutrality of scientists
Neutron bomb, 30, 36
New Brunswick, 74
New England Journal of Medicine, 91, 93, 109
New Left, 4, 150, 172, 189, 246, 253, 258–259, 268
New Look, 14–16, 24, 29, 33, 34, 41
Newman, Stuart, 208
New Republic, 78–79, 137
New Right, 246
Newton, Michael, 110
"New York Anti-War Faculty," 151
New York Review of Books, 90, 167
New York Times, 17, 27, 48, 54–55, 74, 78, 81, 84, 86, 90, 91, 143, 157, 164, 187–188, 201, 208, 211, 238, 240, 258
Nichols, Robert, 90
Nierenberg, William, 117–119, 121, 142, 259
Nitro, West Virginia, 100, 101
Nixon, Richard, and administration: and Manhattan Project veterans, 7–8; and defoliant use, 101, 102, 104, 113, 207; and détente, 102; and Geneva Protocol, 102, 103; Jasons and, 149; Lee DuBridge as science advisor to, 184; and antiwar activism, 200, 205–206; and expansion of war into Cambodia, 210; and presidential science advising, 222, 267; and ABM system, 235–239; and destabilization of old Sputnik order, 235–239; aerospace and defense cuts under, 241; anti-science stance of, 244
Nobel Prize, 18, 21, 55, 57, 132, 152, 199, 261
Noble, David, 10
Nolting, Frederick, 77
"Nonlethal gases," 81
Nonlethal weapons, development of, 144
Nonnuclear tools of war, 6–7
Nonnuclear weapons, 34, 63–87
Nonproliferation efforts, 24–26
Nordheim, L. W., 25
North Atlantic Treaty Organization (NATO), 13, 102
Northeastern University, 242, 261

Northrup, 160
North Vietnam: defoliation in, 75; bombing campaigns against, 115, 121–125, 226, 228; and electronic barrier, 128, 130–131; blocking of military imports into, 225
Nortronics, 160
Novick, Richard, 208
Novick, Sheldon, 95
Noyes, Pierre, 203–206
NSA (National Security Agency), 217
NSC. *See* National Security Council
NSF. *See* National Science Foundation
Nuclear deterrence. *See* Deterrence
Nuclear fallout, 58
Nuclear-freeze movement, 247–248, 266
Nuclear Information, 92
Nuclear Instrument and Chemical Corporation, 203
Nuclear proliferation, prevention of. *See* Nonproliferation efforts
Nuclear Rites (Gusterson), 248
Nuclear test ban, 6, 25–31, 35–38. *See also* Partial Test Ban treaty
Nuclear testing, 24; moratorium on, 27–28, 31, 58; Fisk panel and, 35–38
Nuclear weapons: creation of, 2; 1950s policy, 14–15; tactical, 32, 40, 53, 119, 121, 132–136, 248; and Robert McNamara, 33–34, 67; attitudes of researchers, 44, 60, 173, 248, 258, 264–265; stockpiles under Kennedy, 60; potential use in Vietnam, 91, 133–136, 151; missile defense, 158, 245, 248, 250, 252, 256–257. *See also* Antiballistic missile (ABM) systems; Arms control and disarmament; Hydrogen bomb; Nuclear testing
Nuclear winter, 259
Nuremberg trials, 71

Objectivity, 128, 177, 239, 246, 259, 268
O-chlorobenzylidene malononitrile, 95
Office of Defense Mobilization, 15
Office of Defense Research and Engineering, 105
Office of Naval Research, 156
Office of Science and Technology (OST), 223–224, 239

Office of Science and Technology Policy, 239

Office of Scientific Research and Development, 156

Office of Technology Assessment, 112, 251, 252

"Official Z," 133

Ohio National Guard, 210

"One Man's Meat" (article), 78–79

On-site inspections, 52, 55

Operation Farmgate, 80

"Operation Menu," 210

Operation Ranch Hand, 73–75, 78, 79, 93, 98–100, 106, 112, 113

Operation Rolling Thunder, 85, 121–125, 128

Oppenheimer, Robert, 2, 5, 16, 18, 60, 152

Optical radiation, 158

Orear, Jay, 199, 203, 204, 256–257

Oreskes, Naomi, 259

Orians, Gordon, 95, 110–111

OST (Office of Science and Technology), 223–224, 239

Outsider activism, 58–60, 98, 127, 142, 271

Overkill, 28, 34

Owen, Wilfred, 107

Oxford University, 171

Packard, David, 206

Palmer Physical Laboratory (Princeton University), 144

Panama, 135

Panel on Biological and Chemical Warfare, 63–64

Panel on Research Policy, 23

Panofsky, Wolfgang, 19, 35, 40, 42, 142, 219, 255

Panofsky panel, 40–42

Park, Robert, 1–2, 257

Parnas, David, 266

Partial Test Ban treaty (1963), 6, 56, 58, 60

Paschkis, Victor, 93, 196, 206–207

Patel, Kumar, 260

Pauling, Linus, 27, 43, 48, 49, 58–60, 90

Pennsylvania State University, 21, 191

Pentagon. See Department of Defense

Pentagon Papers, 149–152, 203

"People sniffer," 234

Perl, Martin, 199–202

Pfeiffer, E. W., 94–96, 110–111, 114

Phenoxyacetic acids, 70, 73, 75, 97, 99

Phu Lac Peninsula, Vietnam, 83

Phuoc Long Province, 77–78

Physical Review, 194

Physical Review Letters, 194

Physicians' Committee for Social Responsibility, 92

Physicians for Social Responsibility, 93, 247

The Physicists, 263

Physicists and Public Affairs (symposium), 203

Physics departments, 141, 155

Physics Today, 194, 196–197, 258, 261

Picloram, 99–100

Pimentel, George, 139, 142, 255

Piore, Emanuel, 23, 142, 190

Pitzer, Kenneth, 142

Planetary research, 158

Plant Sciences Laboratories (Fort Detrick), 72, 98

Plowshares program, 54

Podell, Bertram, 112

Poisonous gas, 75–76

Polaris missile, 32, 160

Polaris submarines, 60

Polaroid, 15

Political neutrality of scientists: and American Physical Society, 198–199, 205, 246; other professional societies, 206; John Baldeschwieler, 235; and debates over SDI, 246, 259–262, 268. See also Academic neutrality

Pollard, Ernest, 96

Porter, Edwin, Jr., 176

Poseidon missile, 160, 171, 175–176, 187, 189

Poseidon submarines, 160

Pounds, William, 171, 180, 181, 185–186

Pounds Panel, 182, 183, 185, 188, 189, 191, 192, 219, 220, 263

"Practice Nine," 137

Presidential Politics and Science Policy (Katz), 240

President's Science Advisory Committee (PSAC): creation of, 6; Lyndon Johnson and, 7, 224, 226, 227; Dwight Eisenhower and, 17, 18, 63; initial purview of, 18–19; and ABM systems, 19–20;

science education advocated by, 22–23; and Sputnik, 24, 26; Ad Hoc Panel on Nuclear Testing, 40–41; and Test Ban Treaty, 56–59; John Kennedy and, 65–67; and Sentinel program, 177; and escalation of Vietnam War, 183, 222–224; and Office of Science and Technology, 223–224; military panels of, 225–227; "Vietnam problem" report of, 225–228; and Jasons, 226–228; reservations about Vietnam War among members of, 228; "outsider" status of, 229; Richard Nixon and, 237–239; dissolution of, 246

Press, Frank, 35, 40, 171, 173–174

Primack, Joel, 239

"Princeton in the Nation's Service," 212–213

Princeton University, 7, 21, 144, 149, 193, 210–217, 261

Probstein, Ronald, 169–170

Professional societies, 7, 201, 207, 267

Project AGILE, 65, 69, 72, 117

Project Air War, 150

Project CAM, 170, 186

Project Charles, 157, 162

Project Lincoln, 156

Project Orion, 205

Project Plowshare, 54

Project PRESS, 158, 180

Project THEMIS, 241–242

Propellants, missile, 19

PSAC. See President's Science Advisory Committee

Psywar, 81

Public Health Service, 169

Puerto Rico, 74, 98, 99

Pugwash conferences and movement, 2, 31, 52, 59, 203

Pupin Hall (Columbia University), 151

Purcell, Edward, 18, 195, 200–201

Purdue University, 263

"Pure" research, 196, 215, 217–218

Quakers, 173, 203

Rabi, I. I., 6, 18, 22, 57, 58, 60, 66, 120, 136, 156

Rabinowitch, Eugene, 22, 90

Radar, 158–159, 178

Radiation Laboratory (MIT Rad Lab), 31, 156, 158, 161

Radioactive fallout, 24–25, 47

Radio astronomy, 158

Ramo, Simon, 53

Ranch Hand. See Operation Ranch Hand

RAND Corporation, 15, 37–39, 53, 66, 68, 80, 125, 133, 134, 160, 170, 231

Raskin, Marc, 39–40, 45

Rathjens, George, 120, 170

Rational Approach to Disarmament and Peace, 163

Raytheon, 159, 160, 175, 176, 182

Reagan, Ronald, and administration, 8, 11, 146, 191, 210, 233, 245, 248–249, 254, 255, 258

"Reconversion," 169–170, 189

Reed, Thomas, 240

Republican Party, 234–235, 247, 258–260

Research Analysis Corporation, 133

Research Laboratory of Electronics (RLE) (MIT), 156, 157

Reserve Officer Training Corps (ROTC), 155, 211

RESIST, 163

Resistance, 163

Resnick, Joel, 177

"The Responsibility of Intellectuals," 166, 167

Reviews in Modern Physics, 255

Rice, Berkeley, 240, 241

Rich, Alex, 115

Riot control, 83–87, 91, 102–103

RJ Reynolds, 259

RLE (Research Laboratory of Electronics) (MIT), 156, 157

Roberts, John, 142

Robertson, H. P., 64, 158

Rockefeller, Nelson, 143

Rockefeller Center, 166

Rockefeller Foundation, 155

Rockefeller University, 201

Rodberg, Leonard, 203

Rolling Thunder. See Operation Rolling Thunder

Romania, 59

Rome Air Development Center, 138

Romney, George, 234

Roosevelt, Franklin D., and administration, 76, 84, 156
Rosenberg, David, 14
Ross, Marc, 201
Rossi, Bruno, 89, 163
Rostow, Walt, 65, 66, 69, 75, 81, 124, 125, 130, 142, 227, 228
Rotblat, Joseph, 2
ROTC (Reserve Officer Training Corps), 155, 211
Route 128 (Boston), 156, 191, 240–242, 263
Rubenzahl, Ira, 163
Ruderman, Malvin, 143–144, 151
Ruina, Jack, 118, 138, 142, 161, 179, 188
Rusk, Dean, 35, 39, 43–45, 51, 52, 55, 56, 83
Russell, Bertrand, 78, 79

SABRE ballistic missile program, 160
SAC. See Strategic Air Command
SACC. See Science Action Coordinating Committee
SADEYE "Bomblet Dispenser Weapon," 126–127
Safeguard ABM system, 236, 238
SAGE (Semi-Automatic Ground Environment) system, 158, 162
Saigon, South Vietnam, 69, 72, 83
St. Louis Committee for Nuclear Information, 92
Saletan, Eugene, 198
Salpeter, Edwin, 148–149
Sanctuary, 163
Sandia National Laboratories, 1, 254, 255
SANE. See Committee for a Sane Nuclear Policy
Santa Clara valley, 229
Satellite-based lasers, 251
Satellite communications, 158
Saxenian, AnnaLee, 242
Scheer, Robert, 266
Schelling, Thomas, 166
Schevitz, Jeffrey, 229–230
Schlesinger, Arthur, Jr., 45
Schwartz, Brian, 200, 202
Schwartz, Charles, 7, 151–152, 195–196, 198–202, 204, 209, 256, 272
Schwartz Amendment, 196–197, 199–200

Science (periodical), 28, 95, 96, 98, 99, 104, 106, 107, 110, 143, 147, 150, 151, 209, 241, 264
Science Action Coordinating Committee (SACC), 164–166, 168, 170–174, 188
Science advising, 4, 10, 16, 26, 58, 67, 222, 231, 235, 237, 239–240. See also President's Science Advisory Committee (PSAC)
Science Advisory Committee (Office of Defense Mobilization), 15, 17
"Science against the People" (SESPA report), 151–153
Science and Engineering Education panel, 23
"Science and Technology of Directed Energy Weapons" (study), 255–261
Science education, 20, 22–23
Science for the People, 202, 209
Scientific Advisory Board (SAB). See Air Force Scientific Advisory Board
Scientific American, 236, 248
"Scientific Progress, the Universities, and the Federal Government" (report), 23
Scientist and Citizen, 92–95, 97, 106, 108, 109
Scientists and Engineers against Johnson, 143
Scientists and Engineers for Johnson, 80, 115, 116
Scientists and Engineers for McCarthy, 163
Scientists and Engineers for Social and Political Action (SESPA), 151–154, 200–202, 207–209, 238, 253, 272
Scientists' Committee on Chemical and Biological Warfare, 95–96, 102
Scientists' Institute for Public Information, 92
Scott, Wilbur, 106
Scoville, Herbert, 35, 39
SDI Advisory Committee, 260
SDIO. See Strategic Defense Initiative Organization
SDS. See Students for a Democratic Society
Seaborg, Glenn, 23, 35, 41, 43–45, 50, 54
Security clearances, 5, 160–16, 176
Seismology, 35, 51

Seitz, Frederick, 197, 198, 246, 253, 258–260, 268
Senate, U.S. *See* U.S. Senate
Senate Armed Forces Committee, 174
Senate Armed Services Committee, 251
Senate Committee on Foreign Relations, 56, 102, 108, 123–124
Senate Preparedness Subcommittee, 136–137
Sensors, 65, 67–68, 70, 105, 125–128, 131, 137–139, 141, 144–146, 234
Sentinel ABM system, 177, 236
Serber, Robert, 195
SESPA. *See* Scientists and Engineers for Social and Political Action
Shapiro, Ascher, 189
Shapley, Deborah, 104–105, 150, 151, 241
Shaw, Warren, 73
Showalter, English, 216
Sidel, Victor, 91–93, 108, 109
Silent Spring (Carson), 79, 100
Silicon Valley, 242
Single Integrated Operational Plan (SIOP), 28, 29
SIPRI (Stockholm International Peace Research Institute), 111–112
Sirbu, Marvin, 172–173
Skolnikoff, Eugene, 173
Slayton, Rebecca, 266
Sloan School of Management, 171, 173
Smullin, Louis, 161
Society for Social Responsibility in Science, 89, 163, 203
Software, 247, 252, 266
Sorensen, Ted, 57
South Africa, 135
Southeast Asia, 67
Southern California Federation of Scientists, 265–266
South Vietnam: Diem regime in, 68, 69; Limited War Task Group visit to, 69; CDTC in, 69–70; defoliation efforts in, 72–79; transfer of tear gases to, 82; medical effects of weapons in, 92; ecological destruction in, 103–105, 113; insurgencies in, 122; William Davidon's trip to, 203–204
South Vietnam, barrier into. *See* Electronic barrier

Soviet Union, 13, 17, 26–29, 32, 33, 35, 36, 43, 102, 122, 125, 128, 134, 135, 157, 226, 251, 253, 268. *See also* Sputnik
Space-based lasers, 248
Space-based weapons, 254
Space shuttle, 191
Special Committee on Disarmament Problems, 26
"Special Group," 65
Sperry Gyroscope, 159, 160
Sperry Rand, 160
"Spider bombs," 146
"Spin-offs," 15, 138, 156, 158, 187–188, 191, 205, 220, 221, 240–242
Spratt, John, 260
Sputnik, 6, 17, 22–24, 54, 58, 157, 162, 183
"Sputnik children," 166
SRI (Stanford Research Institute), 149, 218
SST (supersonic transport program), 235–239
Stahl, Franklin, 90
Stalin, Joseph, 260
Stanford Applied Electronics Laboratory, 22, 157
Stanford Research Institute (SRI), 149, 218
Stanford University, 21, 22, 157, 210, 218, 233–234. *See also* Applied Electronics Laboratory
Starbird, Alfred, 28, 35, 129, 130
"Star Wars." *See* Strategic Defense Initiative (SDI)
Stassen, Harold, 26
State Department, 26–28, 30, 31, 66, 76, 80, 100
Steinbeck, John, 55
Steininger, Donald, 223
Stockholm International Peace Research Institute (SIPRI), 111–112
Stone, Jeremy, 206
Strategic Air Command (SAC), 14, 15, 28, 32, 33
Strategic Defense Initiative (SDI, "Star Wars"), 4, 8, 146, 191, 198, 245–269; Reagan's speech announcing, 245, 248–249; Reagan administration studies supporting, 249–250; Richard Garwin's critique of, 251–252; and APS report on directed-energy weapons,

Strategic Defense Initiative *(continued)* 252–261; boycott effort, 261–262; and economics of research funding, 263–264; Roy Woodruff's skepticism toward, 264–266; and scientist activism, 267–269

Strategic Defense Initiative Organization (SDIO), 250, 255–256, 263, 266

Stratton, Julius, 171

Strauss, Lewis, 24, 26

"Strengthening American Science" (report), 23

Strong Program, 11

Strontium 90, 92

Structure of Scientific Revolutions (Kuhn), 212

Students for a Democratic Society (SDS), 155, 162–163, 210

Sub-Limited Warfare Research Project, 65

Sullivan, Edward, 181

Sullivan, Walter, 157

Summerfield, Martin, 212–213, 214

SUNY Buffalo, 263

Supersonic transport program (SST), 235–239

Super tear gas (o-chlorobenzylidenemalononitrile, CS), 81–83, 85–87

"Symmetry," nuclear, 13–14

"The System," 170

Systems Development Corporation, 158, 160

Szent Gyorgyi, Albert, 55, 96, 208

Szilard, Leo, 3, 49

Szilard Award, 266

Tactical nuclear weapons (TNW), 32, 121, 132–136

"Tactical Nuclear Weapons in Southeast Asia" (Jason report), 150, 151

Tam Giang, Vietnam, 83

Taylor, Maxwell, 15, 41–42, 52–56, 81, 147, 148

TCDD. *See* Dioxin

Tear gas (chloromalononitrile, CN), 67, 76, 81–87, 102

Technical assessments, 9, 128, 133–134, 258

Technical Development Corporation, 169

Technological Capabilities Panel, 15–17, 19

Technology, 67, 163, 260

Technology Review, 162, 165, 190

Teledyne, 138

Teller, Edward, 5, 6, 13, 14, 16, 22, 27, 28, 38, 44–45, 53–54, 56, 58, 60, 152, 197–198, 208, 209, 230, 245–249, 264–266, 268, 269, 272

Tennessee Valley Authority, 74

Test ban. *See* Nuclear test ban

Testing moratorium, 27–28, 31, 58

Texas Instruments, 138

Texas Tech, 263

Thailand, 74, 98, 144, 146

Thanh Ham, Vietnam, 83

Thermal tracking, 158

Thia Thien Province, 78

Thor missile, 19, 160

Three Mile Island, 247

Threshold, 34, 38, 49–50, 52, 54, 135

TIBA (2,3,5-triiodobenzoic acid), 88

Time Magazine, 239

TNW. *See* Tactical nuclear weapons

Tordon (picloram), 97

Townes, Charles, 152, 195, 199, 254, 255

Treaty of Versailles, 75

Trident missile, 191

Trieste symposium, 150

Trinity test, 2

Truman, Harry S., and administration, 13–14, 17, 25

Tschirley, Fred, 98–99, 104, 113

Tulane University, 21

Twining Commission, 46, 53, 54

2,3-D, 99–100

2,4,5-T, 72, 95, 101, 102, 106

2,4-dichlorophenoxyacetic acid (2,4-D), 70, 71, 95, 97, 101, 214

2,4,5-trichlorophenoxyacetic acid (2,4,5-D), 70, 71

U-2, 15, 29, 59

UCLA (University of California, Los Angeles), 231

UCS. *See* Union of Concerned Scientists

Udall, Stewart, 207–208

Ulam, Stanislaw, 5–6, 53, 212, 270–271

UN. *See* United Nations

The Uncertain Trumpet (Taylor), 41

Underground testing, 35, 41

Underseas Warfare Technical Center, 225
Union of Concerned Scientists (UCS), 7,
    164–166, 181, 220, 247, 251, 261
United Nations (UN), 25, 42, 95, 102, 144
United Press International, 78
Univac, 160
University of California, 265, 266
University of California, Berkeley: antiwar
    activism at, 150–151, 209, 210; SESPA
    at, 151; and sponsored research, 212,
    217, 242; and SDI, 261. See also
    Lawrence Livermore Laboratory
University of California, Los Angeles
    (UCLA), 231
University of Chicago, 25, 70, 203
University of Hawaii, 74
University of Illinois, 206, 261, 262
University of Maryland, 1
University of Michigan, 201, 218–219, 220
University of Mississippi, 109
University of Nevada, 210
University of Pennsylvania, 166
University of Southern California, 210
University of Wisconsin, 9, 208, 219
Urey, Harold, 48
U.S. Air Force: and nuclear weapons
    policy, 14–15, 28, 32; interservice rivalry,
    15, 28, 32, 105; Air Force Scientific
    Advisory Board, 20, 53, 92, 157, 230,
    240; World War II efficiency studies,
    33; and nuclear testing, 46, 50, 53; and
    chemical/biological weapons, 64, 97;
    and limited war concept, 64–65, 138;
    and defoliation, 71, 73–74, 79–81, 105;
    William McMillan and, 130, 136, 230,
    232; MIT and, 138, 157, 159; Charles
    Schwartz and, 195, 209; Princeton
    contracts with, 215; Donald Hornig and,
    223. See also Operation Ranch Hand
U.S. Army, 215
U.S. Army Chemical Corps, 71
"U.S. Bombing in Vietnam" (report), 122
U.S. Department of Agriculture (USDA),
    72–74, 98, 101, 102
U.S. Disarmament Administration, 26, 31
U.S. Information Agency, 43
U.S. Senate, 55, 56, 59, 61, 102, 125, 236
USDA. See U.S. Department of Agri-
    culture

Utah State University, 263
Utter, Leon, 85, 86

VC. See Viet Cong
VELA, 158
VELA-Hotel, 27
VELA nuclear test detection program, 21,
    26–27, 36, 37, 50, 51, 117, 146
VELA-Sierra, 27
Venezuela, 135
Veteran Administration, 106
Viet Cong (VC), 72, 77, 80–84, 104, 116
Vietnam War: de-escalation, 7, 116,
    119–120, 125, 130–132, 139–142, 146,
    150, 151, 210; "electronic barrier" in, 7,
    116–117, 125–132, 136–139; Manhattan
    Project vs., 7–8, 67–68; loss of
    scientists' prestige during, 8–9; and
    debates on epistemology of science, 10;
    and graduated escalation, 34, 87,
    121–123, 125, 195; Kennedy adminis-
    tration and lead-up to, 64, 65, 67–70,
    73–79; ARPA and, 65; stigma of, 68,
    173–174; as "proving ground" for new
    weapons technologies, 69–70, 81–82,
    107; defoliant use in, 70–78, 80–81,
    88–91, 93–99, 101–107, 109, 111–113,
    207; gas use in, 81–87, 90–92, 103,
    106–110; public opposition to gas and
    defoliant use in, 88–114; possibility of
    nuclear weapon use in, 136; and Jasons,
    147; PSAC involvement and escalation
    of, 222–224
"Vomit gas." See Adamsite
von Hippel, Frank, 235, 239
Vonnegut, Kurt, 200
Von Neumann, John, 15, 16, 26
The Voyage of the Lucky Dragon, 25

Wakefield, Massachusetts, 160
Wald, George, 165
Wall Street Journal, 256, 259, 260
Wang, Zuoyue, 57
War crime accusations, 71, 150, 205
War games, 133–135
Warheads, nuclear, 36, 41, 44, 160, 231,
    251, 254
War Industries Board, 155
Warsaw Pact, 102

Washington Post, 60, 81, 137, 205–206, 209
Waterman, Alan, 22
Watertown Arsenal, 160
Watson, James, 55
Watson, Kenneth, 117–119, 195
Watson Research Center, 199
Weaponized napalm, 215
Weapons categorization, 93
Weapons Systems Evaluation Group, 21
Weed Society of America, 97
Weinberg, Steven, 132–134
Weinberger, Caspar, 248, 265
Weisberg, Barry, 111
Weisskopf, Victor, 2–3, 161, 163, 165–167, 169, 171–173, 195
Weldon, Curt, 260
Wellesley, Massachusetts, 121
Westing, Arthur, 103, 111
Westinghouse, 160, 242
Westmoreland, William, 68, 80, 83, 86, 125, 129, 136, 137, 166, 168, 231–233
"Westmoreland Umbrella," 146
Wheeler, Earle, 86, 137, 146, 227
Wheeler, John, 21, 53, 195
Whirlwind computer, 158, 162
White House Science and Technology Policy Office, 239
White House Science Council, 248–249, 268
White phosphorus, 65, 91
Whittam, Donald, 73
Wiener, Norbert, 3, 5, 156, 271
Wiesner, Jerome: and nuclear arms control, 6, 31–33, 39–51, 54–56, 58–62, 157, 247; and Eisenhower administration, 18; and PSAC, 19, 239;

and Committee of Principals, 28, 35; and limited war, 63, 66; and Cambridge Discussion Group, 116, 120; on federal contracting at MIT, 162, 183, 190; on balancing moral and technical concerns, 192–193; and Safeguard program, 238
Wigner, Eugene, 5, 93, 197
Wilson, E. Bright, 223
Wolfowitz, Jacob, 47
Wood, Lowell, 258–259, 260, 261, 264–266
Woodbridge, Caspar, 139
Woodbury, Roger, 174–175, 177
Woodruff, Roy, 264–266
World Federation of Scientific Workers, 271
World War I, 75, 82, 107, 108, 155
World War II, 2–3, 5, 71, 76, 88, 108, 109, 156, 159, 161
Wright, S. Courtenay, 133

X-ray laser, 264, 265

Yarmolinsky, Adam, 145
Yonas, Gerold, 250, 255
York, Herbert, 18, 20, 21, 28, 29, 35, 57–60, 66, 80, 145, 149, 238, 249, 255
Young, Alvin, 97
Yucca Flats test, 28

Zacharias, Jerrold, 116, 119, 120, 125, 129, 157
Zachariasen, Fredrik, 138–141
Zeus system, 19
Zinn, Howard, 155, 165